© Natalia Jimenez

About the Author

NOGA ARIKHA grew up in Paris and studied in London. She received a doctorate in history at London's Warburg Institute, was a Fellow at the Italian Academy for Advanced Studies at Columbia University, and has taught at Bard College and at the Bard Graduate Center in New York. She lives in New York and London.

Passions *and* Tempers

A History *of the* Humours

Noga Arikha

AN ECCO BOOK

HARPER PERENNIAL

NEW YORK • LONDON • TORONTO • SYDNEY • NEW DELHI • AUCKLAND

HARPER ● PERENNIAL

A hardcover edition of this book was published in 2007 by Ecco,
an imprint of HarperCollins Publishers.

HarperCollins books may be purchased for educational,
business, or sales promotional use.
For information, please e-mail the Special Markets
Department at SPsales@harpercollins.com.

Library of Congress Cataloging-in-Publication Data
is available upon request.

FIRST HARPER PERENNIAL EDITION PUBLISHED 2008.

Designed by Kate Nichols

ISBN 978-0-06-073117-5 (pbk.)

HB 03.28.2022

To my parents, Avigdor Arikha and Anne Atik Arikha,

and

in memory of Miriam Rothschild
(1908–2005)

Acknowledgments

THIS BOOK owes its existence not only to thoughts and stories contained within other books, but also to friends, acquaintances, mentors, and chance encounters. I first discovered many of the books, especially those concerning the classical, medieval, and early modern worlds, in the Warburg Institute in London, where some years ago I conducted under the attentive supervision of Jill Kraye the doctoral work without which I would not have been able to conceive of a history of humours. Although much of this book was written in New York, it remains something of a Warburgian enterprise.

The actual seeds of this book were sown on a cold December day in 2001, in New York, over lunch with the philosopher Avishai Margalit, whose maieutic gifts led me to first formulate the idea. I am deeply grateful to him for that inspiring lunch, and for his subsequent encouragement. The project crystallized during a fellowship in 2002–2003 at the Italian Academy for Advanced Studies in America at Columbia University, whose director, David Freedberg, I thank for his generous hospitality and support, and for his invaluable friendship since then.

I am grateful to Leyla Selmi, who believed in the idea from the very beginning, and who was instrumental in helping me turn it into a book project. It took off thanks to the enthusiasm and hard work of Deirdre Mullane and the trust that Dan Halpern of Ecco put in it, as did Julia Serebrinsky, then editor there. Her successor, Emily Takoudes, has been a demanding, exceptionally attentive editor, and the manuscript was much improved under her guidance. Deirdre Mullane also helped me to undertake revisions, which benefited from the input of Jon Jackson.

Thanks to Alissa Quart for coming up with the title's final version, to

Robert Martensen for his critical and detailed evaluation of the text, and to Edmund Fawcett for some sharp comments; to Elisabetta Mori for sharing her research on the fate of Isabella de' Medici; and to Jay Weissberg for information on Mrs. Carleton. At the Warburg Institute, I owe thanks to its director Charles Hope; to Charles Burnett, the expert on Arabic medicine there, for bibliographical clues and for kindly correcting the chapter on this topic; to Ian Jones, for logistical assistance; and to Rembrandt Duits and Paul Taylor, for help with images.

I was able to discuss the work in progress on a number of occasions. Thanks to Antoine Compagnon for inviting me early on to give a seminar on Renaissance passions to his students of Montaigne at Columbia University; to Rosanna Warren for inviting me to submit some preliminary thoughts to her colleagues at Boston University; to Edna Margalit and Rivka Feldhay for the opportunity to discuss relevant ideas at the Bar Hillel Colloquium in Jerusalem; and to Françoise Longy for inviting me to talk about humours at a seminar on form and function at the CNRS in Paris. Barry and Bobbi Coller's hospitality allowed me to test my thoughts on Rockefeller University clinicians and biologists: I thank them and the participants for the stimulating occasion, as well as my old friend Margrit Wiesendanger for initiating it. I should also like to thank Bard College faculty Helena Gibbs, Richard Gordon, Garry Hagberg, Gregory Moynahan, Kristin Scheible, David Shein, and Marina van Zuylen, for some thought-provoking exchanges; and the students who took my class on the history of medicine and psychiatry in the fall of 2005, especially Emily Brennan, Giulia Carrozzini, and Eleanor Levinson, for their insights.

Almost everyone I encounter has something to say about humours, and many exchanges, even brief ones, have nourished the book's argument. It would be impossible to credit everyone, but I would like to acknowledge a few of those—including good friends old and new—whose insights, thoughts, tips, and clues have left their mark, however tangentially, over the years during which this book was being written. They are, alphabetically: Amir Amirani, Tony Bourne, Guido Branca, Vicky Brandt, Hillel Braude, Winsome Brown, Ian Buruma, Rosanna Camerlingo, Margherita Castellani, Daniele Derossi, Larry Dreyfus, Jennifer Dworkin, Edmund Fawcett and Natalia Jimenez, Roberto Farneti, David Freedberg, Enrico Galliani, Sheila Hale, Oren Harman, Elliott Jurist, Ute and Jonathan Kagan, Danny Katz, Richard Keatley, Paul Keyser and Michele Lowrie,

Claudia La Malfa, Phillis Levin, Rhodri Lewis, Robert Martensen, Zoe Martlew, Jonathan Miller, Peter Miller, Amy Morris, Allegra Mostyn-Owen, Turi Munthe, Keren Osman, Rami Osman, Pia Pera, Tanya Pollard, Yves Pouliquen, Alissa Quart, François Quiviger, Lisa Roscioni, Matthew Rutenberg, Joseph Rykwert, Didier Sicard, Lavinia Snyder, Rosanna Warren, Margrit Wiesendanger, Nick Wilding.

Many ideas were developed in conversation with Gloria Origgi, she and Dan Sperber have been primary interlocutors and supportive friends, energizing and inspiring. The input and friendship of Laura Bossi and Gérard Régnier have also been central to this project; and I am grateful to the latter for inviting me to pursue thoughts on melancholy on the occasion of his exhibition on that topic.

Miriam Rothschild never got to see this book. But over a dozen years of near-monthly visits to her home at Ashton Wold in Northamptonshire, I had the luck of being inspired by her energy and wit, her affection and encouragement, and challenged by her immensely wide-ranging conversation, which fanned my interest in science and nourished my questions regarding its boundaries. I only wish she could have enjoyed this volume.

My family has been ever-present. My sister, Alba Branca, is essential company even when transatlantically so, and, to an extent, I have written this book with her in mind. Our extraordinary parents, Anne and Avigdor Arikha, dedicated readers of their daughters' drafts, believed in this undertaking from the start. They have seen the book take shape; and it is thanks to their unwavering, loving support that I was able at all to take the road that led to its writing.

My husband, Marcello Simonetta, provided key information on the Renaissance, a period central to the history of the humours. He read the chapters as they accumulated, praised and criticized, and kept me going. Without his entreaties for me to sit and write, and without his capacity to humor my tempers, I might never have been able to finish. I thank him for much more than what can be acknowledged in a book's acknowledgments.

Contents

I. FOUNDATIONS: ANCIENT INSIGHTS
(Antiquity: Sixth Century BC to Second Century AD)

II. ESSENCES: THE CLASSICAL TRAIL
(Eastern Middle Ages: Seventh to Twelfth Centuries)

III. REMEDIES: MIRACULOUS MEDICINE
(Western Middle Ages: Fifth to Fourteenth Centuries)

IV. HARMONIES: RENAISSANCE BODIES AND
MELANCHOLY SOULS
(Renaissance: Fifteenth Century to Early Seventeenth Century)

V. NATURE: OF BLOOD, AIRS, AND REASONS
(Scientific Revolution: Seventeenth Century)

VI. BRAIN: PASSIONS AND NERVES
(The Making of Modernity: Seventeenth to Nineteenth Centuries)

VII. SCIENCE: CONTEMPORARY HUMOURS
(Twentieth and Twenty-first Centuries)

List of Illustrations

Prologue

I N 1463, Giovanni de' Medici, son of Cosimo the Elder—the effective founder of the famed Medici dynasty—died in his native Florence, at age forty-two. What killed him was apparently a sudden onset of "phlegmatic complexion." Giovanni had been drinking too much cold water and had not taken enough exercise. Giovanni himself is not a key figure in history: it was under the rule of his famous nephew, Lorenzo "the Magnificent," that humanism and the arts flowered in Florence. Nor, for that matter, is his death of particular consequence. But a long, fascinating story underlies the diagnosis of the illness that killed him.

For centuries before Giovanni's birth, and for centuries after his death, the notion that an excess of water could induce an excess of phlegm was an accepted medical diagnosis. Phlegm was one of the so-called "humours," and the theory of humours was orthodoxy among doctors and patients alike. Humoural theory began in Greece in the fifth century BC with the body of work attributed to the physician Hippocrates. It then continued with Galen, the Roman doctor who adopted the Hippocratic doctrine in the second century AD. For over two thousand years thereafter, humoural theory explained most things about a person's character, psychology, medical history, tastes, appearance, and behavior. Doctors continued to work on the assumption that the body and the mind were intimately connected, that emotions were corporeal, that vapors caused headaches, and that a cold stomach caused indigestion. It was on the authority of humoural theory that they would advocate leechings, bleedings, cataplasms and fomentations as a cure for all ailments, from stomach aches to fevers, from skin rashes to chest pains.

The humours were substances that circulated within the human body, much like water in pipes. A humour is literally a fluid—*humon* in Greek, *(h)umor* in Latin—and bodily humours are fluids within a living organism. In the West, the theory developed that the human body was constituted of four of these humours, all central to its functioning. Phlegm was one them; the three others were yellow bile, black bile, and blood. They were concocted out of the heat of digestive processes in the stomach: food turned into so-called *chyle* in the liver, from where, thanks to the heat produced by these digestive concoctions, particles in the bloodstream called "vital spirits" were expedited to the heart, and from there to the brain. The cerebellum refined some of these spirits into smaller "animal spirits." Heat and cold, dryness and moistness affected the course of the spirits, and determined the effects of each humour on mood, thought, or health. There was thus a continuum between passions and cognition, physiology and psychology, individual and environment.

In turn, individual temperaments were the product of variations in the proportion of each humour in the body. An excess of choler (yellow bile) in the blood produced the choleric temperament; an excess of black bile produced the melancholic; an excess of phlegm, the phlegmatic; an excess of blood, the sanguine. Women tended to be moist, old people dry. Children's brains were moist, and this moistness explained their ability to learn and memorize. Regardless of one's predominant temperament, however, humours shifted according to what one ate and drank, to where one lived, and to climate and season. A sanguine person could suffer a bout of melancholy on one day, and a phlegmatic individual might become choleric on another. An excess of food or drink could cause a humoural imbalance, bringing about illness or, at worst, death.

Had Giovanni de' Medici lived today, no doctor in his right mind would have diagnosed "phlegmatic complexion," just as no serious physician today believes that a good bleeding will reduce a temperature. Once the connection between disease and the existence of germs had been firmly established, about 150 years ago, it was indeed impossible to hold on to the theory of humours. In fact, humours had already begun to lose theoretical credibility by the seventeenth century, when the circulation of the blood was properly understood. But in practice, the theory continued for another couple of centuries to sustain medicine and to offer a general scheme within which anatomy, physiology, and psychology could be made to fit. Some medical manuals were still recommending treatments based on humoural theory as recently as in the early 1900s. The principle underlying humours remained potent in the form of substances, particles, or currents traveling through the blood from limbs and organs to heart and brain, and back. It accounted for health and illness, for all sensation, emotion and cognition. The basic humoural model revealed how the sight of a beautiful maiden could trigger desire, induce a rush of blood in the veins, and increase the heartbeat, or how the excessive ingestion of wine resulted in a wobbly step and an altered mood.

Humours now remain familiar mostly metaphorically: we have all reacted cholerically to an insult, or awakened in a melancholy mood on a dark wintry morning. A mood is an *humeur* in French and an *umore* in Italian; and English-speakers still have to humor the whims of a temperamental colleague, face a Monday with ill-humor, and remain good-humored throughout the week. But humours do not survive just as linguistic habits: this book argues that their explanatory power has actually never gone away. It tells how and why this is, bringing them back to light, delving beneath the names we give to states of mind, to illnesses, and to the invisible world beneath our skin. It shows how humours have been recycled, continually reappearing in new guises, ever-present within evolving scientific systems and medical cultures. By now, the original four humours imagined by the ancients have been multiplied by the hundreds into hormones, enzymes, neurotransmitters, particles, and the like, constituting the hydraulic system that is our body, providing us with a partial picture of what is going on inside our organism.

We usually recognize a physical element in aspects of character or a particular mood: we say that mood swings might be related to hormonal shifts, or that an excitable person might be prone to high blood pressure. This is not very different from humoural thinking, which recognized the interdependence of mind and body. The historical equivalent of depression, for instance, is melancholy, which was understood to be caused by an excess of black bile in the organism; and the search for cures began long before modern antidepressants were conceived.

Natural remedies alone were available until the advent of pharmaceuticals, but herbal concoctions were the pharmaceuticals of yore. Today herbal concoctions and "alternative" medicines are increasingly popular. The mechanistic medicine which we have inherited from the Enlightenment, and according to which our bodies are mechanisms that must be taken apart in order to be understood, dissatisfies many of those who resort to such alternatives. A number of these alternative medical practices derive from Asian traditions, many of which are also humoural, based on the notion of fluids circulating within the body, energy flows, mind-body interaction, and balances between hot and cold, dry and moist. But few people know when, why, or how western medicine became so mechanistic; or that our attraction to Asian medicines might be due to their closeness to the humoural models that prevailed in the West until so recently. In effect, the history of humoural theory is not widely known. There is little awareness that there even is such a history to be told, and that its telling can reveal just how much our present is impregnated with the past.

Some of the remedies whose efficacy was once explained by humoural theory have survived its official demise simply because they have remained efficient. It still seems reasonable to eat root vegetables and "warming" spices like mustard, ginger, pepper, and cloves in the winter and "cooling" foods like green vegetables and lemons in the summer. There once was a humoural rationale to such practices: a winter chill supposedly strengthened the organism's own heat, thereby increasing appetite and the capacity for digestion, while one needed to counteract the heat and dryness of summer weather with cold and moist substances. Cures for mental or bodily ills based on these beliefs may or may not work. Humours no longer account for the cures that do work—not because they are entirely wrong, but because they are based on a largely mistaken picture of the body.

Mistakes, though, are interesting, and necessary for correct theories to exist at all. Today we understand the natural world and the human body in much greater detail than in the past; but there remains much that we do not know. This book concerns itself primarily with our capacity to make mistakes even when our questions are right: its premise is that all theories about how the world works are revealing, in the way that children's questions about the world are revealing. In a sense, we are all children in our relation to scientific information. Whether today, in fifteenth-century Florence, or in fifth-century-BC Greece, we need commonsense explanations, regardless of whether or not they are provable, or true. Even when wrong, a theory can help us understand, if not the world, then perhaps ourselves.

Popular beliefs have always interacted with learned theories about how our minds and bodies function. This history of humours is also a history of this interaction. The book chronicles the fate within the western world of the protean, invisible substances that are humours, from their origin in ancient Greece via the medieval, Renaissance, and modern worlds, up to the present day. It explores the sources of beliefs in the West about the relationship of body and mind, and about the role of humours in binding them together. It looks at the fears and myths that have surrounded these beliefs, at the gaps between medical theory and medical practice, between the visible body and its invisible processes, between clinical care and human pain. It presents a 2,500-year journey inside the scientific mind; but by its end, the distance traveled might reveal itself to have been remarkably short.

A Note on Terminology

THE THEORY of humours as it was developed in Greece and then transmitted to medieval and Renaissance Europe has pride of place in western history, and in the following pages. But etymologically and technically, humours are fluids, and the author has seen fit to use the term in the widest possible sense: humours are here understood to designate a conceptual field as well as a historical moment.

Throughout the following pages, some key words occur—psychology, soul, science—whose meaning changes according to historical context. The science of psychology as we know it today was born in the nineteenth century, but when the term is used within an earlier context, it should be taken to mean the study of the soul. The soul itself is used to designate what we today understand as the mind and its various functions. For over half of the book, science is used in the general sense of technical knowledge about a particular field; only later does it designate the particular set of practices and protocols that the term now denotes. In all cases herein, the context within which the word occurs will cast light on its meaning: history itself can serve as a shield against anachronisms.

This is literally "a," not "the" history of humours in the West. The validity of such a history cannot depend on completeness. It is relatively short, and necessarily selective. Given the chronological and thematic breadth of the terrain covered, it was thought best to favor interpretation over accumulation, and clarity over glut—at the expense of having to omit many figures, texts, references, and episodes that are as relevant to this history as those that have been included.

I

Foundations:
Ancient Insights

(Antiquity: Sixth Century BC to Second Century AD)

> *To begin at the beginning: the elements from which the world is made are air, fire, water and earth; the seasons from which the year is composed are spring, summer, winter and autumn; the humours from which animals and humans are composed are yellow bile, blood, phlegm and black bile.*
>
> GALEN[1]

> *People's characters are not carved on the clay tablets of their natures unalterably.*
>
> ELISABETH YOUNG-BRUEHL[2]

I. Cosmic Elements

BEFORE HUMOURS, there were elements: air, fire, water, earth. The earliest myths of creation—from the biblical apparition of light to the birth of Greek gods out of primordial chaos—tell how life emerged from the heavens and its fiery stars, from the earth and from the sea, created by divine forces. Humours, the liquids that sustained, directed, and defined the human organism, were begotten then too. The origins of humours are enmeshed with the first accounts of the origins of life itself.

Our bodies and their humours are part of the natural world, but they are not always in harmony with it. Nature's elements can destroy the very life they create; floods, fires, earthquakes, heat waves, and stagnant waters are its enemies. Human cultures began with gods and shamans, prayers and priests that were supposed to act as vectors to health and shields against illness and nature's onslaughts. But the history of humours in the West starts in earnest with the first stabs at philosophical wonder, the first efforts at understanding nature in its own terms. This was at a time when the gods were still powerful but no longer sufficed to explain where the world and its elements came from, what it was made of, what made life possible, what caused storms or earthquakes, illness, health, or death.

In the early sixth century BC, many decades after Hesiod had described the genealogy of the gods in his *Theogony*, a number of thinkers in Greek Ionia (the coast of Asia Minor, today's Turkey) began to interrogate nature. They used reason to analyze the universe, matter, the soul, divinity, and eventually thought itself. Thales of Miletus famously claimed that all matter was made of water. But alternatives soon followed. Thales's disciple Anaximander suggested that all qualities had combined to form matter through their mutual conflict: cold and wet formed earth, hot and dry created the fiery elements that had dried the earth, and life emerged out of the combination of hot and wet. It could also be that air was the principle and origin of all things, as Anaximander's own disciple Anaximenes believed—its rarefaction producing fire, its condensation clouds, water, and finally earth. The world breathed, just as we did, and the human soul, or psyche, itself was breath, just as air was the soul of the world. Around the late sixth century BC, Heraclitus of Ephesus proposed that water, air, and strife between opposites did not explain much: instead, fire best symbolized the continual state of flux of all things, and of human life itself. Farther west, in southern Italy, Parmenides of Elea (near today's Salerno), declared that the universe was a single entity and that everything in existence was unchangeable, ungenerated, and indestructible.

The writings of the so-called pre-Socratic philosophers are not extant: we have only a few fragments, reported by later thinkers such as Theophrastus and Diogenes Laertius. But these broad, often divergent claims about the origin, substance, and structure of the universe and of man's place within it mark the beginning of philosophical speculation. It would come to full fruition with Socrates, who was born in 470 BC and whose words Plato recorded for posterity. No one really knows why the Greek peninsula, along with the area of Greater Greece, or Magna Graecia, was such a ferment of innovation. A few of the novel ideas might have grown out of cultural exchanges. Some scholars speculate that Indian sages who traveled westward had fertilized Greek minds with their own cosmology —and also with their own conception of the human body, which bears strong resemblances to that of the Greeks. It is hard to know for sure, although a sect known as the Pythagorean Society did promote beliefs that are not unlike those of Hinduism, such as the centrality of air or wind to life and diseases, the transmigration of souls, and the necessity for vegetarianism.[3] This Society had been founded in southern Italy, in the Calabrian town of Croton, by the thinker Pythagoras. He was born in 569 BC in Ionia, but it was in Croton that he explored connections between mathematics, music, mysticism, and cosmology. He believed that revelation was a suitable medium for the acquisition of knowledge, and he was the first Greek to use the term *philosophia*.

According to the Greek historian Herodotus, Croton also hosted one of the best medical schools of the period. One Crotonese Pythagorean named Alcmaeon, active around 500 BC, has an important place in the history of humours. He was a keen observer of the body who managed to explain aspects of its functioning with impressive accuracy. He discovered the optic nerve, for instance, and differentiated the veins and the arteries. He was also convinced that the brain, rather than the heart, was the seat of perception and intellect. Plato would later present this craniocentric view in the *Timaeus,* a book in which figures an account of the elements and humours, and whose fortune would prove long-lived.

Alcmaeon seems to have agreed with the slightly older Parmenides and his Eleatic cohort that the universe was one indivisible entity. Yet in one of his most important books, the *Metaphysics,* Aristotle referred to Alcmaeon as the possible originator of the fundamental idea that the world was bred of opposites and therefore contained opposite qualities. According to Aristotle, Alcmaeon either had taken over from the Pythagoreans or had himself come up with the notion that "most human affairs go in pairs," such as "white and black, sweet and bitter, good and bad, great and small."[4] Wherever it comes from, it is a powerful thought, and it plays a central role in the genesis of the idea of humours.

Much mystery, if not mystique, remains attached to the Pythagorean movement, whose influence was still alive 1,500 years after its foundation. Empedocles of Acragas (today's Agrigento, in Sicily) was a Pythagorean—he believed in the transmigration of souls and was a vegetarian. He was a self-possessed, charismatic public figure who not only wrote philosophy and poetry but also practiced both politics and medicine. And he was the first of these remarkable new thinkers to claim with great persuasive force that the world was not reducible to any one entity but was instead constituted equally of all four elements. He claimed that all things emerged out of the various combinations of air, earth, fire, and water, and dissolved with the dissociation of these elements. Matter itself was unchanging, but the powers of attraction and repulsion bestowed upon it all its motions and variations. The elements were united by "love" or attraction, and driven apart by "strife" or repulsion.

By the mid-fifth century BC, the Empedoclean version of the doctrine that all matter was divided into four opposing pairs of principles (hot and cold, dry and moist) and four elements (air, earth, fire, water) was established. It is out of this simple scheme that humoural theory grew: the human body was now also built out of the four elements. The microcosm of the body corresponded to the macrocosm of the universe: to each one of the cosmic elements corresponded a bodily humour, and it was thanks to this correspondence that we were able to perceive the world. There were debates as to the exact number of humours, just as the number of qualities themselves was subject to some variations. But the useful symmetry inherent in the fourfold division of the universe into seasons, qualities, and elements was applied to the bodily humours as well. Air corresponded to blood (which was perhaps categorized as a humour in order to preserve this fourfold division); water corresponded to phlegm; fire corresponded to choler, the yellow bile; and earth corresponded to melancholy, the black bile. To each humour corresponded two of the basic qualities associated with its element: blood was hot and moist, phlegm was cold and moist, choler or yellow bile was hot and dry, and melancholy or black bile was cold and dry.

2. Human Elements

MEDICINE AND PHILOSOPHY are twin disciplines, born at the same time. Just as the pre-Socratic philosophers were defining the world in terms of its natural elements, so *physis,* or nature, was now to account for the vagaries of mind and body. In the fifth century BC, doctors were beginning to use systematic reason, just as the philosophers were, to identify symptoms and illnesses on their own terms, without immediate recourse to gods. They were separating themselves from the priesthood and practicing their skills increasingly as professionals: many of them were itinerant, traveling to visit patients (the original meaning of the Greek word *epidemic),* and receiving remuneration for their work.

Hippocrates is known as the first doctor to have engaged in this new, rationalist medicine, and to have helped establish it as a practice in its own right. He is still remembered, in fact, as the father of medicine. Members of the medical corps today continue to swear to a modernized version of the Hippocratic Oath,[5] one of the texts collected within the corpus of Hippocratic writings canonized by the third century BC. Most of the texts attributed to Hippocrates are by other, unknown authors. But he did exist. He was born on the Aegean island of Kos, in about 460 BC, into the so-called Asklepiad family of physicians, an illustrious dynasty which had been practicing in Kos and also in Cnidus, a nearby town on the coast of Asia Minor. He was a famous figure in his lifetime, mentioned by Plato—a younger contemporary—and a few decades later by Aristotle.[6]

Hippocrates was studying and practicing medicine just as it was becoming a proper art—a *tekhnè,* in Greek—that concerned the nature and conditions of human life as a whole. For the Hippocratic physician, diseases were caused by nature rather than by the direct intervention of gods in human affairs, and medicine was a matter of reason and understanding. Doctors in ancient Egypt and Mesopotamia had already developed sophisticated techniques of bandaging and suturing wounds, useful for both mummification and living patients. One basic salve, composed of grease or resins and honey, has since been shown to be an effective bactericide.[7] But empirical medicine did not emerge in a fully-fledged form in Egypt, or in Mesopotamia. Throughout the ancient world, sorcerers and priests tended to view most afflictions of body and soul as initially caused by divinities and other forces external to the body.

Hippocratic medicine instead focused on the organic, humoural processes that caused diseases, that is, on the aetiology (from *aitia*, "cause" in Greek) of illness. Shifting their gaze from the heavens to their patients, Hippocratic doctors began to scrutinize the body with an intensity that had not existed before in this part of the world. They were concerned with the whole of the body because "every part of the body, on becoming ill, immediately produces disease in some other part," as one Hippocratic author put it in a text called *Places in Man*.[8] New skills were necessary for the Hippocratic doctor: relying on a holistic and humoural understanding of the organism, he now paid thorough attention to symptoms, in order then to offer a *diagnosis*, establish a *prognosis*, and present a cure.

Superstition was useless and treatments based on it were wrong, according to the author of an influential Hippocratic treatise known as *The Sacred Disease*. The symptoms of this "sacred disease" would be identified today as epilepsy—and an epileptic fit could easily have seemed to be a case of possession by spirits or demons. The Hippocratic author, however, insisted that a good, virtuous doctor was one who took responsibility for his own failings, and who understood how the body worked, on its own terms. This is why he forbade superstitious practices, such as the use of baths or "the wearing of black (black is the sign of death)," "not to lie on or wear goatskin, nor to put foot on foot or hand on hand." Those who enforced such practices did so, he wrote, "because of the divine origin of the disease, claiming superior knowledge and alleging other causes, so that, should the patient recover, the reputation for cleverness may be theirs; but should he die, they may have a sure fund of excuses, with the defense that they are not at all to blame, but the gods."[9]

The Hippocratic treatise in which the theory of humours is formulated most explicitly and clearly, and in its connection to nature's elements, is known as *Nature of Man*. It also happens to be one of the few treatises whose authorship is precisely attributed—in this case, to Polybus, who was Hippocrates's disciple and son-in-law, so it is likely to bear testimony to the master's beliefs.[10] Here one hears the voice of the rationalist doctor taking issue with the claims of contemporary pre-Socratic philosophers that man was made of either "air, or fire, or water, or earth," and with the claims of contemporary physicians that man was either blood, or bile, or phlegm. The body, wrote Polybus, perhaps echoing Empedocles, had "many constituents, which, by heating, by cooling, by drying, or by wetting one another contrary to nature, engender diseases." Man was not a unity, and was not reducible to one element or one humour: he was "born out of a human being having all these elements." All the elements and substances were present in each individual, from birth to death; all four humours made up "the nature of his body, and through these he feels pain or enjoys health."

Humours generated the parts of the body by nourishing it. Together, as what Aristotle called a *mixis,* they constituted the blood (distinguished from the humour blood) that flowed through the veins, and, along with air—or *pneuma* in Greek—through the arteries. Disease was understood as a state of imbalance or *dyskrasia* between the humours that made up the body's *krasis,* its general constitution or *complexion.* Curing a disease meant rectifying the imbalance within the organism by returning the humours to their proper "mixing," or *eukrasis,* and thereby to a healthy state of balance, or *isonomia.* The process of rebalancing the humours had to take into account both external and internal factors, including the patient's lifestyle and his or her temperament. Temperaments themselves were constituted by the humoural complexion.

These humours were not themselves visible, although they were based on visible substances. Everyone had seen blood; phlegm was apparent in the form of a runny nose, or tears, say; yellow bile appeared in wounds as what we understand to be pus, or within vomit. As for black bile, it might have been inferred from the observation of clotted blood, excrement, and dark vomit; but it had never been isolated as any one substance, and if it was posited at all, that is because it was necessary to the symmetry of the scheme—there had to be one cold, dry humour if all the possible combinations of the four elements and qualities were to be included in the humoural system. Still, black bile represented a real component of the human psyche: it did not need to be observable to have explanatory power.

In fact, none of the humours needed to be visible to exert their hold on the imagination, and to provide a credible, at times effective physiological account of the unseen operations within the body. To a large extent, the physiological and psychological theory of the four humours emerged out of further assumptions about the material basis of our passions and thoughts. In fifth-century BC Greece, consciousness was thought to be located, literally, in the viscera—the *splanchna*—and basic passions were fomented within the guts, in the spleen, liver, gallbladder, and heart.[11] Three centuries before, Homer had already used the term *cholos*, the bile that later Greek thinkers called *chole*. In tragedies of the fifth century BC, *chole* ranged from yellow to black: *cholao* meant "I fill with bile" and *melancholao*, "I fill with black bile" or "I am passionate" or indeed "I am becoming mad" *(melan* means black in Greek).[12] The notorious association of melancholy with madness has an etymological as well as a genealogical ground; but the notion that passions were literally organic events, actions of the body that turned the soul inside out, so to speak, made more than metaphorical sense. The bloody occurrences of Greek tragedy were rooted in a mythical culture where diviners could read meaning into the configuration of potent organs. The science of medicine—of soul and body—would gradually be built upon this culture. The humoural system provided a rational scheme that encompassed passions, illness, blood, and guts, ordering the darkness and disorder of inner life.[13]

3. Types, Temperaments, and Environments

THE HUMOURAL SYSTEM was precise and calibrated, accounting at once for types and for individuals. Crucially, humours were not present in equal quantities in everyone. It was the preponderance of some over others that determined the temperament of each individual. Personality and body types—thin, fat, sallow, nervous—depended on the humoural characteristics present within the individual from birth or even from conception on: "every year participates in every element, the hot, the cold, the dry and the moist," and "if any of these congenital elements were to fail, the man could not live."[14]

The humour blood, *haima*—different from the visible venous blood— was the most neutral. Produced in the liver, it was warm and moist. It was associated with springtime and childhood, with the sanguine temperament; and its prevalence within the organism was generally correlated with health and mental balance, serenity, sensuousness, and optimism.

Phlegm, *phlegmos*—sometimes a by-product of the secretion of blood—could be found anywhere in the body. Phlegmatic temperaments usually had an excess of it in the brain or lungs, and tended to be sluggish in action and reaction. It was cold and moist, associated with the winter and old age; many common illnesses, such as headcolds, were also attributed to its actions.

Yellow bile—*chole*—was the attribute of cholerics, who were quick-tempered, sometimes resentful or envious, generally argumentative. Associated with summer and adolescence or youth, yellow bile was warm and dry, and produced in the gallbladder.

As for black bile—*melaina chole* or melancholy, the most ineffable and most illustrious of all humours—it was concocted, in some versions, out of the yellow bile or out of the blood, but it was stored in the spleen. Cold and dry, it was the humour of maturity, and its season was autumn. Melancholic people tended to be introspective, and so this humour could be useful for creativity, although it was also instrumental in delirium, madness, or conditions associated today with depression.

Some humoural types were more prone to certain diseases than to others; but generally illness was a matter of excess or lack, of the exacerbation of one quality over another, of "the hot being too hot, the cold too cold, the dry too dry and the wet too wet."[15] Not only did the author of *The Sacred Disease* dismiss the notion that this spectacular disease had the supernatural causes implied by its name; he also identified it as, typically, merely a phlegmatic ailment. The brain was its seat, as indeed it was "of other very violent diseases." This was because of the structure of the brain, which was "double; a thin membrane runs down the middle and divides it," and of the veins that connected it to the rest of the body— "two large vessels, one coming from the liver and one from the spleen." A detailed description of the various branches of these veins followed. The author, like his contemporaries, believed that we breathed thanks to these

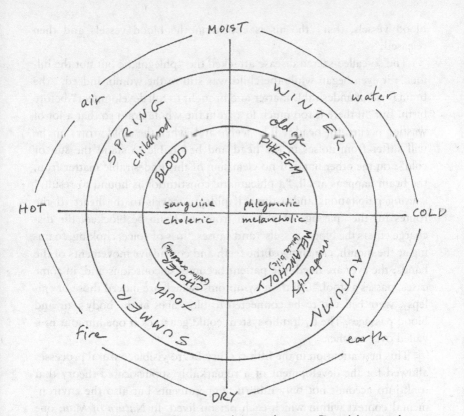

Fig. 1. Diagram of humours, elements, qualities, and seasons. To each humoural category would eventually correspond also a time of day, a color, a taste, a type of fever, a main organ, governing musical modes, a tutelary planet, and a set of astrological signs. *Blood:* morning; red; sweet; continuous fever; heart; lydian and hypolydian modes; Jupiter; Gemini, Libra, Aquarius. *Choler:* midday; yellow; bitter; tertiary fever; spleen; phrygian and hypophrygian modes; Mars; Aries, Leo, Sagittarius. *Melancholy:* afternoon; black; sour; quartan fever; liver; mixolydian and hypomixolydian modes; Saturn; Taurus, Virgo, Capricorn. *Phlegm:* evening; white; salty; quotidian fever; brain; dorian and hypodorian modes; Moon; Cancer, Scorpio, Pisces.

blood vessels, that "the air is cooled in the blood vessels and then released."

The so-called sacred disease attacked the "phlegmatic but not the bilious"; it even began while the child was still in the womb. Indeed, "the brain is rid of undesirable matter and brought to full development" before birth. But "if there is too much lost from the whole brain so that a lot of wasting occurs, the head will be feeble and, when the child grows up, he will suffer from noises in the head and be unable to stand the sun or cold"; on the other hand, if no cleansing of this undesirable matter from the brain happens at all, "a phlegmatic constitution is bound to result," causing palpitations and asthma if phlegm travels to the heart. If the routes for the "passage of phlegm from the brain be blocked, the discharge enters the blood-vessels" and causes "loss of voice, choking, foaming at the mouth, clenching of the teeth and convulsive movements of the hands; the eyes are fixed, the patient becomes unconscious and, in some cases, passes a stool." All these symptoms, which are indeed those of epilepsy, were believed to be connected to blockages in the body's air and blood passages. The hydraulic system could get stuck if one humour prevailed over another.

This new attention in the fifth century BC to visible, natural processes allowed for the development of a remarkably sophisticated theory that took into account not only inborn temperaments but also the environmental context within which each person lived. In *Nature of Man,* one reads that humours varied according to the season—that phlegm increased in winter and decreased in summer, since it was "the coldest constituent of the body"; that blood increased in spring and summer, whereas both types of bile, especially black bile, increased toward autumn. But there was more. In a highly influential treatise called *On Airs, Waters, and Places,* written in about 400 BC, the (unknown) Hippocratic writer made a point of showing how factors such as climate and vegetation shaped the human organism and had a direct effect on its humoural constitution. It is commonsensical enough that the weather—hot or cold, dry or humid— should have an impact on health. But this text went further and described how nature, climate, and therefore temperament determined culture, that is, political and social structures—although culture also affected nature and could compensate for some of its effects.

Winds, elevation, and geographic orientation determined the quality and location of water; water in turn determined vegetation, as well as the general health of the population. Seasonal changes and astrological conditions were crucial factors in determining a population's physiological and psychological tendencies. The Greek author was perhaps writing in a patriotic vein when he claimed that the temperate climate that caused temperance of character also explained the "mental flabbiness and cowardice" of Asians—whom the Greeks had fought, and who were "less warlike than Europeans and tamer of spirit." This was, according to the Hippocratic author, because Asians were not exposed to the changes and "mental stimulation which sharpen tempers and induce recklessness and hot-headedness." An unchanging monarchical rule, he thought, led to "stagnation," while self-rule kept minds and people active.

The temperaments characteristic of a people explained the mores of a given culture, determining, for instance, a certain group's propensity to overeat, or to exercise. The Hippocratic doctor—like Hippocrates himself—was more concerned with observing correlations than with identifying strict causes. Prognosis mattered more urgently than diagnosis (there is a treatise called *Prognostics* in the Hippocratic Corpus), and he was keen to stress that the course of an illness was in fact a process of natural readjustment of the organism to its environment. This is why the author of *Airs, Waters, and Places* pointed out that a competent doctor had to be aware of these environmental conditions: in this way he could recognize which illnesses were endemic—due to local, natural factors—and which were due to a broader range of causes.

For instance, in a city exposed to hot winds, yet protected from southerly currents, it was likely that water would be abundant, albeit shallow, hot in summer, and cold in winter. There, the inhabitants were likely to have "a humid brain, full of phlegm," and their bowels were subject to disruption by the phlegm that descended from the head. They were "neither great eaters nor great drinkers"; indeed, people with a weak brain did not hold their drink too well. In such an environment, there would be a high incidence of miscarriages induced by illness; babies would be prone to frequent convulsions, asthma, and symptoms pertaining to the "sacred disease." Men, for their part, would tend to suffer from "dysentery, diarrhoea, intermittent fevers, prolonged wintry fevers, pustules which would erupt at night, haemorrhoids." On the other hand there would be few cases of "pleurisy, bronchitis, acute fever, or any so-called acute illness," thought to be rare in people with humid bowels. In these cities, inhabitants over the age of fifty were prone to hemiplegia, a condition that resulted from a combination of the phlegm that literally descended into the organism from the brain, and shocks either of heat or cold to the head. The summer could produce "epidemics of dysentery, diarrhoea and long quartan fever"; and if those went on for too long, people with "large, stiff spleens, and hard, thin, hot stomachs," and emaciated "shoulders, collarbones and faces," tended "to develop dropsies that result in death." The simple change of seasons produced a range of illnesses to which these inhabitants were susceptible. The temperaments formed by humours were not fixed entities: they were rather an aspect of nature's variability.

4. Prescriptions and Priests

I T IS INTUITIVELY PLAUSIBLE that the organism should be an integral part of the environment, that complex organisms should be constructed out of simple ones, and that illness should be an imbalance between the elements that constitute the healthy, functioning organism. But it took some further, rational thought to turn these intuitions into a full-blown theory whose main characteristic was its immediate applicability. It was on the basis of this relatively simple theory of humours that, against the overly hot and humid stomach accompanying unstoppable diarrhea, for instance, the doctor would recommend the stomach's cooling and drying, with foods such as barley cakes, boiled fish and birds, vegetables like boiled beets in vinegar, and some dark, dry wine.[16] Similarly, baths helped to cure pneumonia, since they concocted and brought up sputum (which we call phlegm); they were diuretic and moistened the nostrils.[17] Illnesses caused by an excess of exercise could be cured by rest; those due to tension, by relaxation. Prescriptions could be commonsensical, and some of them survive to this day, simply because they work.

In a short Hippocratic treatise entitled *A Regimen for Health,* one may read that "fat people who want to reduce should take their exercise on an empty stomach and sit down to their food out of breath. They should before eating drink some undiluted wine, not too cold, and their meat should be dished up with sesame seeds or seasoning and such-like things. The meat should also be fat as the smallest quantity of this is filling. They should take only one meal a day, go without baths, sleep on hard beds and walk about with as little clothing as may be. Thin people who want to get fat should do exactly the opposite and never take exercise on an empty stomach."[18] The treatise contained a large number of guidelines for the treatment of general ailments; some of the recipes still seem applicable, and others less so. We are told that "in all diseases which affect the lungs and sides, sputum should be brought up early and, in appearance, the yellow matter should be thoroughly mixed with the sputum";[19] and that "to drink a mixture of honey and water throughout an illness caused by an acute disease is generally less suitable for those with bitter bile and enlarged viscera than it is for those who have not these things"; while "softening of the lungs and expectoration of sputum is produced by a greater dilution of honey," although "honey and water is generally acknowledged to enfeeble those who drink it and, for this reason, it has acquired a reputation for hastening death."[20]

Many things, seen and unseen, hastened death. Humoural theory functioned well as a general explanatory framework, but diagnosis was not always clear, prognosis not always hopeful, and treatment rarely effective enough for full recovery. The technique of bleeding, or of inducing violent vomiting, based on the notion that the ill organism should be purged of its noxious humours, occasionally worked. But it could also kill. The treatments tended to be harsh, painful, and dangerous. Illnesses were rarely straightforward or preventable, and few adequate treatments were available for the worst ones. (Medicine has not changed very much in this respect.) In a Hippocratic text called *Epidemics,* one can read case histories, accounts of the day-by-day evolution of a patient's condition. One woman "who suffered from sore throat," whose voice was "becoming indistinct" and whose tongue was "red and parched," for instance, suffered from "shivering; high fever" on the first day of examination. On the third day: "rigor, high fever; a hard reddish swelling on either side of the neck down to the chest, extremities cold and livid, respiration superficial. What she drank was regurgitated through the nostrils and she was unable to swallow. Stools and urine suppressed." On the fourth day, "all symptoms more pronounced." And on the fifth: "died."[21] Many such cases, minutely and impressively described, end with death. Even armed with their humoural theory, physicians often were unable to help. Reason was not all-powerful.

And so the gods were still helping people live and die; when a case was desperate, their effect was more potent than could be that of a doctor. The epidemic that struck Greece in 480 BC forced Xerxes, the king of Persia, to retreat with his men after some 300,000 died of it. Fifty years later, the great Athenian statesman Pericles died in the plague that swept the city. Doctors could do nothing against such events. But priests invoked the deities they served to help those in distress. Animals were slaughtered on altars and offerings brought by the faithful to monumental temples, as they always had been throughout the world.

Asklepios, the Greek god of healing and medicine, was one of the most popular gods in ancient Greece; the Romans would venerate him too, under the name Aesculapius, after Rome was struck with a plague in the late third century BC. He was known as the son of a mortal woman by Apollo, the great god of the sun, poetry, and music, whose lyre-playing charmed all and cured ailing souls. The original Hippocratic Oath, written in the fifth century BC, was explicitly addressed to both Apollo and Asklepios, but it was to the latter that the ill, the lame, and the convalescent flocked in the hope of a rapid cure. No Hippocratic doctor would have denied the god his powers, and indeed, Hippocratic doctors were called Asklepiads because they were healers and thus followers of Asklepios. The entry of Asklepios into the pantheon was gradual: in Homer he was merely a mortal prince, albeit a heroic healer. In Hesiod (probably some decades later), he had become known as Apollo's son and acquired the capacity to resuscitate the dead from Athena's gift to him of the Medusa's life-giving blood. As a result, Zeus had struck him dead with a lightning bolt.[22]

But the caring, healing Asklepios was too uniquely important not to be resuscitated in turn, as a full god; and gods do not die. From at least the sixth century BC to the second century AD, thousands upon thousands were to visit Asklepios's temples, the Asklepieia—especially the one in Epidaurus, his birthplace, which had become the most important healing center in antiquity, well positioned near spring waters. A physician at wit's end could advise the patient to spend a night on the temple's ample, peaceful grounds, in the hope that the hereditary priests might succeed where doctors had failed. The god exerted his power especially during sleep, so visitors who had been allowed in, after a ritual bath, settled for the night in a special sleeping area called the *ábaton*. The lucky ones might be visited in their dreams by the god himself or by his snake, the *drakón*, a long, harmless, tree-climbing constrictor whose tongue was believed to heal wounds. (It still roams Europe as the *Elaphe longissima.*) The luckiest awoke in health. One Anticrates of Cnidos had been blinded by a

Fig. 2. Statue of Asklepios: Roman copy, in Uffizi Gallery, Florence.

spear that remained stuck in his face, and after a nocturnal vision of the god extracting the spear, he was entirely cured when he awoke the next morning. Another man had a wound on his toe healed in his sleep by the *drakón*'s tongue.[23] There was also room on the temple's sacred grounds for water treatments, pharmaceutical herbs, and scalpels; and the priests might prepare medications and perform surgery.

Later, in the first century AD, the Roman polymath Pliny the Elder would report in his *Natural History* that patients who had recovered should "inscribe in the temple of that god an account of the help that they had received, so that afterwards similar remedies might be enjoyed." Pliny went on to mention a rumor according to which Hippocrates copied out these inscriptions and used them to found "that branch of medicine called 'clinical.'" The rumor, which derived from the Roman historian Varro, of the first century BC, was ill-founded: the shrine to Asklepios on Kos does not seem to have existed in Hippocrates's lifetime. Rationalist medicine was not a hoax.[24]

But Pliny himself listed hundreds of magical beliefs in his *Natural History*, and was capable of writing, for instance: "I find that a heavy cold clears up if the sufferer kisses a mule's muzzle."[25] In fact, the line between medical, natural care and miraculous, supernatural cure remained a thin one for a long time. The *drakón,* coiled around a staff, figured as the emblem of Asklepios. Pharmacies today use a similar emblem on their storefronts, though it differs from that of Asklepios in that it consists of a staff around which two serpents are coiled, rather than one. This emblem is called the caduceus, and is the symbol of another deity: Hermes, the youth with winged feet, known by the Romans as Mercury—the god of travelers, messengers, information, and alchemy. By the sixteenth century, physicians, apothecaries, and alchemists had adopted it, replacing the original single-snake emblem of Asklepios with the double-snake caduceus. The roles of Asklepios and Mercury had become conflated. Perhaps that is because doctors helped patients travel from one state to another, from the state of illness to the state of health, or, all too often, from the land of the living to the land of the dead. They dealt with physiological, humoural transformation, of health into illness or illness into health, just as apothecaries and alchemists transformed physical elements into remedies.

The outcome of such transformations was not, and still is not, always predictable; the capacity to cure was never merely a matter of medical expertise, of Hippocratic *technè*. When they are effective, doctors seem priestlike; but when they are not, they reveal themselves to be ordinary humans. When we are physically well, and in control of our rational faculties, we are potential doctors; but it takes little to turn any of us into patients. The best doctors are those who recognize their own fragility, who are closer to fallible, ailing humans than to the gods. Medicine developed as a collective, ordinary effort at gathering and collecting information. It depended on the anonymous messengers who, starting in classical Greece, communicated observations about the complex, mysterious workings of body and mind. The collective nature of the Hippocratic corpus therefore does not detract from the authority of Hippocrates himself. In fact, it is the transmissibility of these early Greek texts that ensured their survival.

5. From Greece to Alexandria

THE THEORY OF the four humours as it appears in the Hippocratic Corpus would be transmitted beyond Greece and beyond the fifth century BC, especially by the physician and scholar Galen of Pergamon. For 1,500 years after his death, western and middle eastern conceptions of the body, anatomy, physiology, illness, and health would remain "Galenic." Galen, a Roman citizen, was born in AD 131 in the refined, cosmopolitan city of Pergamon in Asia Minor. He wrote (in Greek) and practiced on the basis of the great Hippocratic corpus as well as of treatises of Plato and Aristotle, all of whom he adopted, commented on, and revised for his own purposes. But he himself never proved, in the modern sense of the word, the validity of the theory of humours. Nor had the Hippocratics:

although they were remarkably precise, they had not actually witnessed the concoction of humours flowing throughout the body's organs.

For a long time, textual transmission alone ensured the authority of the theory—sometimes against the odds. First conceived as essences, as a meeting point between our bodies and the world, nature, and the universe, humours were a construct rather than the strict outcome of observation. It is probable, though, that the idea of humours first emerged out of the visible bodily fluids: blood, sweat, tears, pus, saliva, bile, urine, milk, and sperm—the only substances in our internal organism that can appear on its surface.

As it happens, most of us would rather not pay attention to these substances. Secretions are, precisely, secret. They seem intimate, not for public view; and their appearance on the surface of the skin at the wrong moment and the wrong place can typically cause embarrassment, unease, revulsion, or even horror. Most of the time, the complex, mucous, bloody mess inside ourselves seems separate from the selves we think ourselves to be. We leave it up to professionals—nurses, physicians, researchers—to treat wounds and to investigate the body's functions. It is true that what is hidden from view fascinates as much as it disquiets. But, however curious we might be, we like our skin best, and we like it smooth. We are removed from our innards, at once conscious of our embodiment and afraid of it. Oddly enough, it is our very rationality—our very capacity to look at ourselves—that makes us forget our own humoural guts.

Humours bring—are—life. The body contains them, enclosed by the skin. When they spill out, either they are inappropriate in a social world where we must hide even our skin; or they spell death. But in life, they have always filled the gap between the visible and the invisible. Dissections that would have allowed for better visibility of the body were forbidden in ancient Greece, insofar as the body was the soul's container: damage to the contents was thought to impede a peaceful death, and so provoke the birth of monsters. Empirical, anatomical research on the human body—the close, detached observation of innards—does require its desacralization. The dissection of animals, on the other hand, was never a problem. The blood of other species is less gory and less repelling, though also less fascinating than that of humans. Aristotle, for whom anatomy was an integral part of philosophy, performed from an early age countless dissections on animals including monkeys. The idea of dissecting a human body for the sake of learning from it did of course occur. The ancient

Egyptians, especially, had practiced a sort of dissection when they em-
balmed their dead, although since embalmers seem to have remained separate from physicians, they drew few medical conclusions from this
treatment of their deceased. But the Hippocratic physicians, who had adopted some of the Egyptian methods of treating a wound (such as versions
of the effective honey salve), managed to make their acute observations
without seeing very far inside. They reconstructed anatomy on the basis
of wounds and operations, and imagined physiology in terms of the humoural framework.

And yet elaborate dissections on the human body would eventually
take place—not on the Greek peninsula but on the Egyptian coast, in the
great Hellenistic city of Alexandria, about 200 years after Hippocrates's
lifetime. Some Greek physicians acquired a reputation as practitioners not
just of the dissection of human corpses but also—according to the account in the first century AD by the Roman polymath Aulus Cornelius
Celsus in his great, and only extant work *De medicina*—as vivisectors
who "laid open men whilst alive."[26]

It was here, too, that the Hippocratic writings were gathered into the
Corpus we still know today. Alexandria was certainly one of the most sophisticated, rich, and cosmopolitan cities in the world at that point. It had
been envisioned by Alexander the Great—Aristotle's most famous pupil—
as his capital, on conquered Persian lands. The city was built on the grid
scheme inaugurated in the fifth century BC by Hippodamus of Miletus
(another Greek innovator, whose idea for urban order survives to this
day), a few years before Alexander's death in Babylon in 323 BC. Happily
positioned at the heart of a fertile land, the city was successful in trade
within the Mediterranean, with the rest of Africa, and with markets as
far-flung as India. It quickly grew into an economic and cultural center,
whose inhabitants included not only Egyptians and Greeks but also Jews,
Arabs, Phoenicians, Nabataeans, and Indians. Its monuments, palaces, and
temples were grandiose. Its famous lighthouse on Pharos, in the city's
harbor, was begun around 290 BC by Alexander's general Ptolemy I Soter,
who, fifteen years earlier, had instated himself as king of Egypt; it was to
become one of the seven wonders of the world, proudly welcoming visitors to the metropolis. (An earthquake would bring it down in 1303.)

But the library, also built by Ptolemy, was especially renowned. It was
a part of the city's *Museion,* a scholarly academy attached to the Royal Pal-
ace and to its temples, led by a priestly director appointed by the Ptolemies,

and conceived very much as are today's centers for advanced academic research. The *Museion* attracted minds of the highest caliber, such as Archimedes and Euclid. Ptolemy had commissioned a fellow pupil of Aristotle, Demetrius of Phaleron, to act as its first librarian, with the mission to transplant to conquered Egypt both the Aristotelian Lyceum and the Platonic Academy, no less. (It was Demetrius, too, who, it is said, commissioned seventy-two translators of the Bible into Greek—thus the Septuagint.)

The Library eventually became an extraordinary collection, one of the most substantial in antiquity. Figures vary, but it is estimated that there were about half a million scrolls, a majority of which contained more than one work. The Ptolemies formed the library by storing, copying, and translating texts from a wide range of provenances—they were written in Greek, Hebrew, Syriac, Egyptian, Sanskrit. Galen tells us that one means of acquisition was the systematic search of all ships arriving in the city harbor: those transporting texts were detained, texts were copied and translated, the originals kept and the copies sometimes (but only sometimes) returned to their owners. Among these works, shelved in those spectacular stacks for use by select students, researchers, and professionals, were the Hippocratic texts. Sixteen of these would be chosen as the core sources for teaching in the school of medicine which, from its basis in the *Museion* and Library, would form physicians for centuries.

6. The Naked Eye

I T IS THE NAMES of Herophilus and Erasistratus, in particular, that are most immediately associated with the so-called Alexandrian school. In contrast to local Egyptian doctors, who tended to be specialists, they were general practitioners whose anatomical research revealed aspects of the body that the Hippocratic corpus had not brought to light. They owed their research in part to Ptolemy I Soter himself, who, as a veritable patron of learning, believed that scholarship required state support and that the political prestige and authority of his reign could only benefit from it. At any rate, it was in the name of scholarship that he authorized the human dissections and vivisections that Celsus reports they practiced. Only condemned criminals could be used in these practices, which were justified by Celsus on the grounds that they alone allowed one to understand how organs and viscera worked, and to establish what were the causes of pain and thus its possible remedies. Vivisection perhaps seemed cruel and gruesome, but, he wondered, was it really so terrible to hurt a small number of criminals for the sake of finding cures for the long term, and for a large number of good people? Whether the Ptolemies used such a justification or not, anatomical knowledge sprouted from the flaying of outlaws.

No firsthand material from the work of either Herophilus or Erasistratus survives. Everything we know about them derives from the details preserved in secondhand accounts. Those of Celsus are crucial, and there are vivid passages about them in texts of the first- and second-century physicians Soranus and Rufus of Ephesus, of Pliny the Elder, and of the Carthaginian Christian Tertullian among others. Galen's accounts, however, are the richest of all, since he used his analysis of their work as fodder for his own thoughts on the body, both visible and invisible. It is impossible to understand Hippocratic and Alexandrian medicine without Galen.

Herophilus was a pupil of Praxagoras, a physician from Kos who had subscribed to humoural theory—although he had believed there were eleven rather than four humours. According to Galen, Herophilus was an Asklepiad, and recorded as an esteemed clinician and writer on medical subjects. As was the norm for physicians from the very birth of medicine as a *tekhnè* on, he probably traveled around the region, selling his expertise. Erasistratus, for his part, was a pupil of the Stoic Chrysippus from Cnidus, across the way from Kos. Both physicians were thus schooled in the Hippocratic tradition. But their emphasis on human anatomy was a notable departure from it. They were interested in structures more than in the diagnosis and classification of disease or in the treatment of patients. Humours were, in fact, of no use to them at all. They were independent-minded; and while they were under the huge influence of the recently deceased Aristotle (he died in 322 BC), they also examined and often dared to contradict his theories.

Aristotle had believed that the seat of sense and intellect was the heart—not the brain, as Plato, his teacher, believed. Herophilus followed Plato in placing the rational soul in the brain. He also described the brain's four ventricles: he thought that each one housed a part of the soul. The rational soul was seated in the posterior, fourth ventricle. Most famously, and to the great satisfaction of Galen, he observed that nerves (*neura* in Greek) were rooted not, as Aristotle had stated, in the heart, but in the cerebellum and spinal cord, and that they differed from arteries and veins.

Nerves transmitted "sensation and voluntary motion to all the limbs of the animal, just as the heart is the source of the arteries," as Galen explained.[27] A simple, rather brutal experiment proved the point: "when the ventricles of the brain have been pressed or wounded, the whole animal immediately becomes stupefied, but neither the motion in the arteries nor that in the heart is destroyed"; and exactly the inverse was true when one compressed the heart. This version of nerve function was evidently closer to the truth than the Aristotelian one; and Galen did manage to prove that the brain, not the heart, controlled both movement and speech. He even performed a public experiment in which he cut the laryngeal nerves of animals who then fell quite silent (and remained alive). Still, nerves remained poorly understood: Galen thought that they were related to ligaments and tendons, while tendons, in his words, were "the nerve-like termination of a muscle, the product of ligament and nerve."[28]

For a very long time, up until the seventeenth century, nerves were believed to be hollow and to contain *pneuma:* the air, breath, spirit, or vital principle which was conceived so early on in the history of science, and which, for Aristotle, was the instrument—*organon*—of the heart-based soul. It was Empedocles who had first envisioned within the body channels or pipes, or *poroi,* through which traveled the blood that was essential to thought along with the *pneuma.*[29] Praxagoras himself was convinced that arteries contained *pneuma,* not blood, and he thought, as Galen recounted, that "as the arteries advance and divide they become constricted and change into nerves."[30] But Herophilus observed anew, and he acutely concluded that in fact there existed two sorts of nerves: motor and sensitive. Only the sensitive nerves, which were responsible for all sense-perception, might, he thought, have contained the vital spirit or *pneuma.* (At any rate the optic nerves—first noticed by Alcmaeon—connecting brain and eye contained it, and Galen, who reports on this in his *On the Usefulness of the Parts of the Body,* assumed that Herophilus meant to generalize from these to all sensory nerves.) Herophilus was also the first to conclude that motor nerves, which originated in the brain and spinal cord and were responsible for all physical motion, were hard and not hollow. Galen, by contrast, was certain that all nerves, not just sensitive nerves, were hollow, and that they harbored canals transporting *pneuma.* The conception of nerves as hollow would actually prove long-lived: it was still current in the seventeenth century.

There is no end to the difficulty of establishing how the conscious will, mental and intangible as it is, produces movement at all. It is therefore understandably tempting to imagine, on the basis of the body's visible channels, how matter travels through them, enabling the communication between key organs and limbs—between thought, intention, or emotion, and word, movement, or action. This is arguably why *pneuma* and humours were outlined in the first place: problematic as it is to pin down the essence of an individual, the unseen, flowing substances at its core did just that. They bridged the soul and the body, the invisible and the visible. But identifying the body's structure with the naked eye is at least as arduous as imagining how the structure actually functions—it was arduous then, and remains so today. Herophilus, in this respect, achieved a great deal: apart from being on record as the first to establish a connection between the heart and the veins and arteries, he was the first to have suggested that the pulse corresponded to the heart. He also suggested that the pulse bore a relation to fever, that most central of symptoms.

7. The Breath of Life

ERASISTRATUS HAD even more of a penchant for speculation than did Herophilus, and he refined the pulse theory, the ventricular theory, and the old doctrine of *pneuma*, uniting them, as far as one can tell, in a rather daring synthesis. We have Galen's detailed, plentiful, and highly critical reports to remind us of his sophistication: he described Erasistratus's view of the heart as "the source of arteries and veins" (though nerves too, he believed, were made of arteries and veins)[31] and as an organ equipped with membranes, some of which drew "matter" in, and others which pushed it out. Erasistratus had taken from Herophilus's master Praxagoras the idea that veins and arteries started in the heart, which pumped blood from the right ventricle into the lungs, via the veins, and vital *pneuma* from the left ventricle into the whole body, via the arteries. (He did not, however, consider Praxagoras's belief that nerves, too, originated in the heart.) Of course, no *pneuma* could be seen when one cut open an artery, as Galen objected: the blood that gushed out actually overflowed from the veins, from which had previously poured out the invisible but dense *pneuma*. This substance, suggested Erasistratus, was breathed in along with the outside air, entering the body through the nose as "vital heat," and traveling to the heart; some of it passed through the brain. So a slower pulse—understood as the beat of the arteries—was due to low air intake; and minimal air intake caused the *pneuma* not to be dense enough to ensure the injection of heat into the organism. The result was death.

Invisible but distributed throughout the body, *pneuma* was the air itself, carried by the arteries via the heart and responsible for the body's innate heat. If the organs were the solid elements of the organism, and humours its liquid element, then the *pneuma* was the gaseous, airy element. It was also the material substance at the basis of perception and sensation, what made consciousness possible at all. It was, as Galen put it, "the soul's first instrument for all the sensations of the animal and for its voluntary motions" as well.[32] It was partly breathed in by arteries that were part of a network of vessels at the base of the brain, the *rete mirabile,* or miraculous network. (Galen had observed this network on dissected animals—probably oxen or sheep—but, as would emerge some 1,400 years later, it did not actually exist in human beings.)

Pneuma has an ancient pedigree. Alcmaeon had been one of the first to suggest that it circulated within the arteries, Empedocles had discussed it, and it was present in Plato's *Timaeus.* But it had generally been considered inherent in, not separate from, the blood. Aristotle was the first to understand it as a fifth element of sorts—literally a *quintessence*—thanks to which the organism was a coherent whole. For him, *pneuma* and humours worked in tandem, within the same channels. The Stoics, for whom pneuma permeated the universe as well as the body, agreed. So, in the first century AD, did the members of the so-called Pneumatic school, of which Rufus of Ephesus was a member.

Galen adopted this notion that *pneuma* was produced by the humours themselves and traveled in the body together with them. He believed that the veins originated not in the heart, as the two Alexandrians had thought, but in the liver; and it was in the liver that the concoction of *pneuma* began. At this stage, it was called "natural" and was the instrument of the so-called appetitive soul. It then coursed through the veins to the heart, and entered the arteries by traversing what Galen thought was a porous septum dividing one heart ventricle from the other. In the arteries, he explained, the "natural" *pneuma* was refined, through "inhalation" and the "vaporization of humours," into the "vital" *pneuma,* which was the instrument of the sensitive soul. Finally, once it reached the brain, the "vital" *pneuma* was further refined into the "psychic" *pneuma,* which coursed

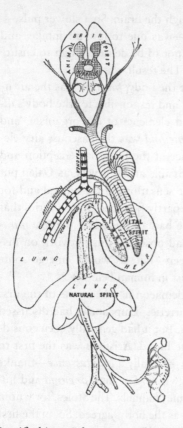

Fig. 3. Diagram showing the course of *pneuma* through the organs. From Charles Singer, *The Evolution of Anatomy* (New York: Knopf, 1925).

through the nerves and was the instrument of the rational soul. All *pneuma* transported particles, or *pneumata*, which were natural, vital, and animal "spirits"—liminal elements that were essential to the humoural, hydraulic metabolism.

Each type of *pneuma*, along with its spirits, thus served a particular soul. The sensitive soul was in charge of regulating the consistency and temperature of *pneuma*, while nerves were the conduits through which the rational soul enacted its commands to the body. Aristotelians and Stoics were wrongheaded, wrote the ever-vehement Galen, in their belief that the heart's "being situated in the middle of the animal" justified it as "the source of everything." There were indeed no "scientific *(epistèmonikos)* or sufficient" premises to suppose that it was "the source of both sensation and voluntary motion." It could only be true that "the seat of the soul's governing part is enclosed in the brain, that of its spirited part in the heart, and that of its desiderative part in the liver."[33] There was a hierarchy of related functions within the soul, and each component of the humoural body played a key part in ensuring their operations.

It was obvious to Galen that humoural categories explained organic and psychological functions, that "bodies act upon and are acted upon by each other as the Warm, Cold, Moist and Dry." Clearly, the activities of "vein, liver, arteries, heart, alimentary canal, or any part" depended "upon the way in which the four qualities are blended."[34] Brain, heart, and liver were the organs—and nerves, arteries, and veins were the channels—of life, movement, sentience, and thought. The innate heat, if it was "well blended and moderately moist," generated the blood (not the humour blood). When the innate heat was not "in proper proportion," then all the other humours were produced. Phlegm arose out of "phlegmatic foods" when they were first "cooked" in the stomach, where it turned into blood. That was why "there exists no specific organ for the purge of phlegm"; whereas the bladder flushed out yellow bile, and the spleen, black bile, both of which had been generated in the liver.[35] But the environment mattered: warm foods generated more bile; cold foods generated more phlegm. It was the same process with warmer or colder "periods of life," with "occupations also, localities and seasons," with diseases, and indeed with individual organisms and temperaments.

Galen devoted a treatise to the discussion of temperaments: *De temperamentis (Peri kraseon* in the Greek). Here, he formalized the scheme of correspondence between types and elements, explaining how the elements and qualities constitutive of humours accounted for fixed temperaments, and not only for emotions or for fluid states of health or illness, as was the case with the Hippocratics. Galen depicted nine types: four corresponding to each element, and four others to each combination of elements; the ninth corresponded to a type in which all elements were equally present.[36] Elsewhere, Galen adopted and developed the Hippocratic idea that local water and food determined to a considerable extent what sorts of temperaments would be most common in a given environment. (He believed that the author *of Airs, Waters, Places* was Hippocrates himself.) In a treatise he wrote quite late in life—sometime after AD 193—called *The Faculties of the Soul,*[37] he described how the combination of the environmental elements inflected the temperaments of local inhabitants. Asians, for instance, had a better character than Europeans because in Asia the seasons varied little and everything there was bigger, softer, and more beautiful; and people who lived under the Bear constellation were of a disposition that was antipodean to that of inhabitants of areas exposed to the sun.

This, he thought, was due to the fact that food—hot, cold, dry, humid—was first absorbed in the stomach, where it underwent an initial transformation into chyle. It then traveled into the veins that connected the liver to the stomach, producing the humours that fed all the organs, including brain, heart, and liver. This was why the faculties of the soul—all housed in the brain—depended on the corporeal humours produced out of the four qualities of hot, cold, dry, or humid. Aristotle himself had used humoural terms, and Galen referred copiously to his accounts[38] of how the composition of the temperaments differed from animal to animal as well as from individual to individual; this variation explained why certain animals were stronger, faster, or cleverer than others. For example, beasts whose temperament consisted of a high proportion of humidity were bound to be rather cowardly, because fear tended to cool down the body and water was hardened by the cold. Those whose brain was humid and cold tended to be lethargic, to have blunted senses, and to catch colds easily. The body's temperature depended on its material constituents, and temperature, in turn, determined which emotion dominated.

The faculties of the soul determined the body in a direct way. Even physiognomical traits—such as skin color, facial features, and sound of voice—were manifestations of the temperaments. The humoural system was therefore broadly deterministic; and it became even more so with Galen. It assumed that all emotional as well as all cognitive faculties had a physical basis; and it purported to explain why animals—and people—behaved as they did. The Alexandrian anatomists' rejection of humours thus did not stick, and many of the theoretical innovations conceived in Alexandria were eventually left behind. Yet the Alexandrian school is important because it produced new methods and assumptions, impressive surgical instruments and procedures (like trepanations), and a new, precise anatomical terminology.

Galen, to be sure, was one of the chief witnesses and students of the Alexandrians, and it is in great part thanks to him that their ideas would be known at all to subsequent generations. Although Galen lived 300 years after they were active, their legacy was very much alive in his mind, and, as an attentive reader, he was in constant dialogue with them. He mixed keen observation with brilliant, sometimes abstract theory, analyzing the functions of minute anatomical parts, smoothly brushing over areas of darkness with ingenious and at times aggressive rhetoric.

Galen was also much more of a clinical practitioner than the Alexandrians ever were; and surgery apart, their theoretical innovations were not always easy to integrate with the ordinary clinical care Hippocratics had cultivated. There was a gap between the two. To imagine essences and humours for the sake of healing was undeniably different from examining the innards of a human being, carved open and laid out to view on an operating table, for the sake of knowledge. A curious anatomist unafraid of opening up bodies was not necessarily the good doctor that a sick person required.

Alongside the new knowledge that emerged out of the desire on the part of a few talented individuals to understand the body's inner structures, there remained the simpler, common need to alleviate anxieties about the body, its frailty, and its mortality; and so long-standing cults whose functions had always included those of ministering to the sick remained in place. During the five centuries or so of the Alexandrian heyday, the local Egyptian worship of Sarapis—invented by Ptolemy I Soter himself—and of Isis, Osiris, Imhotep and Thoth was still strong. Thoth was similar to Hermes, and Imhotep to Asklepios. Knowledge of invisible parts could not kill faith, and the religious imagination remained powerful.

8. Alexandrian Sects and Galenic Travels

I N THE MINDS of the anatomists, however, knowledge and faith had been quite separate. Herophilus believed that medicine must rely on logical deductions from the observation of the visible body. Experience passed on through the generations mattered much less. His approach quickly became associated with that of a new breed of Rationalists. In opposition to them stood the Empirics, headed at first by a pupil of Herophilus called Philinos of Kos. They sprang to the defense of the idea that individual cases—such as those described in the newly gathered Hippocratic corpus—were reliable enough to serve the medical profession. For the medical community, reason was, in effect, of little use. In reaction against this faction, the Rationalists became more extreme; some turned into so-called Dogmatists. Alexandria gradually became a sophisticated ideological battleground between various medical sects, all of which held differing beliefs about what it was possible to find out about the workings of the human body. This was the medical world Galen was contending with in the second century, although all these sects claimed some sort of allegiance to Hippocratic ideas. Later Rationalists—many of them disciples of Herophilus—sought to reconcile Hippocratism with new anatomical knowledge, claiming that some causes of illness were hidden, but that one could discover them through a combination of reason and observation.

The Empirics, on the other hand, were skeptics: they held that the body was fundamentally incomprehensible, that dissections and vivisections led nowhere. Erasistratus was, in Galen's view, such an empiric. Causes of illness, apart from the obvious, commonsense ones (such as excessive cold or heat, fatigue, hunger, or thirst), were ultimately hidden and there was no point for doctors to try to uncover them. An empirical physician's sole responsibility was to treat the illness, which needed to be understood solely as a collection of symptoms: it could be known only through its *historia* (from the Greek *historein,* to observe) and in its resemblance to other recorded histories. For the Empirics, medicine was nothing more than an accumulation of experience: records of attentively transmitted individual cases were collected into a body of case histories or *anamnesis,* that led to a *diagnosis,* which supported the examination, or *autopsy* (literally "seeing with one's own eye" in Greek) of each patient.

The most successful sect of all was that of the Methodists, whose members were much sought after in fashionable Roman society. The Methodists were atomists who believed that the world consisted of particles: they were uninterested in invisible substances like humours, indifferent (like the Empirics) to physiology or speculation, qualities and faculties: health was simply a matter of mechanisms, and of particles. The atomist Asclepiades of Bithynia, in the first century BC, had established that the degree of constriction or relaxation of the body's channels or *poroi* alone determined the state of health of the organism. Other noteworthy members of this sect included Soranus of Ephesus, who, like his predecessors, left humours out of diagnosis and cure; melancholy, for instance, was a disease of the body, but it probably had little to do with black bile and had to be treated with psychological acumen. Galen thoroughly despised the Methodists for refusing to see in nature the structured order of which he was certain; but he reserved his sharpest criticism for the other sects, whose members had thrown away humours so thoughtlessly. Humours were such an evident factor in the body's processes, he wrote, that it was misguided to dismiss them. Galen wondered indignantly how an empiric—meaning Erasistratus—expected to "diagnose or cure diseases if he is entirely ignorant of what they are."[39]

Galen stood apart from each sect, adopting from each what he found valuable—its contributions to anatomy in particular—but never swearing allegiance to any of them. He was well-read, adept at theory as well

as observation, and respectful though openly critical of his great predecessors. But he believed that he should always have the last word. And
indeed it was Galen who would join Hippocrates as one of the two
greatest authorities for late antique, medieval, and Renaissance medicine. Humours would remain within the body, within medical texts, and
within medical practices. The theory would survive the onslaughts of
skeptics.

Still, the lessons of the Greek Alexandrians would not be forgotten.
Galen, their own critic, helped to ensure that their anatomical ventures
would bear fruit. But so did his immediate predecessors and teachers. Rufus of Ephesus, who studied in Alexandria in the middle of the first cen
tury AD, would in fact be the first to pass on the Alexandrian theories
concerning the vascular system, the nerves (which he also called *poroi*
rather than *neura)*, the muscles, and general anatomy. Marinus, the author of a twenty-volume *Anatomy* (now lost), followed. We know of him
only through Galen, who admired him, as well as his pupil Quintus. In
AD 151, the young Galen had in fact left his native Pergamon in search of
Quintus's most illustrious pupil, Numisianus, who was based in Corinth.
Galen's father, a successful architect called Nikon who had died two years
earlier, had literally dreamed of his son's becoming a doctor and had
pushed Galen in the direction he was now taking. Galen first went to
Smyrna, where he studied with another pupil of Numisianus called Pelops,
and set off for Corinth only later, in 152, to find the master. As it turned
out, however, Numisianus had moved to Alexandria. This is how Galen,
too, had sailed to Alexandria.

Galen would not have found the Library of Alexandria intact. Part of
it had been destroyed a good hundred years before, in 48 BC,[40] and a large
number of its substantial holdings may have been burned then. Most
pointedly for a medical student, dissections on humans had ceased to be
allowed when Alexandria had fallen into Roman hands. The physical
body, with its humours and secretions, was no longer accorded the status
of a mere mortal envelope, which anatomists could open up if they wished.
The dead now recovered their right to rest in peace. Greek doctors did
continue to practice and teach here, as in other centers of medical learning
such as Smyrna and Corinth, and comparative anatomy continued to
flourish—but again, with the support of the dissection of animals only.
Human innards could be reconstructed only by analogy with animals, by

inference from the visible body, and by inference from the substances that percolated to its surface from the humours.

When Galen arrived there, however, he still would have had the resources of the *Museion* to draw on. He probably studied the human skeletons that had remained on display since the days of the Ptolemies; and he certainly conducted dissections, many of them public, on a large number of animals—pigs and especially monkeys—the accounts of which fill many of the fifteen volumes of his important *On Anatomical Procedures* (a work that, incidentally, would remain unknown in the West until the sixteenth century). He also traveled around Egypt. When he returned home in 157, still a young man, he was a fully trained anatomist and physician whose reputation allowed him to open a clinical practice and to be appointed physician to the gladiators at the gymnasium attached to the local Asklepieion. Tending to their often severe, bloody wounds gave him plenty of opportunities for in-depth anatomical investigation of the living human body—a compensation of sorts for the impossibility of practicing actual dissections. However, a regional conflict involving Pergamon led to a temporary halt in the competitions for which the gladiators were employed. Galen was then a talented, ambitious thirty-year-old. Greece was impoverished. Rome, the capital and center of an empire at the height of its glory, was the place to be.

Galen proceeded to the imperial capital in about 161. Great fame and recognition had come quickly to him—deservedly so. But these also brought enemies. His difficult character was infamous. His natural haughtiness remains vividly apparent in his writings, and he had a generally choleric temperament, which he himself believed he had inherited from his mother (he tells us that she used to bite her servants) and which usually impelled him to invective against his peers. In 166, perhaps under pressure to lie low, and prompted by the outbreak of a plague in Rome that eventually would spread across the empire—it has since been called Galen's plague—he decided to return home. Again he attended to gladiators, and resumed his practice. A summons from Marcus Aurelius three years later to be physician to the soldiers at the military base in northern Aquileia—attacks against Germanic tribes were in preparation—turned instead, on Galen's own request, into an assignment to care for the emperor's young son Commodus in Rome. The job would last until the boy's adolescence. By now, though, Galen was successfully settled in the capital, and was able to devote much of his time to research and writing. Toward the end of the century, in 192, the corrupt rule of Commodus—he had developed into a brutish, unpopular emperor—came to an abrupt end when he was strangled by his own people. The chaos that ensued may have driven Galen out of town, and he may have spent his last years back in Pergamon. At any rate, he left at his death, aged seventy or so, an extremely large body of work, in Greek.

It was never a given that so many of his writings, or indeed so many other writings of antiquity, would be transmitted to future centuries. By late antiquity, the Library of Alexandria had, in fact, completely disappeared, and had acquired the legendary status often conferred on institutions or individuals celebrated in their lifetime but accidentally lost to posterity. The Roman emperor Caracalla would ransack the *Museion* in the early third century AD and the city was repeatedly the site of rebellions, occupations, and massacres for another hundred years or so. But the *Museion* probably housed its own libraries for use by its fellows, and it did continue to be a reputed center of higher learning well into the late fourth or early fifth century AD. At this point it disappears from the record: it was possibly a casualty of the early Christians' zeal, at a height during the reign of the emperor Theodosius, to destroy all traces of pagan worship.

By then, however, knowledge had already begun spreading elsewhere, and with it, rationalist medicine and philosophy, those twin disciplines whose concern it was to help people lead good, physically and morally healthy lives. As empires gave way to each other, the doctor's recommendations for a proper diet—in Greek, *diaita*, which means "mode of living"—continued to be based on humoural considerations.

9. Primordial Passions

F THE BIRTH of philosophy is contemporaneous with the beginning of medicine, that is in part because philosophers and doctors were both prone to reflect upon what defined a good life. Each was concerned with unraveling the passions, which all too often could disrupt its pursuit, causing illnesses and disrupting rational coherence. After the early, pre-Socratic efforts at understanding the world in terms of its natural components came the Socratic entreaty to "know thyself" for the sake of living a morally and physically good life, voiced just as the new, professional class of Hippocratic doctors was taking over from the priesthood. The story of how we have tried to understand the emotional, humoural, mortal, conscious creatures that we are began when reason became aware of itself. Rationalist doctors were, by definition, among those who made use of reason in order to understand that which escaped its control.

Hippocrates was the prime exemplar of this tradition, of course. There is an anecdote about him in which the king of Macedonia, Perdiccas II, fell grievously ill. The king thought he might be afflicted with phthisis, a popular designation for a set of symptoms generally pertaining to consumption, or tuberculosis. He summoned the famous Hippocrates, who traveled to the Macedonian court and began to observe the royal patient's symptoms. He took Perdiccas's pulse and realized that it accelerated whenever Phila, his late father's consort, entered the room. The conclusion was swift: this, clearly, was not phthisis. Anguished passion for Phila was instead the cause. A substantial narrative tradition based on this story line developed. It carried on as the medieval concept of *aegritudo amoris,* illness of love or erotic malady. In the Renaissance, it was called love-melancholy: black bile was considered to be a powerful factor in this affliction.

One version has Erasistratus as the doctor, summoned by the Syrian king Seleucis I to attend to his ailing son Antiochus. Plutarch recounts this episode in passing, writing that it was immediately clear to Erasistratus that Antiochus was afflicted with the "distemper" of love. In order to discover who was causing the prince's passion, the physician patiently watched over him, attentive to any of "the changes which he knew to be indicative of the inward passions and inclinations of the soul." And indeed, when Seleucis's young consort Stratonice entered the room, Erasistratus immediately noticed on his patient "Sappho's famous symptoms," as Plutarch puts it: "His voice faltered, his face flushed up, his eyes glanced stealthily, a sudden sweat broke out on his skin, the beatings of his heart were irregular and violent, and, unable to support the excess of his passion, he would sink into a state of faintness, prostration, and pallor." Because of the strong bond between himself and his father, Antiochus perceived his love as a curse, impossible and wrong. Afflicted with a combination of unfulfilled desire, shame, and despair, he decided to die, voluntarily starving himself and claiming to all that he was ill. But Seleucis, diplomatically informed of the situation by Erasistratus, ordered the re-marriage of his wife to his son. She became queen, and he king of the provinces of Upper Asia.[41]

But the point here is not the happy ending: it is rather that love could be viewed as an illness especially when it was at odds with reason, and that reason was the very tool which enabled physicians to make inferences from observation to the diagnosis of illnesses of mind and body. Erasistratus was able to conclude, on the basis of observation, that Antiochus's symptoms were caused by emotions which he wished his reason could combat—but which he actually could not control. Insofar as all passions were at odds with the prescriptions of reason, they could be considered diseases of the soul. They were actions of the body upon the passive soul—powerful, humoural events.

Galen, who proudly called himself a philosopher, had sophisticated ideas about the passions, their operations, and the types of behavior that usually accompanied them. In a treatise on the subject, *The Passions and Errors of the Soul,* he showed that in spite of their resemblance to error, passions were actually distinct from it. Error, he wrote, was a false opinion, while passion arose out of an irrational force. And we tended to be blind with regard to our own errors possibly because of self-love—for, as Plato had noticed, one was generally blind with regard to the object of love, whether this object was another or oneself. For Galen, the first step toward a relatively error-free life was to liberate oneself from one's passions—such as rage, anger, fear, grief, envy and excessive desire, love and hate. It was possible to fight off the worst of the most powerful passions, and to abstain from acting on their basis; but it took work. Galen, Stoic-like, recommended a long period of self-conscious, systematic efforts, the help of a wise and disinterested friend capable of giving honest criticism, and the cultivation of a sense of shame. Reason was the instrument at our disposal to fight passions: not to use this instrument, which humans were the only creatures to possess, was to behave like an animal.

Galen recounted how that most Hellenizing of second-century Roman emperors, Hadrian, once gouged out the eye of a servant with a stylus in a fit of rage. When, on realizing what he had done, Hadrian tried to make it up to the servant and asked him what he would wish for compensation, the servant replied that he wanted nothing more than to have his eye

Fig. 4. Jean-Auguste-Dominique Ingres, *Antiochus and Stratonice* (1840),
Musée Condé, Chantilly.

back. Had Hadrian been Greek, he might well have described himself as
invaded by *chole* when he poked out his servant's eye. But that would
have been no excuse. The way to betterment lay in the cultivation of cool-
headed, rational deliberation, which could and should be made to prevail
in the midst of passionate responses and humoural agitation.

It is puzzling, though, that we use our rational faculties both to devise
how to act and to reflect on our frequent inability to act as we intend. For
instance, we might be fully aware that it is wrong and messy to start an
affair with one's best friend's spouse, yet do just that. These are cases of
weakness of the will—for which Greek has one convenient word, *akrasia*.
The temperamental model passed on by Galen actually made room for
this element of human nature: it acknowledged that our behavior could be
a product of actions of the body, actions of the rational soul, or, as was
often the case, of a mixture of the two. Actions of the body—the
passions—were not always willed, because they were the product of physiological processes. They were rarely beneficial, but not necessarily negative. The humoural temperaments responsible for passions were
produced—concocted, as it were—in particular organs, specifically the
liver and the heart. Medieval and Renaissance philosophers would later
contribute their rather complex variations to this scheme.

10. Three Souls

G ALEN HAD BORROWED his psychology—in the sense of a science of the soul, rather than in the modern sense of the word—from Plato and Aristotle, whom he revised for his purposes and integrated with Hippocratic humouralism Aristotle had defined the soul as the form of the body and believed that all living creatures could be placed on a unique "scale of being." All, whether vegetal, animal, or human, shared what he called an appetitive soul, responsible for life's basic functions—breathing, nutrition, and reproduction. The corporeal soul, understood to be present in animals and humans, but not in plants, was responsible for all perception, sensation, imagination, and memory. And topmost in this hierarchy was the rational soul. Aristotle granted a part of the rational soul to beasts. Its highest aspects, however—those that were capable of contemplation, abstract thought, and free will—belonged to humans alone. Aristotle believed that the brain acted merely as the "refrigerator" of the blood, and that the rational soul must be located in the heart, the organ at the center of the body and the first, he thought, to take shape in animal embryos; the senses were connected to the heart, which processed all the perceptual information gathered by them. Galen had instead adopted Plato's craniocentric view, which would prevail in later centuries. But while Plato insisted that the rational soul was immortal and incorporeal, Galen believed that all three souls were material, and as mortal as flesh. This was why they could interact—why, when the body moved, the soul was affected, and inversely, why when the soul produced motion, the body stirred. It was in this way that the temperaments of the body transformed the functions of the soul.

The soul could also be subdivided into as many parts as there were functions. Aristotle had counted three souls; for Plato, there were two: the immortal, rational soul, and the less noble, sensitive, mortal, worldly soul. The latter operated in our emotional life, and Plato described it at great length in the *Timaeus,* which became a crucial source for Galen, consequently exerting its influence for nearly two millennia. There one reads that the sensitive soul was itself made out of two parts: in the heart there was the vital soul, "endowed with courage and passion" and involved in conflicts; in the liver lodged the vegetative soul, which "desires meats and drinks" and all things related to its "bodily nature." The former was akin to Aristotle's corporeal soul, while the latter resembled his appetitive soul.

In the thirteenth century, the great Christian theologian Thomas Aquinas, who adopted Galen and synthesized him with Aristotelian psychology and Christian assumptions about the immortal, immaterial rational soul, would dub the vital soul "irascible" and the vegetative soul "concupiscible." Concupiscence and irascibility each denoted a particular set of passions. The concupiscible power was our tendency to seek the good, the tempting, the beautiful, or the desirable and to avoid the bad, the repugnant, the ugly, and the dangerous; the irascible power was our tendency to fight whatever might block our way to obtaining a good or fleeing an evil. The concupiscible part of the soul was activated when one experienced desire or aversion, and all the emotions in between; and the irascible part was activated when one experienced a hope of obtaining something or anger at losing it, and all the emotions in between.

On this picture of the soul, reason was man's highest good, nobler than—and separate from—the flesh and its desires. Simply, it accounted well for most of the emotional conflicts one was likely to experience, from fear and longing to hope and disappointment; and it also explained why we were capable, or incapable, of judging situations and acting according to our judgments. Central to the so-called faculty psychology that developed during the later Middle Ages and was transmitted to the Renaissance was the idea that these souls of humans were connected to, and therefore influenced, each other. The sight of a beautiful body for example, modulated by the corporeal soul, could activate the appetitive soul—or concupiscible power—while the rational soul could bear on the desire provoked by the wondrous sight and induce the will either to act on it or to control it. This interconnection explained, then, why we were aware of our passions and emotional conflicts, and inversely, why our thought processes were sometimes troubled by the activity of the sensitive soul, thanks to which we were also aware of our sense perceptions. A precise physiology of emotion and cognition, first devised by Galen and refined in subsequent centuries, backed up the notion of the three souls' interdependence and mutual interference—further shaping the doctrine of humours and temperaments.

Humours, in Galen's scheme, relied on the *pneumata* or "spirits," those particles that traveled through the blood, from the liver to the heart via the veins, and from the heart to the brain via the arteries. The initially "natural" spirits, concocted in the liver along with the humours, became "vital" spirits in the heart, themselves refined into "animal" spirits by the cerebellum. At this stage, they were responsible for the transmission of sense perceptions to what Aquinas later called the common sense, separate from reason and will but connected to these highest of faculties. It was a given in faculty psychology that the brain's ventricles were the sites of the higher rational soul's functions: the two anterior lateral ventricles housed sense perception and the imagination (called *fantasia);* the middle ventricle housed reason, or thought; and the posterior ventricle was the seat of memory. The cerebellum, at the back of the brain, was responsible for motor functions. Certainly, this rational soul was the sole antidote to the "wild beast" that was the irrational faculty of the irascible soul. For instance, it was possible to learn how to control one's impulse to violently hit, stab, or—like Galen's choleric mother—bite a servant who had provoked one's anger, and instead to judge, calmly, whether and what sort of punishment was best suited to the wrong the servant might have committed.

But the passions of the irascible soul—of which anger was a central one—could be put to good use and could control the passions of the concupiscible soul. Unlike the irascible soul, which reacted to whatever provoked it, the concupiscible soul tended to abuse of those things—food, drink, sex—by which it was naturally aroused. It could also result in envy or lack of discretion. In any case, it had to be resisted from the onset; and the longer one waited, the worse was the prognosis. But to refrain from acting on impulse was only half the battle: it was a step above wildness but still not the mark of a sensible, good person. The goal was not to experience the passion at all.

The idea that one was better off by overcoming actions induced by the passions was central to the Stoic school of philosophy. Founded in the third century BC by the Greeks Chrysippus and Zeno, this school of thought was taken up once again by the Romans, especially Cicero and Seneca, whose influence would soar during the Renaissance. The Stoics believed that passions were an illness of the soul and that goodness—and therefore happiness—lay in the ability to disregard them entirely. Galen, however, was not suggesting that people should necessarily do so; children, after all, were bundles of passions to begin with, and it was unlikely that an adult should turn out to be the pure embodiment of reason.

The humoural doctrine presupposed that a passionless state would be unreachable in any case: the chemistry of digestion and concoction could

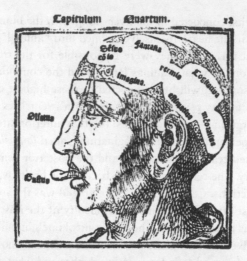

Fig. 5. Faculties of the mind, from *Congestiorum artificiose memorie* by Johann Rombach, 1533, p.12r. Here one can see the brain divided into: common sense *(sensus communis)*, imagination *(imaginatio)*, phantasia *(fantasia)*, judgment *(vis aestimativa)*, reasoning *(vis cogitativa)*, memory *(vis memorativa)*. The organs of smell and taste are also indicated.

not avoid triggering certain physiological and psychological reactions, and individual passions therefore had to be considered in the context of the overall temperament. A bilious person, say, might suffer from heat (yellow bile was a warm humour) and be more prone to anger in the summer than in the winter; and a phlegmatic person might be more slothful in the winter than in the summer (phlegm was a cold humour). Moreover, inherent in the division of appetites into irascible and concupiscible lay a recognition, however implicit, that passionate drives were phenomena distinct from emoted desires, more violent and less manageable. There were degrees of awareness, and degrees to the effectiveness of willpower. But it would have been confusing and extravagantly unrealistic to suppose that rational behavior could simply take the place of impulsive or instinctive action. Galen's ethics were marked by a concern—which had first been Aristotle's—to encourage the cultivation of moderation, at least in those who were so inclined. If each person really was born with a specific temperament, the process of betterment was bound to be difficult. But philosophy helped to identify the pitfalls, accompanying the practitioner of the medical arts as he examined the state of the bodily humours, periodically gazing up at the sky to survey the weather, and, on occasion, the position of the stars as well.

The Hippocratic doctrine of humours, passions, and temperaments survived its earlier passage from pre-Socratic Greece, in the seventh and sixth centuries BC, where reason had a different, less clear-cut role from that prescribed for it by Plato and Aristotle. And the humoural tradition continued to travel well, on the back of the growing, changing Hippocratic corpus that took shape in Alexandria. So did Galen's writings and commentaries on it. These would be translated, glossed, and broadly adopted by the Arabs and Persians of the Middle Ages, who, through the vagaries of conquest, inherited a large part of the classical scientific and philosophical corpus.

II

Essences:
The Classical Trail

(Eastern Middle Ages: Seventh to Twelfth Centuries)

The world was made new to him by a presentiment of endless processes filling the vast spaces planked out of his sight by that wordy ignorance which he had supposed to be knowledge.

GEORGE ELIOT[1]

Look to the body.

TERTULLIAN[2]

1. Byzantium

DESPITE HIS CONSIDERABLE TALENT and achievements, Galen was never able to found a school in Rome or Pergamon. It was later, in the Alexandria of the third and fourth centuries AD, that there emerged a Hippocratic-Galenic school, devoted to the recovery and study of ancient medical sources. About 200 years after Galen, Oribasius, another native of Pergamon, would compile a *Medical Collection* of writings. Those were mostly extracted from the *oeuvre* of Galen—who, to Oribasius, was highly authoritative precisely because of his reliance on the authoritative Hippocrates. Oribasius was celebrated in his lifetime and known as an iatrosophist—a professor of medicine. He had been a personal physician to the pagan Roman emperor Julian the Apostate and subsequently was exiled by the Christians before being called back to Rome because of his self-made reputation as a great doctor. By the fifth and sixth centuries, other iatrosophists in the Byzantine world were still teaching the Hippocratic-Galenic tradition, using the manual of selected extracts, editing and expanding on Hippocratic manuscripts only recently discovered.[3]

Only a few of these scholars, mostly commentators and teachers active in the sixth and seventh centuries, are known. They include figures bearing such geographically determined names as John of Alexandria, Paulus of Aegina, Stephanos of Athens, Aëtius of Amida, and Alexander of Tralles. But the school that had promoted Galenic Hippocratism and humoural medicine had a short life; it was already in slow decline between the fourth and the seventh centuries, and would on the whole not be resurrected for another 500 years. In the sixth century, a few Greek texts, including parts of the Galenic works that Oribasius had compiled, were translated into Latin in northern Italy, notably in Ravenna, the western Byzantine capital. These texts remained fundamental to medical practice, and the school did survive spottily in the Byzantine West, thanks to the few scholars and medical practitioners who knew Greek (many of them were of Greek origin) and to the Latin versions of Greek medicine passed on by Celsus.

The moribund state of the school is easy to glean from surviving documents. For instance, take Agnellus of Ravenna, an iatrosophist who taught in his native town of Ravenna. From his Latin lecture notes on Galen's *On Sects,* one can see that he did little besides restate the basic principles that would come to be associated with Galen and Hippocrates. His utterances concerned the nature of medicine, sects, elements, temperaments and humours, and not much else. He wrote that "the aim of the art of medicine is good health, the end is its attainment," just as Hippocrates had done, and intoned, "We speak of humour because it contains moisture and causes a moistening flow in our bodies. With these remarks ends the sixteenth lecture."[4] No new ideas emerged out of the old humoural fold. The medical student learned the basics: the anatomy and physiology of the healthy body, its pathologies, therapeutic methods, and the principles of dietetics and hygiene that helped prevent illness. But these were established, rather rigid categories that served to explain rather than to observe: they were not tools for the further exploration of the body, or starting points for new theories.

In the Byzantine East, meanwhile, Alexandria remained a famous magnet for scholars of all sorts; but it too was gradually losing its edge. The medical school was still reputed, but it had diminished since its heyday. An influential canon of Galenic and Hippocratic writings, consisting of fifteen or sixteen texts by Galen and about eleven works from the Hippocratic Corpus, had been gathered there by the sixth-century commentators for use in the school's curriculum. But here too, new theories were neither produced nor promoted. Erudite but often barren textual analysis was taking over from original scientific research and fresh philosophical speculation. Scholars were devoting their energies to logical and interpretive problems internal to texts. Exegesis of a broadly Aristotelian kind was the order of the day. The body was now visible only through the mediation of the established authorities.

John of Alexandria, for instance, wrote a *Commentary on Hippocrates's Epidemics VI,* in which he extracted quotation after quotation from the source, then glossing each one. Where the Hippocratic text read, for instance, "Venery helps diseases from phlegm," John explained the passage to his students in these words: "You have learned what venery *[lagneia]* is, namely sexual intercourse *[aphrodisia mixis]*. And it is a matter of dispute whether this intercourse has a warming or cooling effect, but it is accepted by all that it causes drying, seeing that there is a discharge of semen and a dispersion of vital tone. What do we say, then? That it both warms and cools, but it warms with respect to quality. That is why we see with this type of movement and agitation the body becoming warmer and more pungent, and hence also blood is discharged. But it cools with respect to substance, due to the discharge of both vital tone and semen and to the extensive dispersion. Hippocrates, then, had in mind the issue of quality when he said that venery stops diseases arising out of phlegm by making the matter thinner. But sexual intercourse should not be indulged in continually, since it makes the body grow colder and produces more phlegm in addition to what is already present. And Epicurus, dear beginner, rejects all intercourse, just as a philosopher does. However, it should be indulged in at the right time, as Hippocrates says, not when one is either too full or too empty."[5]

The Hippocratic text could certainly be obscure enough at times to warrant clarification and repetition. But critical comments like John's concerned issues of vocabulary and textual meaning exclusively. There was little reference to firsthand observations that might revise the interpretation of the masters' texts—even when, as in the matter of sex, such observations would have been rather easy to obtain. Courses in medicine seemed mainly to consist of exercises of this sort, despite the founding Hippocratics' emphasis on clinical practice. But medical texts were at least being produced, and some were more usable than others. One noteworthy guide was the *Treatise of Medicine*, or *Hypomnema*, by Paulus of Aegina: it was an extensive guide to health and a compendium of diseases, based on Hippocratic and Galenic medicine but rather fresh in tone, and full of dietary advice. Here one could learn, "When one has drunk largely, it is not proper to take much of any other food; but while drinking, one should eat boiled cabbage, and taste some sweetmeat, particularly almonds;"[6] or indeed, in the case of a headache from wine, "If the wine remain undigested, we must procure vomiting, by drinking tepid water; but if the headache remain after digestion, we must use cooling and repellent applications, such as rose-oil alone and with vinegar, or the juice of ivy, or of cabbage."[7]

Paulus probably trained as a doctor in Alexandria, in which case he might well have witnessed the Arab conquest in action. Rather than destroy the classical heritage that had survived in the lands they vanquished, the conquerors would instead take it on board—along with the theory of humours.

2. The Arab Conquest

ALEXANDRIA FELL to the Arabs in 641, barely ten years after the death of Muhammad in 632. The medical school, whose students were attending lectures such as those by John, ceased to function about eighty years later. The Umayyad caliph 'Umar II, in the early eighth century, had it transferred to newly conquered Antioch. (This was the great Syrian city—now in Turkey—founded by Seleucis I, who had called Erasistratus to the side of his son.) Al-Iskandariyah—the Arabic name for Alexandria—continued to be ruled by Greek patriarchs. Even after leaders of the Umayyad dynasty put in place Muslims or Copts, it remained autonomous, and its population multicultural. Throughout the caliphate, many official documents were written in both Greek and Arabic. As the century wore on, scholars gradually returned to the use of their local languages—Coptic in Egypt, Syriac and Aramaic in Syria and Palestine, Pahlavi in Persia—but Greek continued to be the language of instruction, writing, and scholarly work. It would remain the language of administration throughout the whole area, even after the end of the Umayyad reign in 750.

By the time the Abbasid caliphs replaced the Umayyad rulers and, in 756, moved their capital from Damascus to Baghdad, their doctors—most of them Christians, and speakers of Syriac—had at their disposal a rich body of Hellenistic culture. Recognizing its utility and power, the new rulers were eager to make use of it and to fund translations into Arabic. These had in fact begun, though rather faintly, soon after the Arab conquest, under the Umayyads, with the efforts of a few Greek-speaking medical philosophers based in conquered Egypt. The Galenic summary was called 'Jawami in Arabic and would remain an important reference book for medical students. But the Abbasids undoubtedly were responsible for the bulk of the work, translating texts not only from the Greek but also from Pahlavi, Sanskrit, and especially Syriac.

It was especially via Persia that Greek medicine seeped into Arabic and Muslim culture, partly as a result of the internecine conflicts of the early Christian church. Persecuted Nestorian Christians—followers of the bishop of Constantinople Nestorius, whose writings had been ruled heretical and condemned to be burned in 435—fled eastward to Persia, bringing with them their school of science and medicine. By the mid-sixth century, they were settled in the city of Gondeshapur, a now vanished city at the heart of the Sassanian empire (in the region of Khuzestan, which today lies in the southwestern region of Iran). Hippocratic physicians had been present there from the third century, when the Sassanian king Shapur I had brought them along with him when he married the daughter of the defeated Roman emperor.

Under their son Shapur II, Gondeshapur became Persia's cultural and academic capital, and it would remain an important cosmopolitan center for nearly 500 years. Neoplatonists, who were developing a Christianized version of Platonism, fled there when the emperor Justinian closed down their school in Athens in 529. A teaching hospital, perhaps the first ever, was founded there. The city was taken by the Arabs in 638, but Persian and Indian scholars continued to work alongside Jews, Arabs, Christians, and Greeks. The translation of Sanskrit and Greek works into Arabic, Syriac, and Pahlavi proceeded apace, opening the Islamic world to the classical cultures of both India and Greece. Doctors who were trained in Gondeshapur all followed Galenic-Hippocratic medicine, which began to spread throughout the region, ensuring the passage of the Hellenistic medical heritage into Islamic culture. Doctors from the Persian city played a central role in designing and staffing the hospital in Baghdad, which the caliph Harun ar-Rashid had built on the Gondeshapuri model. And members of one Gondeshapuri family (the Bakhtishu) served as court physicians to the Abassids in Baghdad until the tenth century.

By the late eighth century, a sophisticated culture was also growing in Baghdad, around the lavish court of Harun ar-Rashid, which inspired the *Thousand and One Nights*. The caliph founded an impressive library in Baghdad—the Bayt al-Hikma, or "House of Wisdom"—but it was during the reign of his son, al-Ma'mun, in the early ninth century, that Baghdad flowered as a major center of higher learning. The Bayt al-Hikma became the central site for the collection, copying, and translation of Greek scientific works of mathematics, mechanics, geography, astronomy, astrology, alchemy, logic, philosophy, and of course medicine. Pharmacological knowledge developed, with translations of the first century AD *Materia medica* by the doctor and botanist Dioscorides, which described hundreds of botanical and mineral substances in terms of their medical qualities. The Greco-Latin heritage was also enriched by the arrival of Indian scholars, who transmitted to their hosts further botanical, medical, astronomical, astrological, and musical knowledge. The culture was ebullient, and indeed the ninth century is remembered as the golden age of the Abbasid reign.

3. Hunayn ibn-Is'haq and the Translators

CENTRAL to the meandering story of the transmission of Greek medicine to subsequent generations was a ninth-century Persian, Hunayn ibn-Is'haq al-'Ibadi, known as one of the main translators of classical medical sources. Born of Nestorian Christian parents in al-Hirah, located in today's southern Iraq, he traveled to Alexandria, where he learned Greek, and from there went on to Baghdad, where he trained in medicine and would remain for the rest of his life. During his apprenticeship, his knowledge of the ancient sources grew to be formidable. By the time he left Alexandria, he was capable of reading texts in Syriac, Arabic, and Greek, besides Pahlavi. He would eventually know over 100 Galenic texts and, helped by his physician son Is'haq ibn Hunayn, his nephew Hubaysh ibn al-Hasan al-A'sam al-Dimashqi and by a host of assistants—who probably were the authors of translations later attributed to Hunayn himself—he famously tackled a large number of those. He turned some into Syriac before rendering them into Arabic, or translated them straight into Arabic; and he also revised the translations of others, including the Arabic version of Dioscorides's *Materia medica*.[8]

Hunayn's teacher, himself a Nestorian Christian of Gondeshapuri extraction, also looms large in this story. Named Abu Zakariyya' Yuhanna ibn Masawayh, he was not only the director of the Bayt al-Hikma but also a reputed physician to caliphs and a prolific writer, on topics such as melancholy, ophthalmology, and leprosy. He notably wrote a series of *Medical Axioms*, the *Kitab al-Nawadir al-tibbiyya*. These were modeled and built on the Hippocratic *Aphorisms*, but practice mattered to him as much as familiarity with the Greeks and Romans. He is said to have traveled around the Byzantine empire to gather ancient texts, but he is also known to have wanted to perform dissections, just as Galen had. (He is also said to have dissected a large monkey offered by a Nubian prince to the then caliph al-Mu'tasim, in Baghdad, in 836.)

His often wise aphorisms arise out of the firsthand experience that enabled him to revise Hippocratic dicta. He was following the Greek master when he wrote, for instance, that "the treatment of illnesses that are on the surface of the body is more effective in the spring and summer, as opposed to those that are inside the body",[9] but it must have been out of an independent concern with medical ethics that he wrote, "Illiterate doctors, as well as conformist ones, young men, and those who have little empathy and a great many passions, are highly murderous."[10] Masawayh also believed that, although "cold illnesses are difficult to treat in the elderly, and easy to treat in the young,"[11] the doctor had to remember that "illnesses that arise out of a decrease in the quantity and quality of humours are no less frequent than those that arise out of their increase; for this reason, doctors are wrong to rush into purging."[12] A responsible doctor, in other words, would act cautiously, applying recommended treatments only after due consideration of the patient's temperament.

And, crucially, a good doctor must remember that "the soul follows the temperament of the body; if an illness occurs, especially in the main organs, do not forgo to treat the soul through sense, sight, joy and pleasant sound; this constitutes an important part of the treatment."[13] Humoural medicine had been holistic since the very beginning, and it continued to be holistic in the medieval Middle East. Like Galen, Masawayh was much more interested in how humoural concoctions conditioned states of mind than in how states of mind had an impact on humoural concoctions. The doctor, he thought, should be concerned primarily with the body: only secondarily with the soul. Hunayn himself, in fact, had translated a treatise by Galen entitled *That the Powers of the Soul Follow the Body's Mixtures*—devoted to explaining in terms of Hippocratic humours just what the title claimed.

Hunayn also wrote original works, the most noteworthy of which is the *Kitab al-Mudkhal fi t-tibb*. Here he made use of the Alexandrian summary of Galenic texts, synthesizing his extensive knowledge into a manual that later became highly influential in the West, and known in its Latin version as the *Isagoge Johannitii in tegni Galeni*. (Hunayn, at that point, would be renamed Johannitius.) By now, compendia of Hippocratic and Galenic texts like those by Oribasius and Paulus of Aegina were appear-

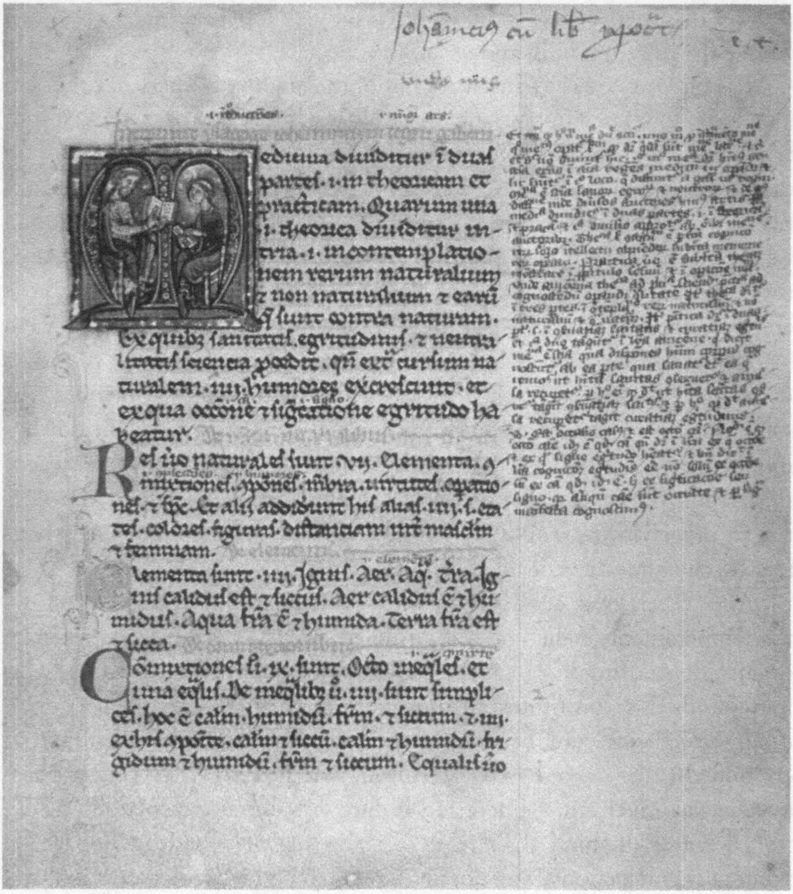

Fig. 6. Hunayn ibn Ishaq al-'Ibadi (Johannitius) manuscript: *Isagoge Johannitii in Tegni Galeni,* Oxford, 13th century. The manuscript contains the "Articella."

ing regularly in simplified Arabic versions; and though they could not replace clinical experience, they did enrich it. By the early tenth century, ancient heritage, current practice, detailed scholarship, and general compendia were all synthesized. This vast scholarly enterprise not only helped ensure the survival of the classical tradition: in its initial phase at least, it was even central to the development of the Arabic language, since, for his translations, Hunayn had created words and forms that did not yet exist in Arabic.

4. Divine Creation and Human Frailty

CLASSICAL LEARNING was adopted in its Arabic guise as a fixed tradition, transmitted throughout an expanding territory that would eventually extend from central Asia to Sicily and Spain, via the Middle East and north Africa. The translation movement thus began as a series of linguistic transfusions. But it also allowed for a transfusion and transformation of ideas from pagan to monotheistic cultures.

Classical medicine was being injected into societies that, like the pre-Hippocratic Greeks, had interpreted illness primarily in terms of divine retribution. Pre-Islamic Arabs had cultivated medical and magical practices similar to those developed in Greater Greece and in the Roman empire, from the use of pharmaceuticals listed in the *materia medica* to the animist attribution of illness to spirits *(jinni)* and the superstitious belief in the "evil eye." For the Jews, illness was more often than not caused by divine wrath rather than natural causes; attention to medicine in the Bible was limited to a concern with prevention, an emphasis on cleanliness, and so the need to establish precise rules of hygiene, which encompassed directives for the proper use of water and the laws of kashruth codifying all aspects of daily life.[14] The Koran, apart from its own emphasis on cleanliness especially as preparation for prayer, has equally little to say either about medicine or about superstitious beliefs. But superstitions would persist among physicians in the Arab world even as they adopted humoural practices like urine analyses, bleedings, and the prescription of pharmaceuticals—just as they had persisted in the Greek and Roman worlds. One tenth-century physician who practiced in Baghdad described how to find out whether a dog who had bitten a patient was infected with rabies: if it was, the patient's urine would contain worms in the shape of dogs, and a rabid patient would urinate blood containing leeches in the shape of feet.[15]

For monotheists, the realm of nature was separate from the realm of God, and nature was no longer inhabited by hidden powers. But nature remained mysterious; its regularities and irregularities still required some explanation. Superstitious traditions were useful for that reason. So was Galenic thinking, which was easy to adapt to the religious prerogatives of monotheists, during a period when both Christianity and Islam were undergoing internal debates and theological consolidation, and when rabbis

Fig. 7. Galen manuscript: *On the Usefulness of the Parts*, end of 14th book; with quotation from summary attributed to Yahya al-Nahwi. Possibly from the 17th century. See also: http://www.nlm.nig.gov/hmd/arabic/galen.html.

were enriching Jewish thought with new commentaries. Galen's teleological thinking, derived from Aristotle's, could be funneled into the vision of a divinely, intelligently designed universe, where all creation partook of a foreordained plan, where the shape of things was perfect, where the body's anatomy alone had determined its functions. This was a perfectly planned biological world where animals, for instance, breathed not in order to bring air into the organism, as we assume today, but rather to fan, cool, and "strengthen" the innate heat. Galen was certain that the motion of breathing dissolved the "smokiness" produced by the "combustion of the blood."[16] The Galenic body functioned as an organic whole according to principles that Aristotle had established: attraction, contraction, propulsion, growth, alteration, elimination, purification, secretion, concoction. Physiology was a matter of temperature, texture, anatomical shape; of canals and containers; of *pneuma*, chyle, blood, and humours. Galen thought it was for the sake of an anatomical part that a function had grown, and that atomists and Methodists were mistaken in their obverse belief that functions developed thanks to the design of anatomical parts. Nature had "artistic skill," he often stated: "There is no part which she [Nature] has not touched, elaborated, and embellished."[17]

This teleological vision presupposed the existence of a supreme being, a divine agency at the origin of the order of the universe that guaranteed this order as well. If the body was created by a supreme being, there was no need for it to be visible. Dissections remained forbidden, but anatomical observation was deemed unnecessary. The elements posed in antiquity worked perfectly well; remedies from the *materia medica*, also transmitted from generation to generation, had passed the test of time. Doctors had everything they needed. Innovation was unwarranted—at least if one stayed within a precise system, such as the scheme of the three Aristotelian souls, that accounted for all appetites and for the human capacity to reason and contemplate God. Meanwhile, apocryphal stories about Galen himself were emerging within the Byzantine world that amounted to his "depaganization"—effectively turning him, by the fourteenth century, into a Christian who died on a pilgrimage to Jerusalem to witness Jesus's miracles, or to Syria to meet his companions. (Some even believed he was the uncle of Saint Paul.)[18]

The Galenic-Hippocratic system allowed, too, for the conjunction of physical with moral health, and for the possibility of channeling passions and correcting excesses through the use of reason and will—all of these central to the ethics developed within the three monotheistic creeds. The control of sexual desire, a crucial element of early Christianity, for instance, was justifiable on the basis of Galenic medicine too (which an abstinence-preaching Church Father like Tertullian read avidly): male vital spirits could be depleted by overly frequent ejaculations. It was best to turn sexual desire into spiritual longing.[19] Medicine continued to include psychology. Whether one lived in Alexandria, Damascus, Antioch, Gondeshapur, or Baghdad, humours were related to states of mind as much as states of body. The concoction of humours within the organs and their circulation within the channels of the organism happened day and night, during the whole of life, in each individual. The body's inner, invisible activities were constants in a changeable world where perceptions, impressions, foods, and climate induced altered states of mind—just as reason, the instrument of divine contemplation, free will, and moral choice, was a stable element that could keep volatile moods and inconstant humours in constant check.

The writers in the Arabic world perpetuated the ancient association of medicine with philosophy and ethics. Their works on the body were, more often than not, studies of the soul, reflections on the nature of its substance and of its relation to God. It is true that neither Muslim, Christian, nor Jewish philosophers quite knew what to make of the often critical revisions of Aristotle's philosophy by the Alexandrians and by Galen himself, since both Galenic medicine and Aristotelian philosophy were being taught in the universities, despite the contradictions between the two systems. Yet Greek and Roman ethics fed into early Christian thought and into the corpus of medical philosophy that Muslim and Jewish thinkers, especially, were developing. Spirituality and medicine were united in the work of some of these scholarly doctors, especially the pious monotheists and rigorous philosophers who pondered the difficulty of rendering visible that which in a human being was invisible—the spirit, and the tripartite soul.

On a more prosaic level, the medical and the legal professions were attractive to Jews and Christians who strove to transcend the second-class status they often fell into in the lands conquered by the Muslims, and to establish themselves in the powerful, prosperous environments that grew around the new courts. This was why so many took up one or both of these professions. Exponents of the three monotheisms, in fact, expended much effort on unifying the philosophical underpinnings of these religions through reflections on the relation of the soul to the body. There are plentiful stories about these philosophical doctors, courted by the powerful, who might then celebrate, or defame them. There was, notably, one Jewish physician from Baghdad, Ishaq ibn 'Imran, who was invited to serve the last ruler of the north African Aghlabid dynasty in al-Qayrawan (Kairouan, in today's Tunisia), but who was eventually put to death by his royal host. Shortly after, the Aghlabids were fleeing before the invading Berbers, in 909.

But tradition transcended the passions and bloodshed of territorial and dynastic politics; and wherever physicians were, it was their spiritual duty to know the body upon which they elaborated their philosophical reflections. Doctors were always needed; and they could achieve glory. Ibn 'Imran, for that matter, had a much luckier disciple, Ishaq ibn Sulayman al Isra'ili—later, in Renaissance Europe, acclaimed as Isaac Israeli or Isaac Judaeus (Latinized forms of his name). He was born in Egypt, and was also called to north Africa, but to serve the founder of the Fatimid dynasty there, 'Ubayd Allah al-Mahdi ibn 'Imran. Unlike the hapless Ishaq ibn 'Imran, he seems to have lived to be 100 years old, dying in 955 or so. During that time al Isra'ili wrote guides on the humoural *regimen* for a healthy life and on the three primary diagnostic tools of the Galenic physician—a *Book of Fevers,* a *Book of Urine,* and a *Book of Pulse.* But he also became reputed and respected as a metaphysician, especially as a Neoplatonist thinker.

For in his view, the humours that transpired in sweat, blood, and urine were just an aspect of nature; and nature itself was nothing less than "a power belonging to the heavenly body which is in human bodies through the mediation of the sphere between the soul and the bodies."[20] He understood the soul as tripartite, in solid Aristotelian fashion, and described death as "the absence of spirit, i.e. the natural warmth, from the heart; we live through the warmth, but die of coldness."[21] The warmth that carried life was at once physical and spiritual—in the same way that emotions, as Galen himself had said, pertained at once to corporeal physiology and to the mind. But he went even farther than Galen in his belief that states of mind, especially those brought on by black bile, must be understood in spiritual terms too. There was a theological point here: there were, he wrote, "Very many holy and pious men who become melancholy owing to their great piety and from fear of God's anger or owing to their great longing for God until this longing masters and overpowers the soul. . . . They fall into melancholy as do lovers and voluptuaries, whereby the abilities of both soul and body are harmed, since the one depends on the other."

5. Religion and Emotion

THIS POWERFUL IDEA—that an excess of religious contemplation could cause melancholy and actually be unhealthy, disabling one from functioning in the real world—would eventually reappear in the Renaissance. It is related to the notion of love-melancholy, and although it might make intuitive sense, it has a specific, textual genesis. It might have arrived in the West through Constantinus Africanus, an eleventh-century monk who translated works from the Arabic corpus into Latin, including texts by Ishaq ibn 'Imran.[22] Ibn 'Imran had based his own thoughts on melancholy on an early-second-century lost treatise *On Melancholy* by Rufus of Ephesus, admiringly referred to by Galen himself. It is possible, too, that the notion, still familiar to us, that melancholy was a noble side effect of creative, active intelligence, was also transmitted by this path. But it could have been a version of a Hippocratic saying, "Fatigue of the soul comes from the soul's thinking," a thought one even finds in Constantinus's Ishaq.[23]

Constantinus also translated a synthetic treatise, the *Kamil al-sina 'a altibbiya,* known mainly as the *Kitab al-Malaki*—"The Royal Book of All

Medicine"—by the celebrated physician and theorist 'Ali ibn al-'Abbas al-Majusi, a tenth-century Persian Zoroastrian who would become known in the West as Haly Abbas. It would make its fortune in the West as the *Liber pantegni* (a later version, by Stephen of Antioch, was known as the *Liber regius*). Al-Majusi's book was also based on Alexandrian collections of Galenic writings, and relied entirely on humoural doctrine. His system was based on texts, not on direct observation, and it was full of Aristotelian distinctions, faculties, and qualities. The body was made up of the four elements, which constituted the four humours; of the "daughters of the elements"; of the organs; and of the mixtures of elements in varying proportions that resulted in the temperament. Humours, as Galen himself had believed, could turn into other humours through the action of heat. To al-Majusi, the humours varied in color and density, and took on various roles accordingly—they helped digestion, nourished organs, caused tumors, and helped blood flow. Blood could be polluted by humoural imbalances and cause illness—so smallpox, for instance, was caused by the failure of the fetus to expel from its organism the menstrual blood that had nourished it. Illnesses that resulted from humoural imbalances could also be transmitted from parents to children, since semen came from the blood; and illnesses could be transmitted to imbalanced organisms via putrid air, that is, air polluted by decaying corpses, rotten foods, and so on.

Physiological processes also accounted for emotions, just as they had for Galen and Hippocrates. Al-Majusi took from the Hippocratic *Epidemics II* the idea that a hot and dry heart combined with a moist brain resulted in the production of black bile from the burning of yellow bile, which was then expedited to the brain. The whole operation induced an onslaught of melancholy. Melancholy was in a class of its own, and a matter of some complexity. It presented a large variety of symptoms, and it could be caused either by internal organs or by external factors. External causes included grief, which led the vital heat to die down; the lovesickness diagnosed as a real disease by Hippocrates, Erasistratus, and others; and passionate love itself, in fact a version of this syndrome. Prescriptions for externally caused melancholy included exercise, moist foods, massages, baths, music, poetry, exemplary tales from the lives of sages, and sexual distraction. All sufferers of melancholy, regardless of its source, should seek out light and gardens, calm and rest, purges and laxatives; inhalations and warm baths with moistening plants, such as nenuphar, vio-

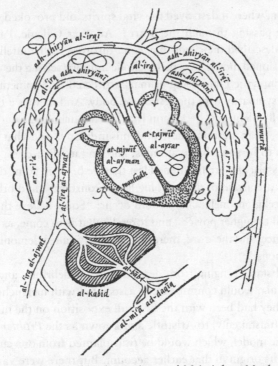

Fig. 8. The movement of the blood according to al-Majusi, from Manfred Ullmann, *Islamic Medicine* (Edinburgh University Press, 1978), p. 66.

lets, and lettuce leaves; and a diet of lamb, lettuce, eggs, fish, and ripe fruit. They should avoid acidic foods like vinegar and mustard, as well as garlic, onion, cabbage, lentils, and red meat. Bleedings helped too, but their duration and intensity should be modulated according to the color, and so the health, of the drawn blood. The use of the root of hellebore had been encouraged since the days of Hippocrates, and it appears in all recipes from that era for the cure of melancholic diseases, among others. (Earlier, Pliny had described black hellebore as "a cure for paralysis, madness, dropsy without fever, chronic gout and diseases of the joints; it draws from the belly bile, phlegms and morbid fluids.")[24]

Melancholy was discussed in many treatises during these medieval centuries. Al-Kaskari wrote an elaborate text on the subject in which he, too, referred at length to Rufus of Ephesus. He explained the phenomenon by following Galen: melancholy consisted in the ascent of black bile to the brain, where it destroyed the vital spirits and provoked sadness and sorrow by passing through the heart.[25] Another Galenic, Paulus of Aegina, had described the syndrome as a "disorder of the intellect without fever, occasioned mostly by a melancholic humour seizing the understanding; sometimes the brain being primarily affected, and sometimes it being altered by sympathy with the rest of the body. And there is a third species called the flatulent and hypochondriac." Melancholics, typically, would experience "fear, despondency, and misanthropy; and that they fancy themselves to be some brute animals, and imitate their cries; and others, earthen-vessels, and are frightened lest they be broken. Some desire death, and others are afraid of dying; some laugh constantly, and others weep." Here too, religious melancholy could occur: "Some believe themselves to be impelled by higher powers, and foretell what is to come, as if under divine influence; and these are, therefore, properly called demoniacs, or possessed demons."[26]

Mental states ranging from despondency to delirium, and from sadness to mania, would continue to be associated with melancholy for centuries, as they had been with the first full exposition on the matter, a text attributed (mistakenly) to Aristotle and known as the *Problemata,* XXX. The Galenic model, which would be transplanted from one culture to the other, had its origin in that earlier account. But there were variations: al-Kaskari departed from the Galenic analysis in thinking that the doctor should pay attention to the patients' sleep patterns, which depended on the warmth of their temperament. Indeed, if a patient had a cold temperament, he or she would develop pleurisy or *birsam,* that is, an inflammation of the chest; if warm, one would see a case of phrenitis, or *sirsam,* an inflammation of the head. The *sirsam* in turn could be either cold or hot: if it was cold and moist, the patient was lethargic and had no fever; if it was hot and dry, there was insomnia—which also occurred if one drank too much wine, since wine heated the brain—and fever. Treatment would have to warm or cool the patient accordingly.[27] No records survive outlining the rate of success of such treatments. That may be because, in the case of melancholic diseases at least, it would be as hard to identify a successful treatment as to define the illness being treated.

6. New Departures

RECORDS of cases are generally scarce, and the available texts, studied by scholars of medicine in the Islamic lands, concern theory rather than clinical practice. But testimonies such as that of al-Kaskari do help reconstruct the state of medicine. Ailments had been described well enough by the Galenics and Hippocratics; few new ones were identified, and the Arabic names corresponded to the symptoms catalogued in the ancient world. The *materia medica* or pharmaceutical corpus derived from Dioscorides and from Pliny the Elder was transplanted to Arabic lands, even though it referred to plants that existed in the Mediterranean rather than in central Asia, Persia, or Syria. (Later on, treatises would begin to include substances from North Africa, India, and even China.) Clinical practice was, on the whole, rationalized, and the body continued to be the site of invisible concoctions and potentially overwhelming secretions.

But in fact there was some room for independent thought, since medicine was intertwined with the philosophy that was developing out of Aristotle and Plato during these centuries, locked in a debate with religion over the role of the rational soul and the order of the body. Doctors were expected by the potentates who employed them to know about dietetics, pharmacology, and astrology, and to be experts on the humours. But they also had to be philosophers in order to be trustworthy: they had to be able to engage in the debate over the nature of medical knowledge and argue whether it was an empirical or a rational practice, based on experience or on logic, on precedent or on reasoning. This tension surfaces in much of the work produced by the great thinkers of the Islamic tradition.

On a continuum with classical thought, one can point to the ninth- and tenth-century figure of Abu-Bakr Muhammad ibn-Zakariyya' ar-Razi, a physician who was born in Rayy (today on the outskirts of Tehran), and became the director of the hospital there after his medical studies in Baghdad. Ar-Razi was prolific and celebrated—notably by Chaucer—and is remembered today for his empirical approach to clinical practice and for his massive treatise, the *Kitab al-Mansuri*, or *Almansor* in Latin.[28] It relied on Hippocrates, Galen, Rufus of Ephesus, Oribasius, Paulus of Aegina, and many others, as well as on Indian sources, but it was a critical work, not only a compilation. The posthumous edition contained the notes ar-Razi had gathered from his clinical experience and acute observations. It was Latinized in the thirteenth century by a Sicilian Jew, and its fortune in the West would resound into the lettered culture of the Renaissance, when its author would be renamed Rhazes.[29]

One can understand the attraction ar-Razi exerted on Christians: his austere, Stoicist-leaning treatise *Spiritual Medicine* outlined the dangers of yielding to the body's call to indulge its need for pleasure and of not resisting its passions. This notion was not novel, of course. Unreciprocated love had always been unhealthy, at the court of Seleucis in the third century BC in Syria as much as in tenth-century Baghdad. Ishaq ibn 'Imran had recommended the moderate practice of sexual intercourse as a good antidote to excessively disruptive symptoms of melancholy, but ar-Razi thought that the key to equanimity was the rational acceptance that, since humans were thinking beings, satisfaction was impossible to achieve. To him, loss was inscribed in possession; attachments and sensual fulfillment accentuated the painful awareness of their fleetingness. A morsel of pleasure and a taste of passion, in the end, led to enduring misery, both in this and in the heavenly world.

But music could help. According to ar-Razi, mental illness could be cured "by means of musical instruments which convey to the soul through the sense of hearing the harmonious sounds which are created by the motions and contacts of the heavenly spheres in their natural motion, which affect the right perception." A further notion was that of the four musical modes devised by the Greeks to classify scales. Each mode—the Arabic word was *maqa*—corresponded to a humour; so a tune, like an Indian *raga*, could be appropriate to the time of day, the season, or the individual's mood. (The idea of musical appropriateness might even have had an Indian origin.)[30] A "cold" *maqam* would not help a melancholic but could be helpful on a hot summer day.

In the ninth century, the philosopher and mathematician al-Kindi—who was born in al-Kufa and studied in nearby Baghdad—went farther and established a correspondence between modes and constellations, effectively arguing that there were twelve musical modes, not four. In the tenth century, ibn Hindu, a Baghdadi physician, would write, "There is a *maqam* which arouses sadness, another which brings joy, one relaxing and tranquilizing, another disquieting and exciting, one which keeps one awake, another which induces sleep. Whenever we order those who suffer from melancholy to be treated with the respective modes, it helps them. The physician need not himself be a performer of the drum, the flute or the dance, just as he does not need to be a pharmacist or a phlebotomist, but rather he employs these people to aid him in his therapy."[31]

The belief in the power of harmony, in the capacity of philosophers to discover it and doctors to deliver it, was connected to a tradition whose Arabic name survives to this day, *al-kimiya*, alchemy. Physicians were expected to know something about alchemy, and Arabic-speaking alchemists were practicing their art well into the fourteenth century. Ar-Razi himself, a firm believer in its virtues, wrote no fewer than three treatises on it. It probably entered Muslim culture just as Galenism had, via Alexandria and Persia. The Arabic word might itself be derived from the Greek *Khemia* (transmutation), or from the ancient Egyptian via Coptic, or, according to some scholars, even from Chinese. At any rate, in its Arabic guise, it favored the development of techniques that would prove useful for the fabrication of pharmaceuticals, since its purpose was the transformation or rather transubstantiation of elements into new substances, and ultimately of base metals into gold.

As an earthbound practice, alchemy sustained the development of techniques familiar to modern chemists: Jabir ibn Hayyan, born in Persia in the eighth century and known in the West as Geber, is often referred to as the father of chemistry. Jabir worked at the court of Harun ar-Rashid and was perhaps the most prolific of the alchemists (he authored twenty-two treatises on alchemy, out of 100 or so on a wide variety of subjects), and was one of the main ideators of processes such as crystallization, extraction, clarification, evaporation, concoction, and especially distillation, which helped create medicines, perfumes, and concentrated plant essences. Alchemists turned "simples" into "complexes" by distilling substances, with the help of a vessel topped by an *al-ambiq*, or alembic. The subtle, volatile parts of a simple substance would be heated in the bottom part of the vessel—the curcubit—and thereby separated from its terrestrial parts, rising toward the alembic, where it began to cool, turning into a liquor, or elixir, that revealed the "real" nature of the substance.

Alchemy was considered by some to be a sacred activity, which revealed hidden, literally occult qualities. Its practitioners were empirics, who thought it crucial to experiment with substances, but their practice partook of an esoteric tradition based on the belief that these occult forces constituted the world and that the search for essences was akin to the search for truth. Alchemy was affiliated with Pythagorean and Neoplatonist beliefs, and paired with astrology—which itself had medical resonances, since temperaments were thought to be determined at least to some extent by one's time and place of birth. The life-giving potions or elixirs prepared by alchemists were medicines, generally based on the four qualities that, in turn, constituted the humoural organism.

In all these ways, the early belief in elemental correspondences between the microcosm and the macrocosm survived the demise of the classical world and became integrated with great force in Muslim culture. And from the belief in correspondences between the earthly and the celestial to the notion that our spiritual health depends on the knowledge of the body, there is but a short step. Religiosity and medicine combined easily, bolstering the confidence of practitioners and the trust of their employers or patients, and this continued to be the case through momentous, bloody dynastic changes. The movements of refugee populations helped to spread, rather than weaken, the classical tradition that provided a training base to clinical practitioners. Medicine had never been a sedentary profession, and the ideas at the heart of the Galenic-Hippocratic system were solid enough to travel with the doctors who used it, as best they could, often in harsh circumstances.

7. Persian Insights

THE STAYING POWER of this system is exemplified by the most illustrious name to have emerged out of the high period of Muslim philosophy: Abu 'Ali al-Husayn ibn 'Abd Allah ibn Sina, or Avicenna. He was born around 980 and was brought up in Bukhara, the capital of the Samanid dynasty of Persia (in today's Uzbekistan). Highly precocious, educated early on in Aristotelian logic, mathematics, natural sciences and metaphysics, he turned to the art of medicine last—aged only thirteen—and, according to his autobiography, "excelled in it in a very short time." He found it rather easy, and was helping to heal the sultan by the age of sixteen. At that point he was also teaching himself jurisprudence, and obtained permission to use the Royal Library, where he opened dusty tomes no one had bothered to look at before. He began to write there, at first focusing on metaphysics and ethics. But in 999, Bukhara fell to the Turkish Qarakhanids and the Samanid reign came to an end. Ibn Sina's father died. The son began a nomadic life, practicing law in the city of Gurganj, and traveling before finally settling in the old Persian city of Hamadan, where he became physician to the court and started work on his highly influential treatise of medicine, *Al Qanun fi al-tibb*, later Latinized as the *Canon*.

Over 400 works by Ibn Sina, covering philosophy, medicine, astronomy, geology, and more, are on record. About half of those survive today, but the influence of the *Canon* outdoes that of all his other books (except for the book on metaphysics known as the *Shifa'*). Ibn Sina brilliantly synthesized Galenism with Aristotelianism, and his interpretation of Aristotelian thinking about logic, metaphysics, and psychology would become established throughout medieval Europe. Although he wrote within a Muslim context, his ideas would be easy to adapt to Christianity. Indeed the Christian church was to be significantly influenced by the Persian thinker and doctor, via its own Thomas Aquinas, who admired him profoundly.

"The Almighty Creator has bestowed upon every animal and every one of its organs the most appropriate and the best adapted temperament for its nature, functions and conditions. Since the verification of this truth is a matter for philosophers and not physicians, we may accept that man has been endowed with the most suitable temperament and the most appropriate faculties for the various actions and reactions of the body. Similarly every organ is endowed with a hot, cold, moist or dry temperament appropriate to its functional requirements."[32] Ibn Sina thought of himself first as a philosopher. The *Canon* was indeed a medical book of a philosophical kind, which delved deep into the hidden structures of the body and soul—not by way of dissection, but through scholarship, observation, reasoning, logical inferences, and rationally based faith. The Hippocratic author of *Regimen I* had written that "men do not understand how to observe the invisible through the visible"; this is perhaps why there still remained a need to define the realm in which the invisible body was observable. It was the religious impulse, rather than the speculative mind, that, perhaps for the Greek of the fifth century BC, and certainly for the Muslim of the tenth century AD, informed the quest for the body's order.

The human organism, visible and invisible, was perfect, and divinely created. The soul was entirely spiritual and increasingly individual as the person aged; it was separate from the body, itself created out of the elements. Ibn Sina, like most of his contemporaries, thought that our "exter-

Fig. 9. Scene in a pharmacy, from manuscript of Avicenna's *Canon Maior*,
15th century (in Hebrew). Bologna, Biblioteca Universitaria.

nal senses" relayed perceptions to the "internal senses" in the brain. Galenics
had identified three of these: the "imagination" at the front of the brain
turned perceptions into information on the "cogitative faculty" in the
middle of the brain, and stored it at the back of the brain, in the "memory." Ibn Sina—along with the other Muslim and Jewish philosophers—
reshuffled these faculties, adding two others, "common sense" and
"estimation". (See Fig. 5.) While Ibn Sina was interested in what we would
call today our cognitive processes, he gave the comtemplative, conscious

rational soul and essential function, separate from the humoural being it served, and in harmonious correspondence with the one, invisible, all-just creator. Theology conditioned psychology. The body was a mysterious—and relatively lowly—outcome of the marriage between God and the elements, although medicine, he famously wrote, was "the knowledge of the states of the human body in health and decline in health: its purpose is to preserve health and endeavor to restore it whenever lost."[33]

Ibn Sina was also one of the few to agree with Aristotle that the heart was the seat of the rational soul. Anatomical evidence was not a way into the invisible. Surely Galen was mistaken: the heart transmitted life, heat, and breath to the brain and the liver. Humours, however, certainly constituted the organism. Heat and cold were instrumental in forming them. Blood formed when heat was moderate; yellow bile when heat was excessive; phlegm when cold was moderate; and black bile when cold was excessive. A combination of heat in the liver, weakness of the spleen, external cold, and a long disease history could lead to a heightened amount of black bile in the organism. A hot temperament was characterized by a powerful voice, rapid speech, anger, vehement gestures, and so on. A cold temperament could be recognized by the exact opposite characteristics. Heat could slow down natural functions and cause insomnia, but sleep was not necessary for health. This was quite a departure from the commonsense guidelines typical of the Hippocratic-Galenic tradition. Ibn Sina took account of both external and internal factors to determine the overall behavior of the bodily structure, trying, as a philosopher, to interpret anew, rather than simply to repeat classical technical knowledge.

But like his predecessors, he also minutely analyzed urine, listed diseases, and recommended regimens and diets according to one's need. So one had to remember that "viscous aliment experiences delay in passing through the intestine"; or that if a tired person ingested heavy foods ("for instance, a dish of rice with soured milk") after a fast, "it will come about that his blood becomes sharp in quality and as if ebullient"; or that "when meat is roasted, and taken with onion and eggs, it is very nutritious; but it is slow in passing through the intestines, and lingers in the caecum." Medicines had to be adapted to the stage of the disease; and so the disease had to be carefully assessed. Melancholy—a disease here as it was in the work of Rufus of Ephesus and al-Majusi—was also recognizable as a set of clear medical symptoms, whether it came from the spleen, the stomach, or the phlegmatic humour: its signs included "bad judgement, fear without cause, quick anger, delight in solitude, shaking, vertigo, inner clamor, tingling, especially in the abdomen." It could also be caused, of course, by love.

Ibn Sina is said to have died while trying to cure himself of a bout of colic that broke out during a military campaign on which he was accompanying, as he often did, the prince he was serving then. (Another version has him return to Hamadan, accompanying the prince from Isfahan who had conquered the city.) He administered a few too many enemas to himself, some with celery seed, and seems to have ended up with perforated intestines. He was given mithridate, and a little too much opium. Death found him at the age of fifty-seven. Holding on to humours could mean holding on to odd, sometimes effective, sometimes useless or outright dangerous solutions. But the outcome did not unsettle the power of his word. His legacy would slowly weave its way into medieval Europe, and by the early Renaissance, the *Canon* was the most oft-printed work in the Western world after the Bible.

8. Out of Spain

BY THE TWELFTH CENTURY, Ibn Sina's *Canon* was already a classic, and the thinkers of the tenth and eleventh centuries constituted a recent but rich heritage upon which further thought would grow, especially in Spain. The Iberian peninsula had been called al-Andalus since its conquest by the Umayyads in 711. Its capital was the great city of Córdoba, which flourished culturally especially during the reign of 'Abd al-Rahman III in the first half of the tenth century, when it was one of the most sophisticated and refined cities in Europe, producing and attracting scientists, philosophers, musicians, and artists of high caliber. Jews and Muslims cohabited here, and texts were translated into Hebrew and into a hybrid language called Judeo-Arabic. But the Umayyad dynasty collapsed in 1031 with the death of its last caliph. The populations were caught up in the political upheavals that ensued, while al-Andalus became a collection of *taifas*, small domains that were often in conflict with one another and were vassals to larger dynasties. The north Africans took over what they could. Some of the Cordoban Jews who fled after 1031, menaced with either conversion or death, moved to Toledo and northern Spain, Italy, and southern France; their descendants eventually forgot Arabic. Others went east.

One eminent refugee from Córdoba was the Jewish physician, rabbi, and philosopher Moses Maimonides (his real name is Abu 'Imran Musa ibn 'Ubayd Allah). Born in 1138, he fled to Fez in Morocco, but as the Jews were persecuted there by the reigning Almohads, he went on to Palestine before settling in al-Fustat—the ancient Cairo—in 1165. He soon acquired a great reputation, both as a doctor and as a leader of the Egyptian Jewish community. He eventually became personal physician to the court of Saladin, and despite the pressures of his office, he produced an impressive number of distinguished works as a rabbi, an expert on Jewish law, a philosopher, and a doctor (he wrote a commentary on the *Aphorisms* of Hippocrates, for instance). The most famous and widely read of his books is the *Guide to the Perplexed*. His medical thought was based on the idea that the body's health served the soul's health. He admired Galen for having stated "that one's moral qualities are also damaged by being habituated to bad things, such as [bad] kinds of food and drink, but that a good regimen greatly improves the moral qualities." Those were "very useful general rules which should be applied in the case of both healthy and sick people."[34] It was the job of philosophers and physicians alike to preserve health and contain illness—by prescription, proscription, and the informed attention to the workings of the soul.

Spain continued to host such philosophers and physicians beyond the collapse of dynasties, changes of regimes, and religious conflict. Caliph Al Ma'mun's Toledo had welcomed the Cordoban refugees, and although the city fell to the Christian king in 1085, cultural exchanges between Jews, Christians, and Muslims remained intense in twelfth-century Toledo. The translation of Arabic works into Latin and Hebrew actively proceeded there. The scholar Gherardo of Cremona traveled to Toledo from his eponymous hometown in Italy in order to work on Ptolemy's *Almagest*, and ended up translating the great ar-Razi, as well as the Alexandrian collection of Galenic writings. And it is to him that we owe the first translation into Latin of Ibn Sina's *Canon*.

Córdoba, however, survived as a *taifa* beyond 1031. A century later it would see the birth of another major philosopher, whom Maimonides would study for his own philosophical purposes: ibn Rushd (the full name is Abu Al-Walid Muhammad ibn Ahmad ibn Muhammad ibn Ahmad ibn Ahmad ibn Rushd), later Latinized as Averroës. Born in 1126, Ibn Rushd came from a family of judges who had served the Almoravids, and later he himself served as a judge, after studying Islamic law, mathematics, philosophy, and medicine. He became well known for his extensive commentaries on numerous works of Aristotle and for his analysis of Plato's *Republic*. But most famously, he argued that philosophy was compatible with religion: it is in this way that he would exert a decisive influence—along with Avicenna, whom he had studied—on the use of Aristotelian thought both by Christians, including Thomas Aquinas, and by Jews from Maimonides onward.

Christian lands in the West would soon welcome the scholarship preserved and translated in the Muslim world, and would inherit the medical philosophy, or philosophical medicine, that had been developed there. Classical medicine had not been passed into these centuries in its original form; but its synthesis and transmission had been ensured, and all along the humours had remained, guiding medical practice and ensuring a cosmic unity in a world that was all but harmonious.

III

Remedies:
Miraculous Medicine

(Western Middle Ages: Fifth to Fourteenth Centuries)

In all this world was none like him to pick
For talk of medicine and surgery;
For he was grounded in astronomy.
He often kept a patient from the pall
By horoscopes and magic natural.
Well could he tell the fortune ascendent
Within the houses for his sick patient.
He knew the cause of every malady,
Were it of hot or cold, of moist or dry,
And where engendered, and of what humour;
He was a very good practitioner.

GEOFFREY CHAUCER[1]

among all my acquaintance, I see no people so soon sick, and
so long before they are well, as those who take much physic;
their very health is altered and corrupted by their frequent
prescriptions. Physicians are not content to deal only with the
sick, but they will moreover corrupt health itself, for fear men
should at any time escape their authority.

MICHEL DE MONTAIGNE[2]

1. Faith and Healing

JUST ABOVE the town of Cassino, midway between Rome and Naples, at the top of a fertile promontory, lies the abbey of Monte Cassino. It remains especially renowned for its role in the passage of the classical heritage from the Middle East to the European Middle Ages. Constantinus Africanus settled here around 1075, moving up from Salerno, where he had lived since 1060 after having traveled widely, studied in Baghdad, and been expelled from his native Carthage. Within these monastic walls, he translated Greek and Latin manuscripts from Arabic—some of which had themselves been translated from Hebrew. He was but one, albeit probably the most illustrious, of the monks in Monte Cassino who were engaged in translating the texts that would eventually spread into the lettered world.

The abbey had always been a place of great sanctity. First built by Saint Benedict in the sixth century near the site of a Roman temple to Apollo, it served as the model for monasteries in the West. It is still a peaceful haven, though it has survived numerous attempts at destruction. Most recently, Allied forces bombarded it in 1944 (in the erroneous belief that the Germans were hiding in it), but it was impressively rebuilt after then. Lombards had destroyed it within fifty years of its foundation in 529 and Saracens had invaded it in the late ninth century. But it was fully functional a few decades later, and by the eleventh century, it was flourishing as a center of learning in the Christian West under its Abbot Desiderius (later to be Pope Victor III). It was to this abbot that Constantinus would dedicate the influential *Liber pantegni,* his rendition of al-Majusi's *Kitab al-Malaki.*

Monasteries—increasingly numerous in western Europe from the time when the early Christians, fleeing persecution, founded the first ones in the Egyptian desert—were appropriate settings for the pursuit of medical enquiry. Considerations on the body's fate, in the early Middle Ages, were intimately connected to religious contemplation. Asklepios had given way to Saints Cosmas and Damian, patrons of all medical practitioners. A combination of fervent Christian faith, sophisticated superstition, and an inherently pagan reliance on the supernatural alleviated the hardship of lives generally cut short by illness, if not by natural disasters or human violence. Illness, malnutrition, and indigence could affect everyone, regardless of class, especially during times of famine. Ergot-ridden grain, notably, caused the so-called "sacred fire," described in the late eleventh century by a French witness as a disease that consumed the innards, burned the legs and arms, caused the skin to darken and the extremities to putrefy.[3] The early Middle Ages—the sixth and seventh centuries in particular—were especially rough. Leprosy was rampant. Beggars were ubiquitous. Blindness was common. Nutritional deficiencies led to rickets and poor teeth: no one knew how to conserve produce, and whatever fresh fruit and vegetables were available often lacked vitamins. Many children were born with defects and deformities; child mortality was high. What we know as diabetes was a common outcome of inconstant diets, overly rich one month and deficient the next.

And so, although the medical rationalism inaugurated at first with the Hippocratics in fifth-century BC Greece was slowly being revived in the later European Middle Ages, it inevitably continued to coexist with the religious impulse to effect miraculous cures. Miracles were often witnessed in the form of the healing of ailing bodies. In that sense religion was itself a form of medicine—just as medicine could be a religious exercise, concerned with the body's passage from life to death. Relics were thought to have miraculous powers, and their presence in a church helped confer on it its sacredness.[4] The course from health to illness, from order to disorder, could be reversed if only one contemplated divine mysteries ardently enough—if only one was a good enough Christian.

Still today, it is not uncommon for the committed faithful to hope for miracles and report their occurrence with zeal. The need for the miraculous—for the advent of a good by means that seem unexplainable without appeal to a divine force—has never died. This need was particularly acute in this early part of the first Christian millennium, when more saintly miracles were invented than new theories of human nature or new dimensions of learning. The realm of the body, too often in pain, was in day-to-day life less urgently an area of study than the preamble to the afterlife. The divine maker of the body mattered more to defining a good life than did the fulfillment of bodily needs. This is partly why the healing powers of Christian saints—and of pagan rituals too—were considered more potent than those of lay physicians. Tales were told of resurrections

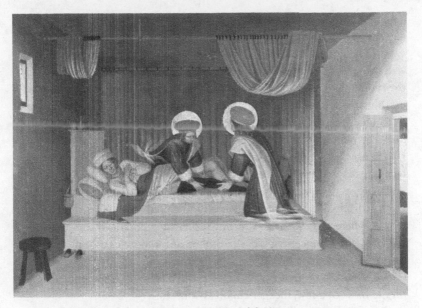

Fig. 10. Fra Angelico, *The Healing of the Deacon Justinian by Saints Cosmas and Damian*, from Pala di San Marco (1440–1442). Florence, Museo San Marco.

of children and lightning cures of paralysis. Saint Augustine described in his *Confessions* how a blind man regained sight after putting to his eyes the handkerchief with which he had touched a bier containing the relics of the martyrs Protasius and Gervasius, brought by Saint Ambrose to the cathedral in Milan for its consecration. Visual contact with remnants of the bodies and belongings of saintly souls, or with iconic representations of the bodily form of holy beings, had at least an apotropaic function. Miracles, moreover, were both physical and spiritual events, exceptions to causal laws over which humans, in any case, had little control. A miracle had no knowable causes, by definition: the result was all that mattered, the end point of all struggle. Human knowledge paled next to the miraculous occurrence. Those who witnessed one were touched with holiness themselves.

Until the twelfth century, lay physicians were not always welcomed by the clergymen for whom the body's health depended on the soul's integrity, rather than the other way around. The combination of monastic learning and piety could on its own amount to a healthy regimen for both soul and body; and not surprisingly, it was in the libraries of monasteries such as Monte Cassino that, in the early Middle Ages, classical knowledge was studied, preserved, translated, and copied by monks eager to emphasize the spiritual dimension of the care one must bestow on the body. At the founding of Monte Cassino in 529, the Rule of Saint Benedict was first established, setting influential guidelines for the regulation of the monk's daily life, of moral and physical conduct, and of the moderate but systematic disciplining of the spirit and the senses. It ensured the monastery's role in providing for and controlling most ordinary human needs, guaranteeing order and the ascetic but humane provision of goods, such as food and clothes. Its central concern was spiritual, social, and corporeal health. Lay medical specialists were generally uncalled for within the monastic world, where a doctor who was not primarily a doctor of the soul was irrelevant, if not objectionable. A number of monks came to possess basic medical skills, which they gleaned from the summaries of Galen and Hippocrates passed on by Constantinus and from treatises assembled locally on the basis of a few, meager classical sources.

But newly compiled knowledge spread beyond monastic walls and into a socially diverse world of secular life, where universities and professional training were to create hierarchies among medical healers. Priests and doctors were professionally close; but it became increasingly possible in the twelfth century to study medicine without belonging to a religious community. The medical school of Salerno had in fact been active from around the ninth century, forming both lay and clerical practitioners on the basis of the fragments from Galen's *oeuvre* that had been extant in the West before the influx of Arabic translations. From the eleventh century, Latin translations from the original Greek texts had begun to appear, such as the translation by a monk at Monte Cassino named Alphanus of a fundamental Galenic tract by the fourth-century Bishop Nemesius of Emesa (Homs, in today's Syria), *On the Nature of Man*. By the twelfth century, Latin translations of the Arabic versions of the classical corpus, including the *Liber pantegni,* were being circulated among wider and wider circles. After the Christian reconquest of the Iberian peninsula, Hebrew translations by Jewish doctors in Spain, Italy, and France were also being translated into Latin. The amount of available information about anatomy, physiology, prognosis, and diagnosis was augmented considerably.

Medical learning was now spreading northward. By the mid-1100s, Montpellier had become an important center, to which began to flock the young men and even some women whose curiosity about the organism was not satiated by monastic culture. Paris, Bologna, and Padua also soon had their established, influential schools. Members of the Salernitan school created a collection of medical writings, known as the *Articella* (the "little art"): by the mid-thirteenth century, this brief, usable compilation was the foundational textbook for most medical teaching in the West. It included the Hippocratic *Aphorisms* and *Prognostics*; Galen's short *Ars parva;* the medically essential and thus ubiquitous treatises *On Pulses* and *On Urines;* and the extensive compendium of Galenic writings by Hunayn ibn Is'haq (Johannitius), the *Isagoge Ioannitii in tegni Galeni,* in the translation by Constantinus Africanus.

Aristotelian natural philosophy was absorbed into the corpus of medical texts known and studied at that point, giving it a strong speculative slant that also transpired in the many commentaries on Avicenna's *Canon,* now circulating widely in a Hebrew translation in both Spain and Italy. Parts of the opus were used in universities—where those whose interest lay in medicine had also to follow a liberal arts curriculum. Other parts were discussed for their philosophical and methodological import rather than for any actual information on anatomy and physiology. Intellectual discourse in general was becoming theoretically sophisticated—and it informed the high end of medical learning. From the twelfth century, alongside the Galenic medical corpus, important works of Aristotle were also being furiously translated into Latin and Hebrew, from the Arabic and from the Greek—endowing the still small, but growing world of lettered people with a rather refined body of thought that encompassed all knowledge about nature, the cosmos, the human body, and the human soul, and the relation between all of these. Aristotle's volumes of natural philosophy were soon in full circulation, as well as his thoughts on logic and on philosophical method. By the middle of the thirteenth century, Thomas Aquinas was studying a complete version of the *Metaphysics,*[5] following in the footsteps of the theologian and natural philosopher Albertus Magnus. Christianity and Judaism were absorbing Greek philosophy, along with the Arabic texts based on it, especially those of Averroës and Avicenna.

In these ways, reason operated hand in hand with faith, although strenuous philosophical reflection was needed to work out the extent to which it could coexist with simple contemplation, the absence of thought and enquiry, the silent acceptance of divine presence, the pure belief in divine power and in the divine word. Rational speculation now reached high levels of abstraction, grounded within Aristotle's rather abstruse methodological framework for the universal classification of knowledge. By the thirteenth century, scholastic thought had reached its apex.

2. Scholasticism and Humoural Care

I T WAS THE VOCATION merely of the educated classes to speculate about the soul, about the nature of matter and its relation to form. Medical practitioners also engaged in these speculations; and some physicians in fourteenth-century Paris went on to take a further degree in theology, smoothly moving from body to soul. But the accumulation of layered texts, the sophisticated scholarly and theological debates—about body and soul, matter and form, universals and particulars, parts and wholes, kinds of causes and kinds of effects, free will and determinism, activity and potentiality, intention and action, space and motion, time and eternity—might seem to have little to do with the actual run of ordinary lives.

In fact, intellectual reflection did not accompany the training of all clinicians. Types of qualification varied enormously, and did not always depend on the formal education provided by the growing number of universities. Lower in the medical hierarchy were empirics—akin to today's specialists—barber-surgeons, apothecaries, and female midwives. They were not always schooled and proficient in Latin, but they were trained and licensed, and they provided essential medical care. The humblest physicians, including village healers, were rarely licensed, although their livelihood depended on their reputation and there was rarely any shortage of clientele. Everyone, in other words, could get some sort of medical attention; and those who were better off were able to afford the services of the most highly trained practitioners.

Yet, in a society where religion governed everyday lives, charitable care was a form of piety; the sophisticated debates could have an urgent ring to them, and provoke real outcomes. Both the Church and religious concerns had a profound impact on medical practice: not all methods of treatment were religiously acceptable, just as not all research methods were compatible with Church doctrines. In 1215, for instance, the fourth Lateran Council—called by Pope Innocent III to gather clergy from around the Christian world—forbade "any physician, under pain of anathema, to prescribe anything for the bodily health of a sick person that may endanger his soul," since the patient's soul was worth much more than the body and "sickness of the body may sometimes be the result of sin." Doctors— "physicians of the body"—were therefore ordered, "when they are called to the sick, to warn and persuade them first of all to call in physicians of the soul so that after their spiritual health has been seen to they may respond better to medicine for their bodies, for when the cause ceases so does the effect."[6] Implicit here was the assumption that doctors were first servants of God, and that the medical arts had therefore little to add to the matter of health. Physicians of the soul—that is, members of the clergy— were more useful than physicians of the body. Regardless of the social status or wealth of the patient, the body was first and foremost the envelope of the soul. And no one would have doubted that.

Of course, the care of the body was a concrete matter: a physician's job was to delve into the organs and sinews, fluids and blood beneath the skin. The actual practice of bloodletting was rather risky, and barber-surgeons soon took on this lowly office exclusively. Physicians' reluctance at engaging directly with the open body compounded their lack of any sense that their old manuals might need updating. Instead, trained doctors became increasingly good at their philosophical speculations about the essential qualities of the humoural sources of life. The availability of texts such as the *Almansur* of Rhazes and Avicenna's *Canon* in particular ensured an ongoing stimulus to engage in discussions that were of rather little use for flesh-and-blood patients.

Rather than stoop to examine real viscera, physicians continued to debate questions about the nature of the mind. It was Avicenna who had revised most radically the "ventricular" theory of the soul put forth in the fifth century by Bishop Nemesius of Emesa. According to this theory, imagination was housed in the anterior ventricle, reason in the middle ventricle, and memory in the posterior one.[7] Avicenna, helping himself to the theory Aristotle had developed in his treatise on the soul, *De anima*, had complicated the picture when he divided the faculties of the soul into five "internal senses": in fact he was partly responsible for bequeathing to the later Middle Ages and early Renaissance a complex, quite fanciful psychology.

These complexities sustained the system that had produced them. When blatant contradictions were noted between texts considered equally authoritative, there arose disputes regarding the right way to interpret them. (One famous debate concerned the reasonableness of supposing that the heart was the seat of the soul, as Avicenna had maintained.) But however the disputes were resolved, there prevailed the belief that truth lay in the words or testimonies of one's predecessors. A learned doctor was one who had primarily read many books, rather than treated many patients.

Certainly the newly translated Arab texts injected a new vitality into medicine, and led to dissections—of condemned criminals only—being resumed in the fourteenth century, probably for the first time since Ptolemaic Alexandria. Dissections became a mandatory part of the medical curriculum in the major medical colleges; but observation was not yet clear enough to displace the authority of the ancients and of their heirs. One needs a map to make sense of bloody, decomposing organs. To view these bodily innards then (and such views, in any case, were still quite rare occurrences) was to see what the existing maps had recorded. Assumptions about the order of souls and the flow of humours were mostly fixed. These assumptions colored vision, and were by now invested with an authority that simply could not be displaced, even by the dissection of human corpses. If anything, anatomical knowledge confirmed humoural flows.

3. Old Convictions

ONE CAN GLEAN how much theory confirmed physical evidence especially from an epochal anatomical text, the *Anatomia,* by Mondino de' Liuzzi, a Bolognese physician who composed this work in 1316. He seems to have intended it for use as a didactic guide to dissections. Mondino entreated medical students to read the ancients' texts while making observations of the body before their eyes; he encouraged in them the independence he himself manifested, and carefully set out the program of dissection for the sake of optimal clarity. He did this according to the

Anathomia Mū dini Emēdata y doctozé melerstat

Fig. 11. Print from Mondino de' Liuzzi's *Anatomia* (Leipzig: Martin Landsberg, c.1495).

rate of decay of each part (abdomen, thorax, head and brain, limbs), in terms that the great anatomist Andreas Vesalius himself would follow over two centuries later.

But this was not a wholesale refutation of the Hippocratic-Galenic superstructure. Mondino read Galen's notions into the bodies he dissected.[8] One of these was the *rete mirabile* that Galen had imagined existed at the base of human brains, although he could have seen it only in the brains of oxen or sheep. The other was the idea, transmitted from the 1200s by a text called *De spermate (On sperm)*, spuriously attributed to Galen,[9] that the uterus had seven "chambers." The right side of the uterus generated males because it was under the influence of the liver, which was hot and moist, and of the gallbladder, which was hot and dry. The left side generated females because it was under the influence of the spleen, which was cold and dry, and of cold and moist phlegm.

There was a physiological correlate to this anatomical picture. If the woman's sperm—as it was then thought to be—was stronger and combined with male sperm on the right side, there would emerge a male with female qualities; and if the man's sperm was stronger and combined with male sperm on the left side, there would emerge a female with male qualities. Sperm was understood to consist mostly of blood (not the humour), mixed in with humours: first the blood became "a hot and moist substance which has the nature of phlegm in arteries and veins to nourish them," then it turned these substances into sperm. In men, sperm descended from the head and could even block the nostrils and ears—otherwise it descended straight into the testicles, through one vein and artery on either side, via "each of those places where the four humours rule and acquires the nature of the humour that dwells there." This was how temperaments were passed on: sperm "acquires the nature of the four humours, and a man made of that sperm retains the nature of those humours, although one of them will prevail." An elaborate theory of embryology and heredity followed, correlating the hour, and hence the predominant humour, of conception with the complexion—temperament and looks—of the resulting child, and factoring into this correlation the relative strength or weakness of the male and female sperm. All along it assumed, as Galen, in Aristotelian vein, had done in his own authentic text *De semine,* that the female semen was indeed weaker and colder than the male semen, although it was active (Aristotle, followed by Aquinas and Albertus Magnus, had presumed it was primarily receptive).

In spite of the regard in which Mondino was held throughout the history of early anatomy, the notion that the uterus had seven chambers would be criticized by the more innovative anatomists in the sixteenth century. But the pseudo-Galenic text was still being referred to in treatises on embryology by then, along with Aristotle's *De generatione animalium.* Even by the seventeenth century, reproduction remained poorly understood, and the inheritance of biological and temperamental traits even less so. There was no reason not to believe, as did the unknown writer of *De spermate,* that if the sperm was "discharged in hours of choler," for instance, the resulting child would have a "red and white complexion, tending towards yellowish brown," that he would suffer "from attacks of fever in his small veins," that his bones would be brittle, and that he would suffer "all over his body from dry cold caused by choler descending from the brain." Humoural determinism was perfectly served by anatomy: the canals through which humours traveled were visible—or seemed to be.

4. The High and the Low

THEORY THUS FLOWED as smoothly as the humours it supported. The body had been mapped and the relations between recognized anatomical parts had been charted, according to the Aristotelian categories of form and matter, activity and receptivity. But however concerned with the finesse of argumentation, learned doctors were not only preoccupied with words. Medicine remained, after all, both a collection of theories and a practice—both a science and an art, whose object was the human organism and whose subject was its state of balance or imbalance. The goal of practical medicine was to help maintain existing health and also, of course, to rectify the imbalances that accrued in illness. Its tools, true to the Hippocratic tradition, were threefold.

The initial step was the establishment of a diagnosis—on the basis first of the pulse and urine, and then of symptoms such as swelling, fever, nausea, vomiting, headache, pus, pain, and breathlessness. In the case of "spleen jaundice," for example, there circulated a description from Galen according to which the symptoms to look out for were: "dryness of the body; the color of the stools is not unusually white; urine tends to be black, indeed the whole body color tends to be black. The whole is followed by twitching, loss of appetite and dislike of sweet foods"; in the bad cases, patients had "sleeplessness, depression, restlessness, and they cannot perspire. Some patients perspire very slightly and lose a little bile with the perspiration. Others perspire many times in the bath. Finally there are those in whom bile comes out in the nasal mucus and in the secretion from the eyes."[10]

After such minute, humourally framed observations, the doctor had to be able not only to prescribe a treatment, but also to elaborate a prognosis, on the basis of further symptoms as well as of external, usually meteorological and astrological conditions or "signs."

The third step was the cure of patients, which could not be planned in the structured way a treatise was. A good doctor, for this reason, was one who could adapt fast to a body's changing states, pay attention to external as well as internal data, and improvise even when the initial prognostic was not panning out. It was hard to keep track of the bounds and leaps the organism made. Doctors could treat only symptoms that corresponded to cases of which records or histories existed already. Other symptoms were made to fit into the body's manifestly hydraulic story, its imagined fluids flowing in and out of veins and arteries, cavities, organs, and senses, responding to heat and causing cold, too acidic or too thick, excessive or insufficient, sluggish or burned.

The humoural scheme, then, was not only theoretical: humours and complexions were to some extent the names for the body's inner world, for the real hydraulic processes that *umor* could possibly refer to—sweat, sputum, pus, bile, blood, urine, semen, milk, and mucus. Humours explained digestive problems and accounted for disruptions in the flow of blood. But humoural hydraulics did not always provide a sound basis for a credible diagnosis, and doctors often needed to resort to intellectual acrobatics for their humoural analyses to make sense of the illness.

One of the most famous (and commensurately expensive) physicians in Bologna during the late 1300s was Taddeo Alderotti.[11] Alderotti was well versed in medical scholarship, and used Aristotelian philosophy as a framework within which to search for the underlying causes of illness. He exchanged with his pupils a series of *quaestiones,* or questions, concerning medicine, and noted a series of *consilia,* advice for the diagnosis and treatment of individual cases, based on evidence gleaned both from consultations and from the theory they all studied so assiduously.

When a doctor set out to understand, for instance, the cause of a "putrid fever," he had to look for a "first principle," as did Pietro Torrigiano, a pupil of Taddeo. This is his paraphrased account: "To demonstrate the cause of a putrid fever, it is sufficient to say that a putrid fever is produced inside the body and heats its container. But if one then inquires why the putrid humour is produced, the answer is because the prohibition of transpiration corrupts the natural heat; the reason for the prohibition of transpiration is the presence of obstruction in the passages of the body. One of the causes of obstruction must therefore be present; and these are either too many humours, or their grossness or viscosity, or excess of coldness or dryness or of dry heat. Whichever of these conditions predominates in any given instance will be the cause in that instance. Suppose that the cause in the case in question is an excess of humours; the cause of this in turn is overeating and the resultant indigestion. This may be due to excessive appetite, which is the result of a faulty constitution [or complexion]; and the last named has, in turn, its own cause. The method of composition, however, requires one begin with the first principles and argue from them to the final manifestation of their activity."[12]

Humoural theory categorized and ordered presumed events in the organism according to (Aristotelian) causes. Intuition and bedside manner aside, it was only by reconstructing the stages in the development of what Torrigiano understood as an "obstruction" that made it possible for the physician to prescribe what he believed to be the correct treatment. Given that the body was intrinsically constituted by its humours, disturbances within the organism were necessarily humoural. Humours, quite simply, bridged medical theory and practice.

But beyond that, establishing the causal history of a disease could be tricky, and the diagnosis of "obstruction" could be used even to obfuscate the physician's own confusion and save face before a terrified, tried patient. The renowned physician, chemist, priest, and astrologer Arnau de Villanova—born in Milan around 1240, and eventually a resident of Barcelona—even admitted so much: if you don't understand the case, he said, just mention obstruction, because "they do not understand what it means, and it helps greatly that a term is not understood by the people."[13] If the doctor seemed knowledgeable, he would be worthy of trust, and his services worth paying for. The mere appearance of authoritativeness could, and indeed still can, reassure the patient.

Patients could rarely hope for much more anyway. On the whole, doctors were much better at obfuscating than at clarifying the nature of illness. It was of course harder to tease out the body's partly visible, partly imagined intricacies than to ascribe its ailments, as the clergy did, to the relation of the wholly invisible soul to God. Doctors were not omniscient. But trust between patient and doctor, then as now, was paramount. Medicine was becoming a structured profession, with new responsibilities; and its authority and autonomy had to be sustained, however tentative was the application of theory to bodily realities. As doctors gathered into new institutions—a College of Physicians was founded in Venice as early as 1316, for instance—the initial involvement of the educated clergy with medical learning diminished. It would never entirely disappear; confraternities and hospices in the fifteenth and sixteenth centuries were still run by religious orders, and nuns continue to tend to the sick in some hospices even today. But by the thirteenth century, physicians already constituted their own hierarchies, with their own priorities, programs of study, qualifications, and clientele.

Those who could afford well-trained physicians—the products of the top medical schools—tended to be privileged themselves. Some of these patients knew the large body of texts underpinning the treatments prescribed for them. Literature, too, testifies to the wide availability of humoural theory as an authoritative account of the workings of body and mind. In *The Nun's Priest's Tale* from the *Canterbury Tales*, Geoffrey Chaucer has Pertelote summarize the medical lore in these colorful terms:

> *Dreams are, God knows, a matter for derision.*
> *Visions are generated by repletions*
> *And vapours and the body's bad secretions*
> *Of humours overabundant in a wight.*
> *Surely this dream, which you have had tonight,*
> *Comes only of the superfluity*
> *Of your bilious irascibility,*
> *Which causes folk to shiver in their dreams*
> *For arrows and for flames with long red gleams,*
> *For great beasts in the fear that they will bite,*
> *For quarrels and for wolf whelps great and slight;*
> *Just as the humour of melancholy*
> *Causes full many a man, in sleep, to cry,*
> *For fear of black bears or of bulls all black,*
> *Or lest black devils put them in a sack.*[14]

These lines by Chaucer convey the extent to which the notion of a complexion and of the humours that constituted it was taken to be a given. Even those patients—the majority—whose medical knowledge was nonexistent held humoural beliefs: humours were accessible to all, easy to understand, and intuitive. They were at once a "folk" and a "high" theory, as essential to self-explanation as genes are today, and in some ways as little understood. It is also true that everyone in this deeply religious world, poor or rich, commoner or aristocrat, learned or illiterate, subscribed to the notion that dreams, visions, and miracles were not merely the outcome of the material body—"a matter for derision," as Chaucer put it, "generated by repletions." They could be experienced spiritually, as messages from God or, by the same token, from the devil. In the popular realm especially, piety could turn into superstition, and into the belief that invisible, supernatural powers operated in the world.

There is a difference between explaining dreams and visions in such supernatural terms and accounting for them in crudely physical terms, merely as the outcome of humours. The fourth-century compiler Oribasius had transmitted the Galenic notion that bad dreams presented nocturnal symptoms that were similar to the diurnal symptoms of epileptics; they were thus precursors of a potentially serious illness, such as apoplexy, mania, or indeed epilepsy, and necessitated a swift purge before things got out of hand. Ingredients that effected such purges included the juice of scammony, a strong-smelling substance such as aniseed or parsley, and the violently poisonous black hellebore.[15] It was also best to stick to a light diet and to avoid foods that induced flatulence. That was medical, not spiritual, advice.

But both supernatural and natural explanations were current; and both required a certain amount of credulousness to pass muster. The body was not just material and humoural: it was fashioned in God's image, a part of the divinely created universe, the place at which spirit and matter met. Remedies were supposed to change the ailing body's condition to some extent and alleviate suffering; but they could be painful, and unintentionally lethal (especially given how often doctors prescribed black hellebore). It was therefore neither physical evidence nor textual authority that sustained the patients' willingness to undergo prescribed treatments, but the belief that these awful, dangerous treatments would work; and the prayers that accompanied the cure multiplied when it failed.

5. Outsiders

THE DOCTOR TO whom the patient bowed in the hope of being saved was not the scholar or priest: he was a technician who could prescribe, concoct, and adjust. When physical intervention was necessary, he was assisted by the less educated midwives or barber-surgeons. But the members of the medical elite were an exclusive guild. They too had to believe in the efficacy of the treatments they prescribed, and they were keen for their professional integrity to be reflected in their ability to cure. Although from the early 1200s on, there actually was a curriculum for the training of both women and barber-surgeons in practical clinical care, female practitioners could not be anything other than midwives. Elite physicians clearly considered women to be more medically effective as saints than as doctors.

Yet there does exist a series of treatises on gynecology and general female matters—including cosmetics—which was gathered into a compendium called the *Trotula*. Some of the texts might have been written by one or a few women; all, in any case, are anonymous.[16] In one of these texts, *On the Conditions of Women* (where one also finds the notion that "women are by nature weaker than men"), we are told that at the beginning of pregnancy, "care ought to be taken that nothing is named in front of her [the woman] which she is not able to have, because if she sets her mind on it and it is not given to her, this occasions miscarriage."[17] The belief that perception had an immediate impact not only on physiology but also on the fetus would endure for centuries. It was still common in the seventeenth century to believe that if a pregnant woman had been frightened by a tiger, say, her baby might emerge covered with stripes.

This idea is perhaps not so far-fetched as might seem: after all, what the mother ingests, feels, or hears affects the fetus, and the belief could easily follow that what she sees also has a powerful impact on the developing child. Pregnant women were considered fragile and impressionable. A woman's physiological life was considered identical to her psychological life. She could be deeply affected by her womb, which was viewed as an animal-like organ prone to "wandering": it impelled her to procreate, lest she become, literally, a hysteric—a word that derives from *hysteron*, "womb" in Greek. Another long-lived notion was Galen's belief that menstruation was a form of purgation, necessary in women, who were cold and wet, as opposed to the hot and dry males. Again, this is not such a far-fetched idea. When the body expels a substance, solid or liquid—of humoural origin—some sort of "purgation" must clearly be going on.

But obstetric and gynecological matters were not high in the hierarchy of medical practice. Nor were the expulsions and purgations of daily life a noble business. Elite physicians were wary of barber-surgeons, those who, like nurses today, had to actually dirty their hands with the body's humoural secretions and excrement. Doctors strove to maintain their social status and did not welcome the forays into medicine of those whose qualifications derived from "folk" callings, or who belonged to the "wrong" social or religious group. A Salernitan poem of the mid-thirteenth century condemns "the unlettered, the empiric, the Jew, the monk, the actor, the barber, the old woman," each one of whom "pretends to be a doctor, as does the alchemist, the maker of cosmetics, the bathkeeper, the forger, the oculist. While they seek profit, the power of medicine suffers."[18]

Surgeons, barbers, and apothecaries—all of whom received their own formal, certified training—were necessary, but the author of these lines clearly wanted to undermine their importance, in order to safeguard the authority that official medicine had built up for itself. Both Jews and women, typically enough, were perceived as threats to the unity of the medical corps; and otherwise it was those whose grasp of clinical care did not depend on theory who were not welcomed by the medical, mostly male, university-trained, Latinate, and profoundly Christian establishment.

In fact, though, throughout the late Middle Ages, unchartered outsiders were caring for the populace and assisting at births. The gap between practice and theory, already present in Galenic times, was wide. Even within the university, each realm had its respective literature and references. One can glean from the medical curriculum at Bologna a couple of centuries later, in 1405,[19] that practical medicine still tended to rely especially on Avicenna's *Canon,* while the readings for theoretical medicine consisted in the main of a variety of Galenic and Hippocratic writings, along with some by Avicenna and Averroës. This gap did not diminish with time. On the contrary, during the centuries to come, the gap between practice and theory would grow.

6. Bloody Treatments

B UT ULTIMATELY, prescription had to be an empirical matter of attention to the individual patient's needs. The remedies prescribed by physicians could take the form of dietary guidance, medicinal and herbal drugs, or surgery. Barber-surgeons who knew their anatomy and physiology were placed on a lower rung of the medical hierarchy, but they could be trusted to properly saw a leg off, operate on a hernia, get a stone out, repair a fracture, bandage a wound, or perform a clean splenectomy. They were also responsible for phlebotomies: common but sometimes spectacular bleedings, considered necessary on the strength of the notion that noxious humours had to be literally expunged from the organism. These bleedings were performed with a selection of somewhat alarming tools, such as lancets and scarificators, as well as cups that, when heated, drew blood to the surface by creating what, in the seventeenth century, would eventually be recognized as a vacuum.[20]

Leeches were famously used for these purposes too. They are a certain type of bloodsucking parasitic worm today known as *Hirudo medicinalis* that breeds in freshwater areas (such as ponds, marshes, and lakes) and is equipped with suckers at both ends of its body. Given its predilection for mammalian blood, a leech needed only a modicum of encouragement to attach itself to its prey's flesh, saw its way through the skin, and forcefully suck out up to ten times its body weight in blood before letting go, satiated. The leech's bite would not have hurt, despite its tripartite jaw and 300 teeth. The animal, it has been found, injects an anesthetic into the skin. Moreover, its saliva contains a blood vessel dilator and anticoagulant (called hirudin), useful for the purpose of phlebotomy since it could ensure a good ten hours of bleeding after the beast had finished its job.

These treatments could be effective; they worked especially well in the cleansing of wounds. We now know that these rather unattractive animals are excellent microsurgeons, or portable pharmacies of sorts, capable of favoring the growth of new blood vessels and so of reducing the swelling of various sorts of wounds and burns. In June 2004, in fact, the Food and Drug Administration reported that it had "for the first time cleared the commercial marketing of leeches for medicinal purposes."[21] Leeches, in other words, are back, prized for skin grafts, for the care of amputations and gangrene, for plastic surgery, and for the reattachment of delicately vasculated organs such as ears, fingers, or toes. And because it is an anticoagulant, hirudin is also effective against strokes and heart attacks.

The power of leeches to cure is not explained today by reference to humours. It is likely that the discovery of these animals' effectiveness was an outcome of empirical trials, rather than of humoural theory. But the justification for all phlebotomies was humoural. The medical goal of the surgical operation of bloodletting was to purge the "bad" humours that might have caused fevers, stomachaches, or headaches. Humoural theory served well as theoretical ballast for time-tested remedies, but practitioners often tended to serve it too well. Bleedings could go too far, and might be stopped just when the patient began to feel faint. It was usual to draw as much as twenty ounces of blood at once, and when this was done to remedy a fever caused by an ailment such as phthisis—identifiable today as tuberculosis—the result was deleterious, if not fatal. Yet bad outcomes did not diminish the explanatory power of the humours.

Physicians, in any case, did not always resort to bleedings for the purgation of humours gone bad. A range of non-life-threatening ailments which were recognizable and controllable, and whose symptoms had been identified and named by the ancients, could call for a variety of treatments other than phlebotomy. These ailments included colds, coughs, benign fevers, headaches, indigestion, muscular cramps, skin rashes, burns, and even insomnia. Recommendations against systematically resorting to bleedings would accompany their effective use throughout the centuries. In the 1450s, Benedetto de' Reguardati da Norcia, physician to Francesco Sforza, Duke of Milan, suggested that one should not resort systematically to phlebotomies, indeed that "we must take care as best we can of blood, which is the treasure of nature."[22]

In general, though, if the humoural aetiology was clear, illnesses caused by blood were best cured by bleedings. Those caused by excessive phlegm might be cured by medicines that induced sweating or expectoration. Ailments due to yellow and black bile were best treated by laxatives and emetics, often potent enough to cause reactions whose violence might override the initial symptoms, to the extent that the patient would probably feel convinced that they were truly effective. And sometimes, of course, they were.

An important tenet for the prescription of treatments was the notion that "contraries," or opposites, attracted and were cured by each other (contraria contrariis curantur in Latin). For instance, a cold disease had better be cured by a hot substance. This was the principle, in modern terms, of allopathy—the word derives from the Greek allos, "different"—which is still the principle sustaining mainstream medicine today. It followed that "like" was repelled by "like": one could find out whether a disease was hot or cold by testing whether the organism drew in or expelled a hot or cold substance. There is an interesting case from the early twelfth century involving the abbot of Cluny, Peter the Venerable. Afflicted with a respiratory disease, Peter was advised by doctors to abstain from his regular bleedings because the loss of heat that accompanied them caused the presence of "sluggish phlegm diffused through the veins and vital channels."[23] His doctors instead recommended moist substances, which included "hyssop, cumin, licorice, or figs steeped in wine and syrups of tragacanth, butter, or ginger."

But Peter was adamant that his disease was itself moist and therefore would be cured by dry substances. So he consulted a learned physician, but this physician's opinion merely confirmed the first one. The doctor tried to persuade his recalcitrant patient that moist remedies were "potentially" dry, recommending "hot baths, inhaling medicated steam, poultices for the chest, lozenges to dissolve in the mouth, gargles, and, for good measure, a laxative."[24] (This is not very different from the treatments we would resort to today to relieve the unpleasant symptoms of a bad cold.) Empirical evidence overrode the explanatory use of the humours. To accord the humoural order with the allopathic principle, the doctors had to invent a new series of explanations to justify their advice. Peter himself, in fact, was convinced that his was a case of "the disease called catarrh"; and it did eventually resolve. The doctor's advice had been good, but he had barely been able to reconcile it with what he knew about the body. Valuable empirical knowledge was often undermined by learned theory.

7. Verbena, Olives, and
Herbal Power

ENEMAS AND CLYSTERS; poultices; vinegar- or water-based syrups with liquorice or endive; potions, powders, pills, and ointments made of such diverse ingredients as mint, chamomile or rose oil, aloe, tamarind, mastic (the resin from the mastic tree), *terra sigillata* (an astringent red clay, in use since antiquity), almonds, saffron, butter, absinth, turpentine, and corals—these were the remedies used, in various forms, until the advent of microbiology and the birth of the pharmaceutical industry in the nineteenth century. Methods of treatment like bleeding and purging were called "drastic"; the softer ones, which included fomentations, inhalations, infusions, and so on, were "non-drastics." We would call them home remedies today, and many were indeed composed of ingredients that are surprisingly familiar to us. Aloe vera, for instance, commonly used today for digestive and dermatological purposes, was also frequently used as a digestive in medieval and Renaissance medicine. Numerous compounds, however, have disappeared from even the most obscure health shops.

One of these is theriaca, a spice-based stimulant composed of dozens of ingredients—up to seventy—including those bought at great cost in eastern markets such as cardamom, ginger, pepper, saffron, frankincense, and myrrh, but also viper flesh and opium. Galen had popularized the prescription of theriacas, used at first as antidotes against venom and poisons (the word comes from the Greek *theriakas,* of dangerous beasts) and even wrote a volume on these reputedly potent antidotes. Over time, they had become extraordinarily popular and were used for countless conditions; the more esoteric the ingredients, the more expensive and sought-after they were.[25] They were probably not very effective—but that never detracted from their perceived value.

All the formulas that composed "drastic" and "non-drastic" medications were based on the old pharmacopoeia, or *materia medica*—the classical, authoritative set of sources of botanical knowledge. These included Hippocrates and Galen, as well as Dioscorides, who, in the middle of the first century AD, had cataloged some 600 plants, roots, trees, minerals, essences, and balms in his *Materia Medica*. It had been rendered from the Greek into Arabic in the ninth century in Abbasid Baghdad (Rhazes and Avicenna had referred to it copiously) but a Latin translation existed in the Byzantine West as early as the sixth century. Copies of it were as precious and valuable as the information it contained. Throughout the medieval and Renaissance periods, Dioscorides remained the central source for herbal remedies.

There were other ancient texts on "simples," the basic natural components. The great Theophrastus's *Enquiry into Plants* would be only first published in a Latin version in 1483, but Pliny the Elder and Galen were available early on.[26] Along with Arabic sources, these were recycled into the herbals that circulated in medieval Europe. In the thirteenth century there appeared in northern Italy a *Tacuinum sanitatis,* translated from Arabic and heavily illustrated, as were by now the other medieval herbals.[27] One collection of simples called the *Liber simplici medicina,* collated in twelfth-century Salerno from both Latin and Arabic sources, would become known as the *Circa instans.* This book would circulate widely, in numerous translations.[28]

All these treatises delivered an elaborate and detailed sum of botanical knowledge, and transmitted from generation to generation empirically established information on the medicinal properties of plants, along with specialized advice on their use. The botanical illustrations—present already in some early, fourth-century AD manuscripts, but increasingly found in the Middle Ages—were necessary for the correct identification of the plant, and they would become increasingly sophisticated. But on the whole, knowledge was transmitted via books at least as much as via the plants themselves: each new herbal was based on previous ones. A work that was popular in seventeenth-century England, for instance, was the *Herball or Generall Historie of Plants* by John Gerard, a sixteenth-century surgeon and gardener. It had been revised into a lavish, comprehensive and still useful edition by a later apothecary and botanist, Thomas Johnson. But it was itself based on a Flemish work, whose illustrations were mostly recycled from those found in the other great botanical books printed throughout Europe in the sixteenth century, by the likes of Leonhart Fuchs in Germany, or Pierandrea Mattioli in Italy.[29]

In the Gerard-Johnson book and in its predecessors, herbs and spices were described in botanical terms, but also in terms of their effects on the humoural body and their chemical characteristics, the two of which were sometimes connected. The authors fully adopted the medical apparatus transmitted from antiquity; and since fevers were the primary clinical preoccupation, they often specified what sort of fever each plant was good at fighting. Indeed, fevers had been classified since Galenic times as simple or quotidian if they were due to phlegm; tertian if they were due to choler; quartan if melancholy was the cause; and continuous when due to blood. They could also be acute, or hot.

The *Circa instans* is a remarkable specimen of the *Materia Medica* manual. It was not strictly a pure herbal, since it also included minerals and metals—such as asphalt and hematite, ammoniac and turpentine, storax, adraganth and bdellium, sulfur, honey and soap, glass, gold, alum, coral, and mercury. But all these, along with plants and spices, were considered for their medicinal properties. They could be cold or hot, and recipes for their use depended on their further classification into degrees.[30] One reads in the *Circa instans* that the herb called *semper viva*, for instance, was cold "to the third degree" and dry "to the first," and was efficient against headaches when its juice was mixed with rose oil; and that when chopped it worked well against red eyes, burning swellings, or hot gout. One also reads that valerian was diuretic and effective against dysuria (painful urination) when cooked in wine with fennel seeds and mastic, or, as a decoction, against a cold-induced obstruction of the liver and of the spleen.

Sage cooked in wine was deemed a good remedy for paralysis, contusions, and epilepsy. Black olives (which are ripe) were hot, a little dry, but also moist—and here the *Circa instans* refers to both Dioscorides, who claimed they were dry, and Galen, who claimed they were moist. They softened the stomach; their digestion was delayed because they oiled the stomach before descending into the intestine, and so were not as healthful as green (unripe) olives. Moreover, they engendered a blood of bad quality; and if they were pickled, they turned into a choleric humour. All olives, in fact, produced a humour of corresponding tint: red ones produced red humours (a version of yellow bile), black ones produced black humours. Red olives contained less oil and so were more healthful; they even comforted the stomach.

But the *Circa instans* was also replete with recipes of a more superstitious nature, that partook of magic rather more than of medicine. The entry on verbena (also known as lemon verbena, now used as a digestive, a relaxing herbal tea with a lemony taste) is a case in point. The guide tells us that the plant is cold and dry, that its juice is a universal antidote, that water in which a few leaves have macerated is good against tertian or quartan fevers. But it then assures us that four leaves and four roots of the plant cooked in wine, then sprinkled over the dining area, will ensure the good spirits of one's guests. The leap from the medical to the magical is strikingly smooth, the information delivered with a constant, unblinking authority.

Such instances occur elsewhere, too. Recipes for the fabrication and use of remedies such as warm, plant-based poultices for the skin could often be followed by further directions: the treatment should be accompanied by the performance of incantations, the sprinkling of plant solutions, the use of charms and amulets, fires, incense, or special robes. Galen himself recommended such practices,[31] and Avicenna believed that a complex pharmaceutical substance was greater than the sum of its parts: once combined, ingredients could give rise to all sorts of curative powers. The existence of occult, hidden properties was a given.

8. Apothecaries, Alchemists, and
Amulets

SUCH A BLEND of superstition with medical rationalism may be surprising, but perhaps that is because our own culture of expertise and
specialism tends to divide "folk" knowledge from learned, "high" knowledge. The status of botanical knowledge, though, has always been ambiguous, since it relied on the authority of the past while remaining separate
from the study or development of anatomy and physiology. It was connected to humoural theory, but nevertheless stood apart from it. It also
shared some terrain with alchemy, on the edge of the world of chemical
transformation and thus of magical, rather than miraculous power.

The great names of alchemy in the West, however, are often the same
ones we associate with the history of "official" medicine and of early
modern natural philosophy. In its theories about the transubstantiation of
metals and the fabrication of elixirs, tinctures, and ferments—some of
which were indeed medicines—alchemy certainly was an empirical practice. It was a given of alchemy that the four elements could change into
one another, and ultimately that metals, all of which were the product of
the combination of mercury and sulfur, could eventually be turned into
gold. Alchemy, in the late medieval West, was a rather peculiar form of
chemistry, and its practitioners liked to mention—not always accurately—
Rhazes, Avicenna, Albertus Magnus, Arnau de Villanova, and the Catalan
poet and mystic Ramon Llull as their references. It was rooted in the notion of an ultimate, of a basic essence or authority; and the more ancient
its source, the stronger the claim it could make to harbor truths, however
"occult" these might be. For this reason it was enmeshed with astrology,
physiognomy and magic, with divination and interpretations of the Book
of Daniel and the Book of Revelation.

And together with humoural medicine, it was central to the *Secretum
secretorum,* or *Secret of Secrets,* a book which circulated widely during
the Middle Ages and well into the seventeenth century, in which were also
gathered entries on moral philosophy and political thought. The book
was spuriously attributed to Aristotle—it was thought to be a letter from
Aristotle to Alexander the Great—but it was in fact a version of a tenth-
century Arabic text (the *Kitab Sirr al-'asrar)* that had been translated from
the Greek via Syriac. Again, ultimate origins were murky, and so the

Secretum seemed all the more venerable to its many readers throughout Europe and the Middle East.

Whether physicians then subscribed to superstitious practices and occult theories, and whether they were concerned with the connection between the elements, and between the cosmos and the human body, they certainly all believed that the mind and body were powerfully conjoined. They understood the organism holistically, as we would say today; and the humoural system functioned on the basis of this holism, which also helped to make sense of the nature of an ill according to its location and color, and to prescribe the suitable medication accordingly.

But the fabrication of remedies relied on traditions that physicians did not necessarily control, and, as with the case of leeches, doctors had little to do with their efficiency. Those remedies that had a theoretical story to explicate their workings cohered with the rationalist Hippocratic-Galenic tradition to which the medical elite claimed allegiance. And so, the dogmatic resort to methods prescribed by previous authorities, as a result of prejudice rather than of attention to each individual case, tended to perpetuate ignorance, and might well have killed more patients than would have died without any treatment. Other "folk" remedies that had been designed over time through trial and error, however, often seemed to work for unexplainable reasons and therefore had an aura of danger. Apothecaries who knew these remedies could also be useful for their ability to create poisons, the obverse of medications—and frequently used in the Renaissance world as political "remedies." Apothecaries had at their disposal deadly recipes, of the sort that Romeo and Juliet took: in Shakespeare's words, "Within the infant rind of this weak flower / Poison hath residence and medicine power."[32]

This is one reason why physicians' guilds sometimes took action against the excesses, misplaced zeal, or even charlatanry of apothecaries who abused their powers and concocted secret, illicit, or uncontrolled substances. Apothecaries were necessary, of course, but they could be condemned as "empirics" whose investigations of the secrets of nature could be perceived as dangerously related to magic, and whose products were sometimes of doubtful efficacy. Apothecaries were in a sense more powerful than practitioners of the medical art liked to admit—not always at odds with this art, but sometimes a little too invasive, too eager to prescribe and treat.

Those apothecaries who dabbled in alchemy remained on the fringe of

Fig. 12. Preparation of medicine, from *Opera Nova intitolata Difficio de ricette* (Venice, 1529).

elite medical practice. They were understandably perceived as liminal figures, not only because their extensive grasp of botany and chemical concoctions lent them the power to kill as well as to cure, but also because their knowledge of nature's occult powers—of magic—carried its own authority. But such traditions could not be eradicated; and "old wives' remedies," which could be made of such repugnant items as excrement, viper poison, bone powder, and blood, always coexisted with the studied regimens that had evolved within the rationalist, Hippocratic-Galenic tradition, indeed, just as the recourse to shrines, amulets, and prayer always had.

To an extent, though, the authority of humoural theory functioned in a similar way to that of the many "folk" remedies that were available. Its age justified the belief that it worked, and all treatments were by definition empirical, whether justifiable by theory or not. And in general, the good doctor, then as now, was one who knew where rationalism—that is, medical theory—and empiricism met, and who could recognize a match between the patient's story and the textual story.

9. Airs, Waters, Places, Diets

WHAT PHYSICIANS were especially good at was the art of prevention, crucial in a world where the outcome of illness was unpredictable at best. The humoural being's life, from conception to death, remained a balancing act. Sexuality, appetites, passions, and perceptions, all changes and crises, and all responses to the environment required a readjustment of the humours and of the body's secretions. Even spiritual life was sustained by physiology, and the line between illness and health was not necessarily clear.

This is why physicians were valued for their ability to prescribe the so-called *regimen sanitatis* that would favor the consolidation and preservation of health. Based mainly on the dietary and hygienic advice first mooted by the Hippocratics and by Galen, it was for the most part extracted from Avicenna's *Canon* as well as from works by Rhazes, Johannitius, and other writers in Arabic. But the line between professionally sanctioned theory and empirical practice based on "folk" belief blurs here, because the *regimen* had itself evolved from a combination of both of these.

All remedies and "receipts," or recipes, had a clear purpose: they were supposed to reestablish a balanced *krasis* in the organism and restore it to its optimal state. Doctors had to take careful account of the body's innate physiology, what Galen had called the "naturals"; but they also had to pay attention to the environmental and hygienic factors that affected its well-being, what Galen had called the "non-naturals." The Galenic naturals encompassed the body's constituents: the four elements, the four complexions or temperaments, the four humours (in their normal, nonpathogenic state), the faculties of the mind, the virtues or moral qualities, and the *pneuma*. The non-naturals encompassed the variables that could affect the body's state: air, food and drink, wakefulness or slumber, the processes of excretion and retention, and, importantly, the passions, or states of mind. Galen also included in this group the humours in their altered, pathogenic state. These categories had been firmly established and transmitted throughout medieval Europe via the *Canon* of Avicenna, the *Isagoge* of Johannitius, and the translation by Constantinus of al-Majusi's (Holy Abbas's) *Kitab al-Malaki* into the *Liber pantegni;* and they would last even beyond the official demise of Galenic medicine.[33]

The Hippocratic texts called *Regimen* had been well known in Byzantium and in the Islamic world. Veritable health guides had emerged out of the combination of *Airs, Waters, Places* with a selection of Galenic writings, and they would remain authoritative throughout the centuries. Many of the prescriptions for the maintenance of health that became so popular for so long would strike any modern reader as familiar and commonsensi-cal: take moderate exercise, do not overeat, adjust your diet to the amount of exercise you take and vice versa; make sure you get sufficient sleep. The advice to practice intercourse with moderation might seem outdated, and other prescriptions are certainly less self-evident for our modern, urban age: one had to live with clean, eastern or northern air, away from ani-mals, not too close to the ground floor, replete as it was with contami-nated waste. Excessive cold and humidity were deleterious; excessive heat and dryness had to be avoided as well. This was particularly true during outbreaks of illness.

For the maintenance of humoural balance, emetics should be taken regularly—about once a fortnight "during the six winter months," ac-cording to the Hippocratic *A Regimen for Health,* "as this is the phleg-matic time of year and diseases are centered around the head and the chest." In the summer, by contrast, it was best to use clysters, or "ene-mata," since the body would be "more bilious and heaviness occurs in the loins and knees, when there are fevers and colic in the belly. It is necessary therefore to cool the body, and draw downwards the matter [the humours] surrounding those regions."[34] The rules of the *Regimen* required one to pay close attention to the variations of season, geography, and location, and they were specific to one's general state of health, weight, gender, and age. The amount and kind of exercise recommended depended on indi-vidual types, and the supposed qualities of foods, evidently enough, de-pended on their provenance.

The notions contained within the *Regimen* were widely diffused throughout Europe, but also in the Middle East. A central source for its guidelines was the medieval text known as the *Regimen sanitatis Salerni-tatum,* which had been rendered into Latin from the Arabic in the twelfth or thirteenth century. According to legend, it was written for Robert of Normandy, the son of William the Conqueror: its Salernitan origins might therefore be spurious. But its efficacy was real, and its popularity was per-haps due to its having been partly transmitted in verse. Versions of this text circulated widely for centuries. Some 150 printed editions appeared in the late fifteenth century, in various languages. In the first printed edi-tion, published in 1480 in Montpellier by Arnau de Villanova, we learn, for example, how to adapt our habits to the seasons in good Hippocratic fashion:

Fasting in summertime dries out the body.

Vomiting is profitable in every month, for it purges harmful
humours, and it washes the circuits of all the stomach.

Spring, summer, autumn, and winter are the seasons of the year.

In springtime the air is warm and humid,

And no time is better for phlebotomy.

In spring lovemaking is beneficial to man in moderation,

As are exercises, laxatives, sweating, and baths.

In that season the body should be purged with medicines.

Summer is usually hot, and is known as a dry season.

The summer encourages the occurrences of red choler.

In summer food of cold and humid qualities should be served, and
lovemaking should be avoided;

Baths are not good then, and phlebotomy should be rare.

Rest is useful, and drink is good in moderation.

Because of its potentially radical effects, phlebotomy was counter-
indicated in those for whom a "cool constitution, a cold region, great
pain" applied; or in those who had just bathed, or had sexual intercourse,
or were young, or old; or if one was in the midst of a "long illness," had
been drinking heavily, and had been eating a lot. One could "take twice
as much blood in spring, but only the normal amount in other seasons,"
and generally practice phlebotomy "at the beginning of acute and very
acute illnesses." Note, moreover, that:

Phlebotomy is scarcely needed before a person is seventeen.

The more productive spirit will escape with your blood during
phlebotomy,

But these spirits will soon be replaced by drinking wine, and

Any harm done by the humors will be gradually repaired by
food.

Phlebotomy clears your eyes, freshens your

Mind and brain, makes your marrow warm,

Purges your bowels and restrains your stomach and belly from
vomiting or menstruation;

It purifies the senses, brings on sleep, takes away weariness;

It cultivates and improves hearing, speech, and strength.

Some advice is more familiar to us: the text entreats us to wash our hands often and generally to wash after a meal; to eat less in the summer, but "as much as you like in the winter"; to take small drinks during meals; and not to heed "ignorant doctors" who are wary of cheese and instead to take cheese at the end of meals because "it brings help to a weak stomach." We are also told that:

Beer nourishes thick humours, gives strength,
Fattens the flesh, produces blood,
Provokes urine, has a laxative effect, causes gas,
And has a cooling effect. Vinegar has more of a drying effect:
It cools, makes a man thin, induces melancholy, decreases the
    number of sperm,
Harms those of dry humour, and dries up the nerve of the fats.[35]

Whether readers and users of this simplified version of the rich Hippocratic corpus knew it or not, a complex theory of digestive and metabolic processes underpinned these rules. For example, in the Hippocratic *Regimen,* we find that moist foods and drinks "warm, moisten and pass by stool better than things that are dry; for being more nourishing to the body they cause a revulsion to the belly, and, moistening, pass readily by stool," whereas warm and dry foods and drinks, "producing neither spittle nor urine nor stools, dry the body" because when the body grows warm, it "is emptied of its moisture, partly by the foods themselves, while part is consumed in giving nourishment to the warmth of the soul, while yet another part, growing warm and thin, forces its way through the skin." As for "things sweet, or fat, or oily," they "are filling, because though of small bulk they are capable of wide diffusion. Growing warm and melting they fill up the warmth in the body and make it calm."[36] We are also told that

a walk after dinner dries the belly and body; it prevents the stomach becoming fat for the following reasons. As the man moves, the food and his body grow warm. So the flesh draws the moisture, and prevents it accumulating about the belly. So the body is filled while the belly grows thin. The drying is caused thus. As the body moves and grows warm, the finest part of the nourishment is either consumed by the innate heat, or secreted out with the breath or by

REGIMEN SANITATIS
SALERNITANUM.

ANGLÓRUM Regi scripsit schola tota Salerni.
Si vis incolumem, si vis te reddere sanum,
Curas tolle graves, irasci crede profanum,
Parce mero, cœnato parùm, non sit tibi vanum
Surgere post epulas, somnum fuge meridianum,   5
Non mictum retine, nec comprime fortitèr anum:
Hæc benè si serves, tu longo tempore vives.
Si tibi deficiant medici, medici tibi fiant
Hæc tria, mens læta, requies, moderata diæta.   9
Lumina manè manus surgens gelidâ lavet aquâ,
Hâc illâc modicùm pergat, modicùmque sua membra
Extendat, crines pectat, dentes fricet. Ista
Confortant cerebrum, confortant cætera membra.
Lote, cale: sta, pranso, vel i; frigesce, minute.
Sit brevis aut nullus tibi somnus meridianus.   15
Febris, pigrities, capitis dolor, atque catarrhus,
Hæc tibi proveniunt ex somno meridiano.
Quatuor ex vento veniunt in ventre retento,

The Banquet
Ex magnâ cœnâ stomacho fit maxima pœna.

Fig. 13. Print from a late English edition of the *Regimen Sanitatis Salernitanum:* a poem on the preservation of health in rhyming Latin verse. Addressed by the school of Salerno to Robert of Normandy with an ancient translation (Oxford: D.A. Talboys, 1830).

the urine. What is left behind in the body is the driest part from the food, so that the belly and the flesh dry up. Early-morning walks too reduce [the body], and render the parts about the head light, bright and of good hearing, while they relax the bowels.

One might want to bear in mind, finally, that "walks after gymnastics render the body pure and thin, prevent the flesh melted by exercise from collecting together, and purge it away."[37]

The art of medicine was really the art of maintaining health, and of regulating the excesses and deficiencies that threw the organism off-kilter, such as wrong foods at the wrong time, wrong exercise at the wrong time, an excess of the one and lack of the other, insufficient or excessive intercourse, and an excess or lack of sleep. The Hippocratic *Regimen* hence gave "advice to the great mass of mankind, who of necessity live a haphazard life without the chance of neglecting everything to concentrate on taking care of their health."[38] There are ways of avoiding dramatic damage: just follow medical advice.

Of course medical advice then, and for centuries after, would hardly correspond to our notion of what constitutes a healthy life. Nutrition and hygiene were poorly understood, as was the rationale for medical advice, whether humoural or not. Theory might have sustained humoural practice to some degree, but practice tended to justify itself, just as prayer could. The instinct to deliver and listen to medical advice concerning diet and lifestyle has remained intact to this day. What has changed most fundamentally is that the authority of public health advisories derives in part from our faith in the newest science, and no longer from faith in established tradition, as was the case, for better and often for worse, well into the seventeenth century.

10. Fearful Epidemics

THE EDIFICE OF BELIEF was self-sustaining for a long time, and it seemed to justify itself so long as the air was healthy, not putrid—when even epidemics could be controlled to some extent. Localized outbreaks of plague or of illnesses sometimes called plague—typhus, malaria, influenza—would recur over and over again throughout history, and it was clear that in such cases illnesses were being transmitted in some way. But the notion of contagion remained fuzzy. In a treatise on fevers, Galen had observed that proximity to those afflicted with such illnesses as plague or phthisis, or with some eye infections, tended to result in transmission of the disease.[39] Medical writers in Arabic, and indeed early Muslim communities, had not doubted that illnesses could be passed on—from camel to camel as much as from ill person to healthy person, or from father to son. (It was a process they generally called i' da'.) Epidemics were clearly a function of noxious, putrid air. Al-Majusi, who referred to Galen's treatise on fevers, had told of the dangers to the organism of exposure to putrefying garbage, dead bodies, and stagnant waters—all of which emitted potentially lethal vapours, especially in excessively variable weather.

Debates about the nature of contagion would continue until the nineteenth century, when the microorganisms that cause diseases began to be understood. Innovative ideas did percolate earlier, however: the sixteenth-century physician, astronomer and poet Giromalo Fracastoro already speculated that contagion was a phenomenon in which "seeds" were transmitted.[40] But he did not doubt that transmission, as Galen himself had ascertained, was favored by the combination of putrid or pestilential air and filthy living conditions with certain astronomical and meteorological conditions. Fracastoro would play an important role in initiating a debate about the modes of contagion. But even in his theory, Galenic "miasmas"—the noxious effluvia that polluted the air—would continue to flow, and prevention would continue to be a matter of control over external atmosphere, humours, diet, and astronomical information. One historian recorded the statements of physicians in 1611 who were unhesitatingly ascribing the deaths of forty or so adults and twenty children in one month—in a small Tuscan town—to phlegmatic humours exacerbated by the cold winter, the sudden advent of spring, and other such environmental factors. (The adults were afflicted with respiratory diseases, and the children with variola.)[41] Causes could be found, however wrong they might aver themselves to be. Medicine, although powerless in such cases, could at least identify the boundaries beyond which it was powerless, and leave the rest to the will of God; while within these boundaries, health could be monitored, a regimen prescribed, and drugs administered.

But when the Black Death struck in the mid-fourteenth century, and the epidemic got horridly out of hand, the preservation of health, let alone survival, was no longer determined very much by self-control or by obedience to a daily regimen. Boccaccio famously begins the "first day" in his *Decameron* with a graphic description of the ravages wrought in Florence by the plague that eventually wiped out a third of Europe's population (probably some 50 million people). Lasting from 1347 to 1351, it first broke out in China in 1334. By the end of the century and after repeated outbreaks of the epidemic, the Chinese population had fallen from an estimated 125 million to 90 million. Trade, especially the spice trade, was largely responsible for the westward progression of the disease. From China, it reached the Black Sea via caravans traveling through Mongolia. It traveled by sea, on ships sailing across the Indian Ocean to the Persian Gulf; and then overland, with the caravans destined for northern Arabia and the Levant, or for southern Arabia, from where they reached Yemen, the Red Sea, and Gaza or the Nile Delta. Rival Genoese,

Venetian, and Pisan merchants, positioned at key points along these routes, ensured the transportation of goods to the western Mediterranean region—Italy, southern France, and Spain—and from there to northern Europe.[42]

Commerce was a global affair. Those very spices that were enriching merchants, turning bland meals into gastronomic events, and being used as medicines came from the farthest reaches of the world and were an international commodity. But rodents liked the dark corners of ships; and the bacterium *Yersinia pestis* was traveling on their backs, within the bodies of rat fleas.

By 1347, the bacterium had hit Constantinople and much of the Byzantine world. Cairo, Rhodes, the Aegean Islands followed. It was in Alexandria in the autumn of that year—in its pneumonic form, which is airborne and thus much more lethal than the bubonic form, transmitted exclusively from flea to host. Gaza, Antioch, Damascus were struck in 1348. North Africa was next: Tunis was caught in the spring. Probably half of the Muslim populations of these areas fell victim to the Black Death by 1349. It was generally believed that the plague was caused by impure air that corrupted the organism, and that measures to maintain cleanliness might hold it at bay. Internal cleanliness was a matter of diet; external purity might be ensured by fumigations with incense, myrrh, and the like, and by the appropriate ventilation of rooms.

Some people, such as Ibn-al-Wardi, a resident of Aleppo, would have none of this, claiming instead that the plague was the will of God and that to die of it was to become a martyr and end up in heaven. After writing a treatise on the matter, al-Wardi succumbed to the plague himself.[43] This theological line would, by the next century, become an important line in medical thinking within Islam.

Sicily was the plague's port of entry into Italy; the bacterium came on a Genoese ship arriving in Messina, from the East, in 1347. The southern Italian peninsula followed. Genoa itself was infested in late 1347. So was Tuscany, via ships docking at the port of Pisa. The plague entered Prato, Lucca, Pistoia, Orvieto, Siena. Umbria followed in the spring of 1348. Death tolls everywhere were in the hundreds of thousands. Boccaccio's fellow Florentines were struck by 1347, and the city was infested well into the following year. The Florentine witness and chronicler Marchione di Coppo Stefani reported:

In the year of the Lord 1348 there was a very great pestilence in the city and district of Florence. It was of such a fury and so tempestuous that in houses in which it took hold previously healthy servants who took care of the ill died of the same illness. Almost none of the ill survived past the fourth day. Neither physicians nor medicines were effective. Whether because these illnesses were previously unknown or because physicians had not previously studied them, there seemed to be no cure. There was such a fear that no one seemed to know what to do. When it took hold in a house it often happened that no one remained who had not died. And it was not just that men and women died, but even sentient animals died. Dogs, cats, chickens, oxen, donkeys, sheep showed the same symptoms and died of the same disease. And almost none, or very few, who showed these symptoms, were cured. The symptoms were the following: a bubo in the groin, where the thigh meets the trunk; or a small swelling under the armpit [these swellings would be recognized today as inflammations of the lymphatic nodes]; sudden fever; spitting blood and saliva (and no one who spit blood survived it). It was such a frightful thing that when it got into a house, as was said, no one remained. Frightened people abandoned the house and fled to another. Those in town fled to villages. Physicians could not be found because they had died like the others. . . . Child abandoned the father, husband the wife, wife the husband, one brother the other, one sister the other. In all the city there was nothing to do but to carry the dead to a burial.[44]

Boccaccio too describes how "in men and women alike" the plague "first betrayed itself by the emergence of certain tumors in the groin or the armpits, some of which grew as large as a common apple, others as an egg, some more, some less, which the common folk called *gavoccioli.*" This swelling, he writes, "soon began to propagate and spread itself in all directions indifferently; after which the form of the malady began to change, black spots or livid making their appearance in many cases on the arm or the thigh or elsewhere, now few and large, now minute and numerous."

The black spots manifested what is now understood as the septicemic version of the disease, in which bleeding occurred within the skin and organs. Death usually ensued within a few days of the onset of symptoms. Those who were not yet ill and tried to continue living normally would carry around, wrote Boccaccio, "in their hands flowers or fragrant herbs or divers sorts of spices, which they frequently raised to their noses, deeming it an excellent thing thus to comfort the brain with such perfumes, because the air seemed to be everywhere laden and reeking with the stench emitted by the dead and the dying, and the odours of drugs."

The disease was in Avignon, then the residence of Pope Clement VI, in 1348. It then struck Paris and London. It was an inexorable traveler, reaching Russia by 1352. The only way to avoid pestilence was to close doors and windows or, better still, to run away, if there were only a place to run away to. The pope resorted to a self-imposed quarantine, confining himself inside his residence at Avignon for four years, surrounded by fires, studiously avoiding any contact with the external world and its corrupted air. (He would, in fact, survive, though he died a few years later, in 1352.) But few people had the means to do this. Flagellants, believing also that the calamity was a result of sin, roamed the city streets. Putrefaction was everywhere; bodies piled up in overcrowded towns where filth, excrement, garbage, and rotting food were already a breeding ground for all sorts of illnesses. It was thought that cleaning the streets would spread the disease, and that washing oneself was also dangerous, since water would open the pores to the corrupt miasmas.

Some remedies were tried, such as butterbur—a potentially toxic plant that has actually been found today to have antihistamine properties. Some people, as Boccaccio noted, walked with herbs under their noses; some carried bells to alert passersby. Gentile da Foligno, who taught medicine in Perugia, recommended the pursuit of joyful, pleasurable things and melodies and the avoidance of sad things as a protection against the disease. But that was wishful thinking. Illness bred more illness. Rodents thrived, and although they had been associated with pestilential environments and noxious miasmas even in Roman antiquity, no one could imagine the central role they played in the transmission of the plague. Superstitions and fears were rife, society was in chaos, unrest and distrust prevailed. Scapegoats were needed in such calamitous circumstances; anti-Judaism grew, as it often did in times of crisis, along with the fervent mistrust of those perceived as outsiders. The belief spread that the plague was caused by poison disseminated by the Jews in a conspiracy against Christians. Many Jews were forced to "confess," often under torture. In some places, most notoriously in Strasbourg and near Lake Geneva, they were burned, stabbed, drowned.

Purgings and bleedings, aimed either at restoring humoural balance or at purifying the organism of the disease that inhabited it, were useless. They even seemed nefarious, since they weakened the organism. Nor did prayer and self-flagellation save anyone. Worthy lives ended brutally. Christians were dying without a chance to receive the last rites. To them, God seemed to have deserted the human community of sinners.

11. Life After Death

THE EPIDEMIC did eventually wane, after killing its millions of victims. By the time of Alderotti, at the end of the century, it was gone. Local outbreaks would occur regularly, in various places. But for now, the air was no longer putrid. The human organism was no longer at war with the natural elements, or, for that matter, with itself. The skies and the stars offered better portents. Control over the flow of innards, over the composition of blood, phlegm, and bile, and over moods and needs, could resume.

But the Black Plague had changed Europe radically. The demographic devastation was enormous. Fortunes and estates had been lost or redistributed. The workforce was frightfully diminished; wages were at an all-time low. Impoverished peasants revolted and moved into deserted castles, taking over property, land, and power. Meanwhile, newly founded banks, especially in Italy and the Low Countries, were further transforming economic and social structures, concentrating and investing capital.

The disaster left its mark on cultural traditions, on thought and faith. God had forsaken all, including physicians. Doctors and priests had neither protected nor been protected. Trust in their powers ebbed, and to some extent local governments took over the role previously ascribed to doctors of soul and body, now enacting public health policies on their own authority. The accessibility of the health regimens to all became a priority even for the clergy: the need to maintain good hygienic standards and pure air was becoming painfully clear.

The medical profession actually evolved out of this disaster. It was partly to repair its damaged reputation that its exponents began to seek new grounds for knowledge; it also became more hierarchical, more rigidly institutional, than it had been, and therefore more powerful. Now that the Church had lost some ground to nature's brutal ways, professorial chairs became valuable in a world where universities were gaining in prestige and influence. Lay education was spreading and, to an extent, becoming a handmaiden to religion. The era that, in the nineteenth century, would be labeled the "Renaissance"—French for "rebirth"—was beginning to unfold.

From 1456, once Gutenberg printed the first Bible using the technique of movable type,[45] texts would circulate more widely, in printed form. Medical treatises, such as the 1480 Montpellier edition of the *Regimen*, started appearing in vernacular languages, as well as in the usual Latin. Only a few decades previously, in 1453, Constantinople had fallen to the Muslims. This also remains a key date in the history of the Renaissance. In the late fourteenth century, new editions, readings of, and commentaries on newfound classical sources had already been transforming the very fabric of culture. After the fall of the Byzantine empire, the transformation accelerated: Greeks fled to western Europe, especially to Italy, bringing with them original Greek texts, many if not most of which had been known only in Arabic translations. They also brought the linguistic expertise to study these texts, and they passed their expertise on to their learned, theretofore mainly Latinist colleagues. The notion of textual authority was now to be debated even more: what became known as humanism came into its own.

Among the newfound sources were Galen's original writings. The long-circulating compilations began to be examined afresh. Slowly, anatomy enriched itself with new, empirical questions, embarrassed itself with philological discrepancies between the truths contained in the derived sources and those contained in the originals. Debates erupted between those—the Arabists—who continued to refer to the former and those—the Galenists—who denigrated them as distortions by barbaric pagans. Humours, of course, continued to flow. But what, slowly, began to crack was the certainty that had until then accompanied all belief, all knowledge, and all ignorance. The early modern era was approaching; and the old accounts of human nature were recycled in complex, imaginative ways, for intensive use in a rapidly changing, expanding world.

IV

Harmonies: Renaissance Bodies and Melancholy Souls

*(Renaissance: Fifteenth Century to
Early Seventeenth Century)*

*There was one so Melancholicke that hee confidently
did affirme, his whole body was made of butter,
wherefore hee never durst come neere any fire,
least the heat should have melted him.*

THOMAS WALKINGTON[1]

*you shall see how all the arteries of his brains are stretched
forth and bent like the string of a crossbow, the more promptly,
dexterously, and copiously to suppeditate, furnish, and supply
him with store of spirits sufficient to replenish and fill up the
ventricles, seats, tunnels, mansions, receptacles, and cellules of
the common sense,—of the imagination, apprehension, and
fancy,—of the ratiocination, arguing, and resolution,—as
likewise of the memory, recordation, and remembrance*

FRANÇOIS RABELAIS[2]

I. Hypochondria at Court

ON MARCH 8, 1466, Francesco Sforza, the Duke of Milan, died after a prolonged, intermittent illness. According to the medical records, the Duke had started to experience severe symptoms—acute flatulence, thirst, pains throughout the body, hot sweats, and fever—a few years earlier. Doctor Benedetto de' Reguardati approved of the treatment prescribed by the other doctors in charge of the Duke: an eighteen-ounce emetic, a few clysters, some mallow. The patient exhibited signs of peritoneal dropsy: fluid was accumulating in his liver, and he had little appetite.[3] On September 20, 1461, the ambassador to Milan of the Gonzaga court in Mantua reported in a letter to the Marquis Ludovico Gonzaga that the doctors feared Sforza would develop a more severe dropsy. As it was, his dropsy was still "airy," but it risked becoming humid, the ambassador dutifully wrote, because a phlegmatic fever was easier to cure at its early stage in a young man than in an older man. He might only survive a maximum of two years, and then only if he took his remedies.

Other doctors thought that he might get well—if he followed his regimen, which included lettuce in broth, egg in broth, or sweetened game.[4] But the fevers continued. The Duke was advised to take therapeutic baths and eat rhubarb and fennel, given more clysters, and advised to follow a carefully controlled diet, which nevertheless made room for the occasional indulgence in delicacies Francesco would not have wanted to give up altogether, such as snails, or roasted thrushes. In 1462, Francesco's wife even asked the Pope to give Francesco a dispensation from Lent: he needed to eat meat, in order to "fortify and restore" his weakened organism.

Although the Duke lived a little longer than the two years predicted by the more pessimist prognosis, the illness did kill him eventually. Perhaps no full cure would have been possible. But from the perspective of his doctors, Francesco was also a bad patient, who did not follow medical advice very carefully, and who often refused to take his medications. He trespassed beyond the regimen prescribed for him, and this was always dangerous. Over time Francesco had been getting visibly thinner; and by early 1466 his legs had swollen beyond measure. The liver area was taut. The whole region between the belly button and the diaphragm—called the hypochondria—seemed to be in trouble. Today, a doctor might have diagnosed a case of hepatic cirrhosis, possibly associated with pancreatitis.[5] But at the time it was clear that the Duke's humours were seriously "corrupt." Yellow bile, secreted by the liver, was at that point turning into black bile—normally stored in the spleen—and diffusing throughout the hypochondria. Despite his evident discomfort, though, the Duke did not give up politics—or womanizing: according to the doctors, his recurring fevers were caused by his overindulgence in carnal pleasures.

Hypochondriac diseases were due to excessive black bile and were, by definition, melancholic, not phlegmatic. (Today, to be a "hypochondriac" is to imagine one is ill—and this obsession was considered a form of melancholy.) Not all melancholic conditions were pathogenic; but this one most definitely was, and, although phlegm might indeed also have been involved in Francesco's disease, there seems to have developed a dyscrasia such that the production of blood was overtaken by a surfeit of yellow bile concocted into black bile. This, at any rate, might be a sound humoural diagnosis of Francesco's illness.

Given the Duke's role in ensuring political stability during this restless period of Italian history, his condition worried his allies, especially the Medici, rulers of Florence and financial backers of the Sforza. Cosimo de' Medici was the founding head of the family whose patronage of scholarship and the arts over some 350 years still remains legendary. He had, notably, amassed an important collection of Greek manuscripts of Plato's works. Marsilio Ficino, a physician, scholar, and musician, was the head of the Platonic Academy founded by Cosimo in 1462. His knowledge of Greek was outstanding, and on Cosimo's request, he set out to translate these manuscripts into Latin. He was also attracted by Plato's views of the soul; convinced that they were profoundly compatible with Christianity, he went on to translate Neoplatonic texts and to compose treatises that set out his own, Neoplatonist views.

Cosimo died in 1464, two years before the death of the Sforza Duke, whose illness at that point was in remission. Cosimo's younger son, Giovanni de' Medici, had died of "phlegmatic disease" the year before, when Giovanni's brother Piero ("the Gouty," as he was called), had become chief Justice of the city of Florence. It was Piero who would father

Lorenzo, under whose rule humanism and the arts decisively bloomed. And it was during Lorenzo's reign, probably in 1489, that Ficino finished his epochal *De vita triplici,* or *Three Books on Life.*

Although Francesco Sforza had been dead for almost a quarter of a century by the time this hugely influential treatise was published, Ficino would probably have prescribed the same sort of remedies that Francesco's doctors did. He would have confirmed the presence in Francesco's organism of a surfeit of black bile, which he thought was due to the excessive dryness of the body, itself likely to ensue from "long wakefulness or much agitation of mind, or worry, or frequent sexual intercourse"—all applicable to the Duke—"and the use of things which are very hot and dry, or the result of any immoderate flux and purgation, or strenuous exercise, or fasting, or thirst, or heat, or a too dry hot wind, or cold."[6]

2. The Black Sun

THE EXCESS of black bile in someone like Francesco, whose temperament did not tend to melancholy—he was clearly a sanguine type—could have only deleterious consequences on physical health. Melancholy literally means black bile in Greek: it is the name of the humour itself. In those who had a predominantly melancholic temperament, its effects were not just physiological. It could produce extreme mental states and psychic disorders, from despondency to madness; in its most pathogenic manifestations, black bile gave birth to *adust,* or burnt, melancholy. When the bile was overcooked, so to speak, it could become what was called *acedia,* an especially acidic, corrosive, nefarious type of black bile. Then it gave way to a despondent, wasteful slothfulness, a profound unwillingness to act or speak, to escape the malaise that it produced. For the Christian Church from the sixth century on, acedia was a deadly sin. Dante had its victims dwell in the fifth circle of Hell:

". . . Lodged in the slime they say: 'Once we were grim
      And sullen in the sweet air above, that took

A further gladness from the play of sun;
      Inside us, we bore acedia's dismal smoke.

We have this black mire now to be sullen in.'
   This canticle they gargle from the craw,
   Unable to speak whole words." We travelled on

Through a great arc of swamp between that slough
   And the dry bank—all the while with eyes
   Turned toward those who swallow the muck below. [7]

Acedics were sinful because they let themselves be overwhelmed by the fumes of the black humour; they gave themselves up to dejection, unable to heed and obey the word of God. Their souls, in other words, were weakened before the onslaughts of the body. Acedia was in this way as powerful and dangerous a passion as anger, or indeed lust, or envy, other deadly sins. [8]

Not all melancholics were sinners, of course; nor were all melancholics ill. Unlike an excess of phlegm, or yellow bile, an excess of black bile did not necessarily give rise to identifiable illnesses. Melancholy could also be natural, nonpathogenic, and, at best, a sign of inspiration. The rich cultural history of melancholy encompasses this ambivalence, and begins even before the advent of the Greek medical theory that gave it its humoural name. Biblical figures such as Job describe melancholic states, as does Ecclesiastes and the refrain of one who sees "nothing new under the sun." Homer's Achilles was a melancholic; and Greek tragedy has its share of melancholics, but these are often more extreme cases. The Hippocratic corpus contains references to melancholy as one of the humours: in the Hippocratic *Aphorisms,* for instance, one reads, "Fear or depression that is prolonged means melancholy"; and, "In melancholic affections the melancholy humour is likely to be determined in the following ways: apoplexy of the whole body, convulsions, madness or blindness." [9]

But the Hippocratic accounts did not make much more of melancholy. The enduring medical reference on the matter, which would turn it into something of greater human significance, would come from another source: the *Problemata,* XXX. Although associated with Aristotle, this text was probably by Aristotle's pupil Theophrastus. It seems to have integrated the notion of a "divine frenzy" that Plato had written about in reference to the passions depicted in tragedies, in particular Sophocles's *Ajax.* Here, what had been a humoural phenomenon took on a broader, even spiritual tincture. The influence of the *Problem XXX,* as it remains known, was limited at first, though it transpired in the first century in the writings on melancholy by Rufus of Ephesus—picked up by Galen—and by Rufus's contemporary Aretaeus of Cappadocia. The Arabic tradition bears traces of this text: Ishaq ibn 'Imran, notably, had inspired himself from Rufus's treatise on melancholy. But after the original *Problem XXX* had been translated into Latin by Bartolomeo da Messina in the mid-thirteenth century, the ideas it contained would become a focal point in the history of western culture.[10]

It is perhaps a peculiar feature of western culture that dejection should have been the focus of so much philosophical, medical, and artistic attention. Everyone everywhere in the world knows what sadness feels like; but melancholy is more complex than sadness, and emerges out of a culturally produced recognition of a spiritual, metaphysical suffering. It seems, more often than not, an outcome of a particular, refined sensitivity—an acute awareness of the passage of time, a deep capacity for the appreciation of life and its fragility, along with a sense of the ultimately paradoxical vanity of all that matters most in it. The *Problem XXX* itself begins with a question that still resists easy answers: why is it, wonders the Aristotelian author, that creative, talented people like artists, statesmen, and the like often seem particularly prone to melancholy? Why indeed do the capacity for insight and a melancholy disposition seem naturally suited to each other?

We tend in our day to confuse melancholy with depression, insofar as we believe that happiness—broadly, vaguely defined—is a universally desirable and attainable state of mind. Melancholics, though, are not necessarily unhappy. Many would claim, unlike depressives, that they are happier in their state. On the other hand, an unremitting, paralyzing obsession with the bleak side of things must be a form of adust melancholy—excessive and ultimately unhealthful, close in meaning to what we call depression today, in its various forms. The two concepts in fact overlap in places: people were once just as keen to turn melancholics to the light as we are today to lift depressives out of their pit, or to sober up manics. The awareness that our states of mind are sensitive to external factors and pharmacological substances is not new. It has always been the case that melancholics needed, and could benefit from physical treatment; and it has always been clear that our minds and bodies influence each other. Yet no one has ever known why, for instance, a happy song might relieve adust melancholy, or why, inversely, the sight of illness might cause its symptoms. Today one may investigate neurotransmitters, enzymes, hormones, and ultimately genes; but while humours reigned, they provided a solid account of our profoundly psychosomatic nature.

The physiology of pathological states, however, does not seem at the outset to tell the whole story of the more common melancholy states. The concept of melancholy described a state of mind, an aspect of human nature rather than one precise pathology. According to an astrological tradition that may have begun with the Arabs, it was governed by the planet Saturn. Sanguines were governed by Jupiter, cholerics by Mars, and phlegmatics by Venus or by the moon. Since Saturn—sometimes identified with the Greek god Chronos, who devoured his own children—was perceived as darker and colder than any of the other planets, it made sense to connect it to black bile. Saturnines tended to be troubled and in pain. In the medieval world, each planet governed one of the seven liberal arts; Saturn had been assigned to astronomy and then geometry. The creative underside of the melancholic state was thus always apparent, and it was emerging as a powerful force during the Renaissance.

The saturnine, enigmatic figure, both female and angel, depicted by Albrecht Dürer in his engraving *Melencolia I*—dated 1514—is still celebrated as a representation of the contemplative, inspired melancholic in the Renaissance. Saturated with symbols and rich with historical, psychological, and metaphysical layers, the engraving is a striking allegory of the human mind at work upon itself, and of the artist representing visible realities. The seated figure is surrounded by objects such as a magic square, a compass, a sphere, and a polyhedron, and seems connected to the world by the alchemical and mathematical relations that these objects symbolize. But the figure is also isolated, seemingly mortal, and endowed with limited faculties; and the work seems to point to the ultimately vain nature of even the most ambitious and advanced scientific knowledge.

Ficino's *De vita* was published in Germany in the late 1400s, and his letters were printed in 1497 by Dürer's godfather, Anton Koberger. This kinship might not mean much, but it is possible to argue that Dürer's vision of melancholy as creative contemplation owes something to Ficino. Much has also been made of Dürer's use of the work of an occult follower of Ficino, Agrippa.[11] In any case, the melancholy they each depicted remains a broad, useful concept that actually differs from its contemporary equivalents. Not all soulful, contemplative inspiration is dejection; not all figures of melancholy, represented seated, cheek resting against hand, gaze turned down and inward-looking, are necessarily depressed in today's sense of the term. The physiological causality of this condition is no simple matter.

In fact, there was room within humoural explanation for the soul's

Fig. 14. Dürer, *Melencolia I*. (1514)

subtle sufferings. The Aristotelian author of the *Problem XXX* was aware of the gradations offered by the black bile, and he endeavored to describe the physiological, temperamental, humoural constitution that underlay the melancholy tendency of those inclined to metaphysical, poetic, philosophical, or psychological insight. He had noted early on in his text that the effects of the melancholic humour were similar to those of wine: both contained an "air"[12] that, given shifts in the organism's temperature, would produce the sort of hypochondriac, physiological disease Francesco Sforza was afflicted with.

Because in this early account, the melancholic humour was at once hot and cold, and thus inconstant, its effects included a wide range of disturbances, ranging from apoplexy to torpor if the subject had a generally cold temperament, and from elation to singing, excessive enthusiasm, and even madness if the subject had a generally hot temperament. Ordinary, occasional dejection was also the effect of cold and dry black bile, but it was on the whole the lot of those in whom temperature was better regulated—in whom, so to speak, the physiological thermostat was in order, usually in those with a sanguine temperament, like Francesco Sforza. The humour caused a pathogenic imbalance, a dyscrasia, only in those whose thermostat was already running amok. Francesco was therefore probably not afflicted with a full-blown melancholic disease, because his ailment was physical, and clearly did not involve the mind. In some sense, he did not let it affect him, continuing his life of intrigue and conquest, separate from the ailing hypochondria.

Galen had thought that the melancholic humours were produced by the liver—which was itself connected to the heart, the source of innate heat—and were attracted to the spleen "by means of a bilious vessel like a canal."[13] The spleen then processed them, while they nourished it. Those humours that could not "be changed into the nature of thin, useful blood" were discharged into the stomach "through another venous canal," and could be experienced as acidity in the stomach. Galen's understanding of the body's physiology—of the secretion of the humours—was related to his understanding of the anatomy of the organs involved: he had lucidly explained the visible through a series of brilliant, logical conjectures. It was impossible not to be convinced by this impressively coherent system.

3. Mind, Matter, and Metaphysics

A GALENIC DIAGNOSIS of melancholy can still make sense to a modern reader. But Dürer's melancholic seems to have little to do with Galen's; it is a poetic figure rather than a product of humoural medicine, the image of a contemplative human rather than an embodiment of the melancholic humour. And indeed, to anyone who has experienced melancholy, it may seem all too easy to analyze our bodies from the outside as Galenic medicine does, to separate ourselves and our lives from the goings-on in our innards, to treat our illnesses and analyze our emotions as medical histories that are separate from our deeper selves, or from the social world we live in.

But the humoural tradition did not only underpin the medical arts: it was a form of self-understanding in a broader sense. Poetry and black bile were compatible, without being reducible to each other. The *Regimen,* informed as it was by the notion of humoural balance, had established what it took for this balance to be sustained, and presupposed a constant correlation between lifestyle and physical health. It also presupposed that we were transparent to ourselves—that, by virtue of our ability to live balanced lives, we could control our passions, our bodies, and hence our fates. The *Regimen* offered moral, not just medical guidance. Voluntary self-restraint—of the sort that Francesco was not very good at—was the mark of virtue, and its reward was physical health, the counterpart of moral vigor.

But a given temperament, or *krasis,* influenced the very capacity to inflect its corresponding physiological type. Francesco's sanguine, hot and moist temperament was such that he could forget physical pain and energetically get on with his life, but also such that he would not pay attention to his hypochondria and adopt a healthier lifestyle. Those who naturally tended to be carried away by humoural passions were naturally less amenable to self-control; and those whose passions were less bold could ponder their reactions and fully exert free will. Humoural temperament determined how strong free will was in the face of appetite or passion. We had no immediate control over the physiology that enabled us to exert self-control: we were born with it, under a particular zodiac.

All we could do was to exert our reason, to understand this physiology and thereby avoid dyscrasia. A human being who was not constituted of humours would have been unthinkable, just as someone who is not constituted of cells is inconceivable today. There was a limit to this knowledge too, however: humours necessarily had to be properties, not substances, since the soul was a substance whose qualities humours defined and embodied. Anyone who would have suggested instead that we were fully reducible to our physiology would, fatally, have sounded like a materialist, denying the reality of spiritual matters, of the immaterial rational soul humans alone were endowed with, and ultimately of God. The limit beyond which humours ceased to be relevant was drawn by God, by metaphysics, by the philosopher-doctors who had shaped the authoritative corpus that sustained humoural theory, and by the authority of religion itself.

It was certainly impossible still in the late fifteenth century or the early sixteenth to step beyond these limits. The power of humoural theory was a given. Ficino, for one, understood it well. But he went farther than just partaking in the general assent or summarizing the classics. He broadened the realm within which the humoural system determined us, and he transformed the way in which it accounted for temperaments and the vagaries of moods. He did this by building an elaborate metaphysics in which humans were at one with the whole of nature and with the stars. Microcosm and macrocosm were in correspondence, just as were the soul of man and the "World-soul"; stars had spirits too, that were relevant to our lives. Saturn was no longer the dark planet whose influence, to be positive, had to be counterbalanced by the force of other stars, especially Jupiter.

All this sustained a serious conception of medicine explicitly derived from Galen but enriched by Plato and by the Neoplatonists that Ficino also translated (notably the third-century Plotinus and the fifth-century Proclus). Our bodies were part of the universe, synchronized with the celestial spheres, and our bodies were spiritual, just as the animal spirits that traveled through them were the messengers of sorts between matter and mind. Others earlier on had already understood our states of mind in spiritual terms: soul and body, as Avicenna had written, depended on each other, and when one broke down, so did the other. Ficino went further than this, however, integrating both magic and astrology into his scheme.

Medicine and astrology had been connected from early on, even in the Hippocratic *Airs, Waters, Places;* and a good doctor, in the western medieval and Renaissance world, would have been expected to master at least a smattering of the art. There had been some advocates of the practice in medieval Islam, too—notably the tenth-century astrologer al-Qabisi, or Alcabitius[14]—although Avicenna, for one, had rejected its use and its relevance for the medical arts, preferring meteorology.[15] Ficino, on the other hand, had no hesitation in integrating it fully within his system. It was on a par with magic, whose effectiveness he considered to be the inevitable outcome of the interrelation of all things. He elaborated on notions such as the world's souls and the chain of being of which we were a part; he took for granted the reality of preestablished harmony and of sympathetic vibrations between the lower spheres we inhabited and the heavenly ones above our heads. Miracles and magic helped to explain why a material entity like a humour or animal spirit had an impact on our immaterial soul.

Although Ficino's guide was explicitly a contribution to "high," humanist culture, it actually merged with "folk" belief. Within this newly painted starry realm, it made sense to recommend that old people should drink the blood of "willing, healthy, happy, temperate" youths—but only when "hungry and thirsty and when the Moon is waxing." The operation required that the elder "suck, therefore, like leeches, an ounce or two from a scarcely-opened vein of the left arm" and "immediately take an equal amount of sugar and wine."[16] (One could alternatively cook the blood with sugar if digestion of the raw stuff proved difficult, or replace it with pig blood, effective, apparently, "in warming the stomach. A sponge soaked with hot wine should absorb this blood flowing from the pig's vein and, while hot, should immediately be applied to the stomach.") Humours were instrumental in the impact of planets and stars on the organism; "to stimulate the bowels with solid medicines, look to Pisces; with liquid medicines, Scorpio; and with something in between, Cancer." It was also best to "avoid the malefic aspect of Saturn and Mars towards the Moon, for the former disturbs the stomach, the latter loosens the intestines."[17]

Yet, the intellect alone was capable of apprehending the truth: it was "wrong to cherish only the slave of the soul, the body, and to neglect the soul, the lord and ruler of the body, especially since the Magi and Plato assert that the entire body depends upon the soul in such a way that if the soul is not well, the body cannot be well." And this was why "Apollo, the founder of medicine, decided the wisest man was not Hippocrates, though born of his own race, but Socrates, since, while Hippocrates strove to heal the body, Socrates strove to heal the soul. But Christ alone accomplished what those men attempted."[18]

In this potent, syncretistic vision, all aspects of the universe and of the complex cultural heritage into which Ficino had been born made sense in relation to one another. It was in part, too, a product of the "Hermetic" tradition, based on the *Corpus Hermeticum* that Ficino translated as well—a body of texts circulating in the West from the fall of Constantinople on and ascribed to Hermes Trismegistus, a mythical visionary in ancient Egypt whose pronouncements were thought to synthesize Plato and Christ, anticipating the one and divinely echoed in the other. (The corpus in fact consisted of early Neoplatonist texts that had been rather more prosaically compiled in the second century, but this would not be found out for another 200 years.) Knowledge of mystical realities was knowledge of the truth; and although such knowledge was meant for the educated few, contemplation of the heavens helped us understand our innards.

4. Cosmic Attunement

HUMOURS, in Ficino's scheme, were just one other aspect of the attuned universe; and attunement was the goal for a person thrown into dyscrasia, whose humoural constitution was out of tune, so to speak. A humourally balanced state was one in which we did not need to think about the body—in which we could forget its rumblings and pay attention to the soul's higher, nobler concerns.

Certainly, no one likes to feel the body's presence in the form of a hangover, an acidic stomach, or jet lag—let alone a bad cough, a headache, a sprained neck, or an inflamed wound. Most of us take good health for granted when we have it, because we live through, not for, our bodies. But Ficino believed that if one was afflicted with an ailment, it would have been unwise, un-Christian, and anti-Hippocratic not to try at least to drive oneself back to the state of harmony. One had to pay attention to the necessary relation between our soul and the World Soul. Health was a spiritual good, and its maintenance was a spiritual duty. A balanced organism was morally sound, partly because it meant that one was available for intellectual work, which Ficino believed was weakened by a less disciplined lifestyle.

Excessive sexual intercourse could undermine the intellect: it "drains the spirits, especially the more subtle ones, it weakens the brain, and it ruins the stomach and the heart." Hippocrates had been right to claim it was like epilepsy; and Avicenna was right to say, "If any sperm should flow away through intercourse beyond that which nature tolerates, it is more harmful than if forty times as much blood should pour forth." Excessive wine and food were unhealthful too: they filled "the head with humours and very bad fumes," drunkenness caused men to be "insane," and too much food in the stomach turned "the power of nature" away from the head. Staying up too late was unadvisable for those with intellectual ambition, in part because melancholy and phlegm dominated at night, whereas the Sun, Venus, and Mercury favored "reflection and eloquence." Negative melancholy, in other words, could be avoided if one followed the *Regimen,* took medical preparations based on *materia medica,* and used the right talismans to channel the effects of the planets.

Indeed, even though we were born under a zodiac that determined our complexion, and even though planets governed each stage of our lives in a geocentric universe, Ficino believed that it was up to us to use astral influences properly and for the sake of living optimally (in this he was following Plotinus). We were conditioned, rather than determined by the "celestial goods," in the sense that, as Ficino put it, "gifts from the celestial bodies come into our bodies through our rightly-prepared spirit" and that "the goods of the celestial souls partly leap forth into this our spirit through rays." Our exposure to these souls happened not naturally but "by the election of free will or affection."[19] Here was a solid foundation for a good life in a Platonic sense: "consider that those who by prayer, by study, by manner of life, and by conduct imitate the beneficence, action, and order of the celestials, since they are more similar to the gods, receive fuller gifts from them."

These lines could seem somewhat confusing to a modern reader. They strike one as pagan yet deeply Christian, superstitious yet spiritual, abstractedly Platonist yet down-to-earth. Throughout the three parts of *De vita,* in fact, Ficino was never short of practical advice to help us achieve a good life through physical means, offering a plethora of references to *materia medica* and writing, for instance, that theriaca should be taken regularly. Theriaca was especially effective when taken with a pill based on wine (or, in the summer, on rose water) and made of chebule myrobolan, purple roses, red sandal, emblic myrobolan, cinnamon, saffron, citron peel, ben, melissa, and the ever-useful aloe. It should be taken "twice a week during winter and fall, but only once a week during summer and spring," in order to "either keep whole or restore the forces of the stomach, heart, brain, spirits, and intelligence"; and it was excellent when "phlegm or black bile becomes excessive or nausea threatens."[20] Certain foods could lead to an increase in the worst, most pathogenic sort of black bile. Ficino's list of such foods has its source in a series of dietary recommendations by Constantinus Africanus, and includes "heavy and thick wine, especially if it is dark food which is hard, dry, salted, bitter, sharp, stale, burnt, roasted, or fried; beef and the meat of the hare, old cheese, food pickled in brine, vegetables (especially the broad-bean, the lentil, the eggplant, the colewort, cabbage, mustard, the radish, the garlic, the onion, the leek, the black medic, and carrots), and whatever causes warmth or cold, and likewise dryness and everything that is black."[21]

Visceral realities were never far from the sublime, and the scholar's advice was never far from the popular, unexamined belief that similars attracted and begat each other—that black foods could produce black moods, for instance, according to the rule of "like by like" which then justified the humoural notion that diseases had to be cured by their contraries. Emotional states were envisioned in terms of the adjectives they were matched with, as if metaphors were substances: alchemy and magic were based on this literalness, and humoural theory could also be based on it. Fully engaged in both, Ficino was quite easily able to accommodate the solid tradition of humours and complexions to his ethereal vision, which he combined with the originally Hippocratic correspondences between matter and mind, macrocosm and microcosm. And he held on to the view of the body as a container of vessels in which circulated the substances thanks to which we were alive, sensing, feeling, and literally aesthetic creatures.

5. Musical Therapy

IT WAS BY VIRTUE of this correspondence between microcosm and macrocosm that humoural theory also accounted well for the potentially curative effects of music on body and soul. Ficino described music as the sound of cosmic harmony, capable of quieting our appetites or exciting our senses. Here he was tapping into another long tradition: the Pythagorean notion of the music of the planetary spheres, also described by Plato in the myth of Er that ends his *Republic*,[22] and in the *Timaeus*, where he compared the "revolutions" of the soul to harmonies.[23] The Pythagoreans had believed that numbers expressed musical scales. As Aristotle later reported, they saw "the whole heaven" as "a musical scale and a number."[24]

In Ficino and in related literature, this idea was mixed in with a dose of Stoic metaphysics here, a pinch of Hermeticism there, and Christian

Platonism everywhere. Mathematics expressed the hidden but all-encompassing harmonies of which the human soul and the cosmos were made, the one in correspondence with the other. Music produced the harmony of the soul, which was an echo of the harmony of the spheres; and proportion prevailed within this unified universe. Just as we exposed "the body seasonably to the light and heat of the Sun through its daily harmony," wrote Ficino, so we exposed "our spirit in order to obtain the occult forces of the stars through a similar harmony of its own, obtained by images," as well as medicines and "odors harmonically composed."[25] The therapeutic effects on our bodies of music, medicines, odors, and images were due to planetary influences to which we had to expose ourselves. Their powers participated in the same phenomena as did those of amulets and incantations. Medicine was just one form of magic—one that Ficino insisted cohered perfectly well with Christianity.

Beyond this particular, rather idiosyncratic vision of a world governed by harmony and by divine sound lay the simple notion that music was a remedy, just as were baths, talismans, and balms. Prescriptions for sleep, sex, and song accompanied dietary advice especially for those afflicted with melancholy, who were advised to seek all calm sources of joy. Much earlier, Theophrastus had recommended music against sickness of soul. The tradition never died. Maimonides had prescribed songs and melodies as remedies against excessive black bile—along with sleep and wine to help induce somnolence and disperse the vapors produced by melancholy. It had long been assumed that music was an excellent cure for the melancholy caused by love, too—through its actions on the "spirit," or that "certain vapor, very thin and clear, produced by the heat of the heart from the thinnest part of the blood," as Ficino would put it in the commentary he wrote on Plato's *Symposium, De amore*.[26] The Ficinian "spirit" was the messenger between soul and body, the substance through which the soul could receive sense-perceptions of the world, "pure" versions of which it then conceived "by its own power," specifically the power of the imagination, or *phantasia*.

And so, it was the contemplative, intuitive mind—not imagination—that was placed under the sign of Saturn. This was why melancholy could be a creative force: at its best, it was called "generous melancholy." Dürer, in *Melencolia I*, was perhaps representing his enigmatic figure in the inspired state associated with it—claiming for the creative artist the power that Ficino had claimed for the philosopher.[27] On this interpretation of the engraving, the artist was capable of insight thanks to the imagination, all too aware as he was of the limitations of the mathematics and geometry with which he strove to read the world (Dürer practiced both with great zeal), but endowed with the artistic genius and inspired imagination that enabled him to melancholically grasp the starry realm of the Neoplatonists.

Ficino actually placed imagination low in the hierarchy of faculties; its astral influence was Mars or the Sun, while reason tended toward Jupiter. But he also believed that music acted precisely on that inferior faculty. In his letters, he wrote that "sound and song easily arouse the fantasy, affect the heart and reach the inmost recesses of the mind; they still, and also set in motion, the humours and the limbs of the body."[28] Music partook of physiology rather than of divine contemplation. The consolation of the soul was a physical matter, and the planets—Jupiter, Venus, Mercury, Mars, and Saturn—emitted the sounds that affected us, via the sun, from where music originated. Music acted on our basic physiology, and stirred the passions above which the contemplative mind was supposed to rise: this is why it had the power to act on our physiology, as a cure for these passions. A musician and melancholic himself, Ficino had a strong sense of what this meant. He claimed to play his lyre and sing when he needed consolation, but also in order "to avoid other sensual pleasures entirely. I do it also to banish vexations of both soul and body, and to raise the mind to the highest considerations and to God as much as I may."[29]

"That old and antique song we heard last night: / Methought it did relieve my passion much," says Duke Orsino in Shakespeare's *Twelfth Night*.[30] Later, in his 1687 *Song for Saint Cecilia's Day* (Cecilia is the patron saint of music and musicians), the poet John Dryden wondered, "What passion cannot music raise or quell?" The power of music on our moods, or souls, or physiology is supreme. It is immediately stirring; the mood of the melody becomes one's own mood. Music consoles and exhilarates. If its effects are physiological, they are experienced as spiritual. It carries us beyond our ordinary emotional states and beyond ourselves. It taps our most visceral—humoural—responses. Religious ceremonies are empowered and deepened by song. The most disparate group of people can be united in chant. Musical sound can seem to embody the divine.

And it has always been so. The biblical story of King David's use of the lyre to soothe a troubled Saul—"it came to pass, when the evil spirit from God was upon Saul, that David took a lyre, and played with his hand: so Saul was refreshed, and was well, and the evil spirit departed from him"[31]—appears in Ficino's *De vita*,[32] and remains as famous as the lyre that Orpheus used to coax the guardians of the underworld into letting him enter the place where no living being had ever been allowed.

Martianus Capella—a Roman lawyer of the fifth and sixth centuries AD who had been born in Carthage and was a pagan Neoplatonist during the zenith of the Roman empire—wrote a book called *The Marriage of Philology and Mercury*, much diffused in early medieval Europe. Here he has Harmony herself report that many Greek cities recited laws and public edicts to the sound of the lyre.[33] Harmony tells us that she "frequently sang a song in order to cure the disturbances of souls and illnesses of bodies," and "healed those suffering from phrensy with a symphony, [a method that] Asclepius the doctor imitated." She recounts how "the ancients used to cure fever and wounds with a song," how Asclepius was able to heal the deaf with a trumpet, and how Theophrastus used flutes to soothe afflictions of the soul. And wasn't everyone aware "that sciatica is driven out by the sweetness of the flute?"

These medical claims of Theophrastus and Asclepius were cited again and again in antiquity, by figures such as Aulus Gellius and Plutarch. Reliance on classical anecdotes and examples was so strong in the Renaissance that we find, for instance, this preceding passage from Capella along with the phrases that follow repeated word for word in an early-sixteenth-century treatise, *On Music and Poetry*, by a Neapolitan humanist called Raffaele Brandolini.[34] A technical work rather emblematic of its time, it recalled, in a wholesale appropriation of Capella, how Thales restored "health to those consumed by sickness and plague with the sweetness of his cithara; Xenocrates restored mind and soul to the insane with tunes on the organ; Theophrastus used the tibia to cure various afflictions." Music, Brandolini went on, in his own words now and in a fashion typical for his day, was "so closely linked to the knowledge of divine matters that not only did their theologians, who dwelt in great and universal darkness, have an extraordinary knowledge of it, but also our own theologians, illuminated by the light of Christian truth, know it well and attentively respect it."[35] Medicine and religion met within music, which countered the power of the passions. Each note had an effect on the humoural constitution. And so humoural temperaments, musical modes, planets, and muses were all in elaborate, precise correspondence.[36] Schemes of this sort, in which microcosm and macrocosm were integrally connected, would remain meaningful well into the following century.

Fig. 15. Woodcut from Isidore of Seville (560–636), *De natura rerum* (Augsburg, 1472). It accompanies a quote from Saint Ambrose's *Hexameron* and depicts "Mundus, Annus, Homo": world, year, man, that is, the elements and qualities, seasons, and humours.

6. Artistic Astrology

ASTROLOGY WAS INHERENT in the interrelation of the heavens with the elements and the human realm; it was a rather ordinary practice in the Renaissance, not just an abstruse, occult tradition for the initiated. The stars were signs that helped direct and regulate one's life, and it was important to many of the rich and powerful of the time to establish some ways of predicting their health, relying on the connection between their humoural selves and the celestial bodies.

This connection had already been established during the Middle Ages through the "zodiac man," of which countless examples were in circulation throughout Europe, representing the relation that each body part was thought to bear to a respective constellation and planet. Until anatomy began its rebirth in the late Middle Ages, this way of relating anatomical parts to external, astrological constructs—known as "melothesia"—might have seemed a precise enough prognostic tool. After then, it remained an attractive instrument. Astrology worked as a series of complex calculations relating planetary spheres to elements, seasons, qualities, and humours: the constellations were factored back into this earthbound realm, and therefore seemed to be legitimate parts of the equation. The demonstrations were circular, of course, but they fooled many.

But why not be fooled? Astrology was a reassuring way of finding meaning and order where neither was easily available. It was also, moreover, a source of entertainment, fun, and beauty. Agostino Chigi, a banker from Siena who eventually became treasurer to Pope Julius II, commissioned for himself a villa to be built in what were then the outskirts of Rome (today's fashionable Trastevere area). It was designed by a fellow Sienese, the architect and painter Baldassare Peruzzi, between 1506 and 1510. In 1534, it was bought by Cardinal Alessandro Farnese, and so it remains known today as the Villa Farnesina. Chigi was a man of means and a *bon vivant*, endowed with a refined taste and an exquisite sensuality. In his bedroom, he had the artist (nicknamed) Il Sodoma paint erotic panels depicting the *Marriage of Alexander and Roxanne*, and is reputed to have hosted fabulous parties in the villa and its gardens. The parties are no more, but the villa is still one of Rome's most delectable treasures.

In 1517, Chigi had the good idea to commission Raphael and other artists, including Raphael's pupil Giulio Romano, to paint the frescoes in what is known as the Loggia di Psiche, which connects the ground-floor entrance to the garden. Here one can admire Cupid staring at the sensuous Three Graces while pointing toward the room below. The ceiling of one of the rooms upstairs, known as the *Sala della Galatea*, is decorated with the exquisite *Triumph of Galatea*, which Raphael himself executed in 1511. There too the theme is love: Galatea was a beautiful Nereid, in love with the handsome shepherd Acis (a half-god, son of Pan the flute-player), but also admired by the cyclops Polyphemus, who eventually killed Acis in a fit of jealousy. In the erotically charged picture she is riding a wave in the company of nymphs and sea creatures, standing on a shell-shaped chariot driven by dolphins not unlike the shell from which Aphrodite emerges in Botticelli's *Birth of Venus* (both pictures were loosely based on the same verse from a poem by the humanist Angelo Poliziano). She is staring at three putti armed with arrows directed at her heart. Polyphemus himself is offstage, as it were, depicted in another fresco nearby by Sebastiano del Piombo.

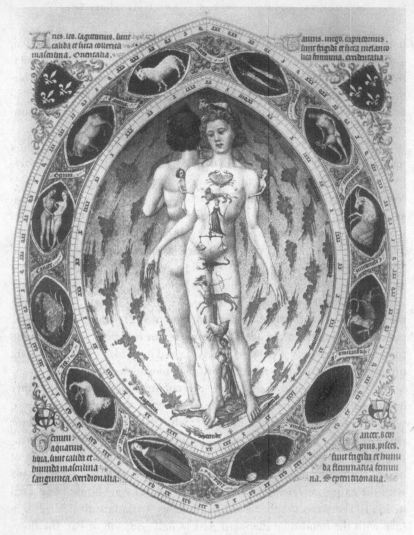

Fig. 16. *Astrological man*, from *Les Très Riches Heures du Duc de Berry*, p. 61 (published in *Journal of the Warburg and Courtauld Institutes*, 1954).

But this heady celebration of love, beauty, and eroticism through the artistic depiction of inspiring mythology is not all one finds in Chigi's villa: astrology is central to the decoration, cheekily connected to these various mythological, celebratory works. Around the *Galatea* herself, Baldassare Peruzzi composed twenty-six frescoes, each set within its own frame, and each depicting signs of the zodiac, gods associated with planets, and various other constellations. What first emerged in the 1920s from the study of this ceiling by the visionary, maverick scholar Aby Warburg was that the disposition of these frescoes was such that it must have an astrological meaning—and, presumably, depict the natal chart of Chigi himself. Chigi's stars, so to speak, seemed to be writ large upon the ceiling of his villa. Warburg died in 1929; but over the decades, scholars continued their calculations, and eventually found out that Warburg's hunch had been accurate.[37] The analysis was complex, especially since the frescoes depicted also, juxtaposed with Chigi's horoscope, mythological figures that did not belong in the zodiac: one can see the goddess Fortuna next to Aquarius, or Perseus next to the Bear. These must have symbolized in some way Chigi's power and fortune—the signs under which he was born, and under which he lived.[38] He was presumably a sanguine man, perhaps not so different, in this way, from Francesco Sforza. The ceiling could only have amused and impressed the guests at Chigi's banquets, and it still speaks its astrological language to those who can decipher it—in a form that is exuberantly pleasing to the eye.

Astrology was therefore not a mere predictive tool. Although it arguably still appeals today to the enduring human need to believe in order, in a world we are not able to control and predict much better now than we could five centuries ago, it was used much more imaginatively then, with a sophistication that is hard for us to match. It was not only a playful, but also a serious, scholarly, activity that many people during the Renaissance believed crucial to self-understanding—an ingredient in the "high" culture of the sixteenth century.

Those who partook of this "high" culture, however, were also beginning to reap the philosophical fruits of humanist philology and to manipulate the pagan heritage of allusion, myth, metaphor, and emblem. It was possible by the early and mid-sixteenth century to rethink old assumptions about the order of the cosmos, world, soul, and body. It was also possible by then to experiment with novel ideas regarding the nature of artistic representation and what a human body should look like, whether it be depicted on canvas or paper, in marble or bronze. The realms of anatomy and astronomy were both evolving. Gradually but surely, the old relation between the humoural microcosm of the body and the macrocosm or universe in which humours made sense was changing.

Fig. 17. *Septem planetae,* Temperaments and elements, Gerard de Jode (after Martin de Vos), 1581. The caption at the center reads:

SEVEN PLANETS
corresponding to the Seven Ages of Man
namely

1. Infancy
2. Childhood
3. Adolescence
4. Youth
5. Maturity
6. Old age
7. Decrepitude

illustrated with very elegant figures for each of these operations and effects.

Man that is born of a woman is of few days and full of trouble.
Job 14

The days of our years in them are threescore and ten years
Psalm 89

Engraved by Gerard de Jode
in the year 1581,
designed by Martin de Vos.

7. Paracelsus and the
Magic of Nature

NOT ALL SCHOLARS and doctors, in fact, held on to humours. A parallel tradition was emerging—another system, which focused on natural magic, rather than on the rationalist tradition of Galenic medicine. But it, too, grew out of the recovery of ancient texts, and especially out of those of the *Corpus Hermeticum,* translated by Ficino. At its core were the alchemical correspondences between microcosm and macrocosm, between body and world—which humouralists themselves had held on to since Hippocrates and Democritus.

The sixteenth-century figure of Theophrastus Paracelsus von Hohenheim, known as Paracelsus, was instrumental in the development of this parallel tradition. He was a largely self-taught doctor who left home early, eventually worked as a surgeon with military troops, and wandered around Europe. He acquired the title of Doctor of Medicine in Verona before becoming professor of medicine and physician in 1526 at the university of Basel, where he treated Erasmus and other renowned patients. But Paracelsus was fiercely independent, and unafraid to express his strong aversion to the Galenic tradition that was being taught in Basel as elsewhere. His relations with Basel authorities deteriorated fast. One day in 1527—it was St. John's Day—in a bold, if rash gesture of protest, he threw a copy of Avicenna's *Canon* into the celebratory bonfire.

The gesture is legendary, and there are various accounts of it. At any rate, it did not make him many friends. He fled soon after, and resumed his wandering life; but along the way he began to gather disciples. In 1536 he had published—in the vernacular German rather than Latin—his immediately popular and impressive *Große Wundartzney,* in which he recounted his care of wounds, notably those caused by gunshots. He had seen many wounds during his military excursions, and that experience formed him better than any university training. He was an expert on the matter. His hands-on approach to medicine earned him enough respect for him to continue seeing patients wherever he went, prolific in spite of his wandering life, writing treatises on syphilis and on diseases caused by mining, for instance.

Paracelsus was thus a medical empiric. But he was mostly a mystic, or visionary Christian on a spiritual mission in a world that was being transformed by the Protestant Reformation. Burning Avicenna was just part of this mission, and alchemy, to which Avicenna had been opposed, was central to it. Nature, Paracelsus believed, was a vital force, and our bodies were a part of nature, a reflection of the cosmos. Our organs, functions, and dysfunctions corresponded to cosmic bodies: our bodies were the "signatures" of the God-created universe. As he wrote, "no brain can fully encompass the structure of man's body and the extent of his virtues; he can be understood only as an image of the macrocosm, of the Great Creature. Only then does it become manifest what is in him."[39] Our soul was in touch with the world's soul—the *astrum,* as he called it—and miracles could happen because souls survived the body's death in the form of astral bodies, the *evestrum.* The world itself was made of forces, embodied in the metals that corresponded to the stars. There was a symbolic import to the three base metals of the alchemists—mercury, salt, and sulfur—and to the alchemical recipes that had been handed down from Arabic sources. Ficino had assumed something similar when he explained that the exposure to all things black was to be avoided by those who suffered from excessive black bile, the begetter of black moods: the metaphorical names given to states of mind were conflated, as ever, with the literal qualities denoted by those names. The whole of life was an alchemical process, united by what Paracelsus called an *archeus;* alchemists, experts at drawing tinctures out of metals, were participating in the divine alchemy that was life and its processes of digestion, putrefaction, and transmutation.

But this was also why, for Paracelsus, diseases had to be cured by their analogue, rather than by their contraries, and why a black mood, for instance, would actually require treatment with a black substance. Paracelsus did acknowledge that "man consists of the four elements," but that was "not only—as some hold—because he has four tempers, but also because he partakes of the nature, essence, and properties of these elements."[40] The tempers, moreover, were associated for him with the body's "four kinds of taste—the sour, the sweet, the bitter, and the salty," where the sour was melancholic, the sweet was phlegmatic ("for everything sweet is cold and moist"), the bitter was choleric and the salty was sanguine.[41] And whereas humoural theory called for the allopathic use of contraries, "like by like" was the principle of the alchemical medicine he favored, which is not unrelated to homeopathy today. Mainstream humouralists like Ficino had been capable of marrying this sort of magical thinking (that a black mood is sensitive to the color black) with the notion that ills required allopathic cures (that a black mood will lift in the presence of white). Paracelsus instead integrated magical thinking with the alchemical principle that an ill must be cured by its likeness.

Paracelsus was rather unique. He was a radical figure—even nick-named by his contemporaries "Lutherus medicorum," the Luther of physicians—with an all-embracing vision, and a man who would inspire future generations. But one did not have to give up humours to advocate the centrality of magic in the divine world. One of the best-sellers of the second half of the sixteenth century was a treatise by Giambattista della Porta, *Magiae naturalis,* published in Naples, in Latin, in 1558. Vernacular editions followed quickly.[42] Della Porta was a prominent scholar and natural philosopher, who wrote about such topics as distillation, optics, refraction, cryptography, and, famously, physiognomy. (He also founded a research academy, the Accademia dei Segreti, which was eventually shut down by the Inquisition.) But the *Magiae naturalis* was highly accessible, and much in demand because it was practical and satisfying. The elixir of life was one of the main goals of alchemical research, and della Porta's book was a contribution to this goal. It combined alchemy, including recipes for obtaining the key tincture of gold, with Hermeticism, astrology, and the sort of magical "folk" and humoural beliefs catalogued over the centuries in works like the *Circa instans.*

Gold was here listed as healthy, a remedy against melancholy because "it contends with the *Sun* in beams, brightness and glory." Not only that, but "the elements are so proportionally mixed in the composition of it, so put and compacted, that they account it a most exactly tempered body, and free from corruption" and when ingested, "it must needs reduce the elements and *Humours* into a right temper. Allay the excessive, and supply the defective, take away all putrefaction. Refresh the natural heat. Purge the blood and increase it. And not only cure all sicknesses, but make us healthy, long lived, and almost immortal."[43]

Less ambitiously, della Porta advised people to eat balm, bugloss (a herb long used for its medicinal properties), and poplar or poplar oil to have pleasant dreams. By contrast, ingredients such as beans, onions, garlic, and leeks tended to produce nasty dreams. In general, dreams were produced like this:

The meat in concoction must be corrupted (this must be taken for granted) and turned into vapors; which, being hot and light, will naturally ascend, and creep through the veins into the brain, which being always cold, condenses them into moisture, as we see clouds generated in the greater world. So by an inward reciprocation, they fall down again upon the heart, the principal seat of the senses. In the meanwhile, the head grows full and heavy, and is overwhelmed in a deep sleep. When it comes to pass, the species descending, meet and mix with other vapors, which make them appear preposterous and monstrous, especially, in the quiet of the night. But in the morning, when the excrement and foul blood is separated from the pure and good, and becomes cool and allayed, then pure, and unmixed, and pleasant visions appear. Wherefore I thought it not irrational, when a man is overcome with drink, that vapors should arise participating, as well of the nature of what he has drank or eaten, as of the humours which abound in his body, that in his sleep he should rejoice or be much troubled. That fires and darkness, hail and putrefactions, should proceed from choler, melancholy, cold and putrid humours. So to dream of killing anyone, or being besmeared with blood, shows an abundance of blood. And Hippocrates and Galen say, we may judge a man to be of a sanguine complexion by it. Therefore, those who eat windy meats, by reason thereof, have rough and monstrous dreams. Meats of thin and small vapors, exhilarate the mind with pleasant phantasms.[44]

We do tend to have "heavier" dreams after a heavy meal. Della Porta's description, in effect, partook of standard humouralism, put in the service of magic. Within such a rich world of metaphorical and literal transformation, there was no reason to exclude alchemy and astrology. One did not have to be a partisan of the humours to discard magic, and despite the rejection by Paracelsus of humoural physiology in favor of unalloyed, exclusive, and universal chemistry, even the latter's system found its advocates among the very Galenists he so despised. Johan Gunther von Andernach, an influential professor of medicine in Paris, thought it worthwhile to accommodate the two, even publishing a treatise called *Of the Ancient and New Physick*. Natural magic would become entwined with much mainstream natural philosophy over the course of the seventeenth century.

8. Corpses, Books, and Reputations

DESPITE ITS CULTURAL and artistic power, however, the notion of a connection between microcosm and macrocosm was becoming more difficult to sustain during the sixteenth century. The world was opening to view. The old, all-encompassing schemes that had filled out the blanks of scientific knowledge began to look obsolete. The year 1543, in particular, was momentous: two very different but equally epochal works were published that would each eventually weaken further the connection between microcosm and macrocosm. *De revolutionibus orbium* displaced the earth from its position at the center of the universe in favor of the sun. It appeared as its author, the Polish scholar Nicolaus Copernicus, lay dying. He had been a practitioner of astronomy but also of medicine, law, and finance, as well as a church and government administrator. His astronomical conclusions arose out of mathematical calculations, not observations: he had pursued a hypothesis, without purporting to demonstrate empirical facts. Nevertheless, in 1616, *De revolutionibus* would end up on the Church Index of Prohibited Books.

The human body was also examined afresh, and with less grievous consequences, with the publication of *De humani corporis fabrica libri septem*. Far from being on his deathbed, its author, a twenty-eight-year-old prodigy named Andreas Vesalius, was at the start of a brilliant career. Vesalius's most important activities are usually associated with the Venetian city of Padua, although he was born in Brussels (as André Wesele Crabbe), the son of an apothecary to the Holy Roman Emperor Charles V. He studied medicine in Louvain (Leuwen) and Paris before going to Padua, where he gained his degree with distinction and swiftly became lecturer of anatomy and surgery. Here, and in Bologna too, Vesalius conducted his famous demonstrations of the falsity of some of Galen's anatomical constructions, dissecting in front of hundreds of fascinated spectators the corpses of condemned criminals. He compared skeletons, studied the newfound editions of Galen in Greek, observed carcasses in butcher shops.[45] He would soon serve the Emperor Charles V himself, as a physician, dedicating the *Fabrica* to Charles; and after a period in successful private medical practice in his native Brussels he ended up in the service of King Philip II. Vesalius would die on his way back to Padua from a pilgrimage he made to Jerusalem.[46]

The *Fabrica* was by no means an isolated phenomenon. Vesalius had studied in Paris in the 1530s, when the teaching of medicine there was changing under the impact of humanism, just as it had changed earlier, in the 1520s, in northern Italian towns like Ferrara, Bologna, Pavia, Venice, and indeed Padua. His professors in Paris included the anatomist Jacques Dubois—also known as Jacob Sylvius—as well as the notable Johan Gunther von Andernach. Both were translators of Galenic texts, and both saw a need to base anatomy directly on dissection and on the newly available Greek sources. Along with a number of their contemporaries, they were keen to correct what they considered the errors of the medieval Arabic past, and to create textbooks that would replace Mondino's *Anatomia* of 1316 in medical schools.

In 1521, there appeared for the first time a Latin translation of the short book by Galen that focused on the determination of temperaments by the four elements, *De temperamentis et de inaequali intemperie*. The translator, Thomas Linacre, was a famous English physician who had studied medicine at Padua and had been instrumental in founding the London College of Physicians a few years earlier, in 1518.[47] The complete works of Galen began to appear in the original Greek in 1525, printed by the great Venetian publisher Aldo Manuzio. They contained important new material. That same year, a Latin translation of Hippocratic works was published for the first time,[48] and the next year they appeared in the original Greek. Translations into Latin of works of Galen continued to appear in those years.[49] The authority of the older, medieval translations from Arabic was rapidly fading, just as Galen's own Greek words were regaining prominence over the Arabic treatises, and over Mondino himself.

To be sure, not all the humanists and forward-looking physicians were successful in renewing anatomy. Jacopo Berengario da Carpi is a case in point. He was the ambitious son of a barber-surgeon, a noted physician, and, from 1502, a professor of surgery and anatomy at Bologna; and he had conceived his own *Isagogae breves*, published in 1522, in exactly that spirit of innovation. But its illustrations were stylized and bore no explicit relation to the text, so the book was not practical enough to be of much use. Berengario, as it happens, was also involved in revising translations of Galen's anatomical work into Latin, and he was the author of an influential commentary on Mondino.

In this commentary, Berengario claimed to have learned from the medieval Mondino the "superiority of seeing and touching over hearing and listening," convinced that medicine required observation, not books or reliance on tradition. Surgery, he wrote, took place on the "uncertain border between life and death"; and because he had examined the corpses of two plague-stricken Spaniards, in order to understand how the disease afflicted the living, as he put it, he was deprived of his professorship in 1527. It must be said that during his tenure, Berengario made some important anatomical discoveries—all indeed based on the dissection of humans. In his commentary on Mondino, he claimed, for instance, that he had seen neither the seven-celled uterus Mondino had believed did exist, nor the *rete mirabile* at the base of the cranium, which everyone since Galen had insisted was instrumental in delivering blood to the brain of humans.

Berengario's reputation was justifiably based on his anatomical skills. But as had been the case with the early Alexandrians, anatomical genius had little to do with clinical proficiency; and although anatomical theory did serve surgical practice to some extent, ordinary illnesses continued to be treated with the same methods as before—some bizarre, others frightening, some commonsensical and straightforward, and others infuriatingly inefficient. We have the testimony of the sculptor and writer Benvenuto Cellini, who met Berengario in Rome, when the physician was visiting during a local outbreak of the plague. In his autobiography, Cellini tells us with customary glee that the great doctor, "in the course of his other practice, undertook the most desperate cases of the so-called French disease. In Rome this kind of illness is very partial to the priests, and especially to the richest of them. When, therefore, Maestro Giacomo had made his talents known, he professed to work miracles in the treatment of such cases by means of certain fumigations; but he only undertook a cure after stipulating for his fees, which he reckoned not by tens, but by hundreds of crowns." [50]

Cellini went on to recount how, as a "great connoisseur in the arts of design," Berengario, passing by Cellini's shop, saw there sketches "for little vases in a capricious style, which I had sketched for my amusement," and ordered from Cellini silver vases based on these sketches. The doctor left Rome shortly after having shown the vases to the Pope, and Cellini tells us that Berengario did "wisely to get out of Rome; for not many months afterwards, all the patients he had treated grew so ill that they

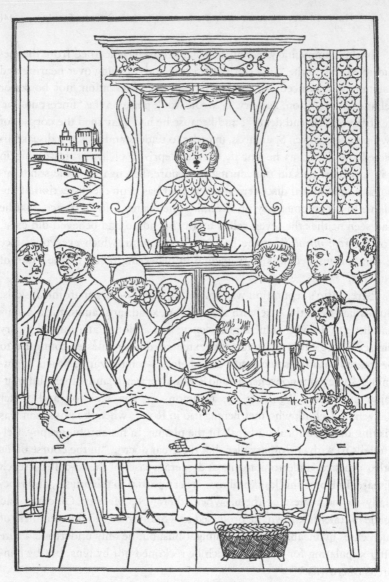

Fig. 18. Johannes de Ketham, *Fascicolo de medicina* (Venice, S & G de Gregoriis 5 February 1493), Essling cat. II, 56. From Jacopo Berengario da Carpi, *Isagogae breves* (Bologna, 1522). Berengario's commentary on Mondino was often found bound along with other short, practical anatomical treatises like the popular, late-fifteenth-century *Fasciculus medicinae* by Johannes de Ketham, published in Venice in 1495. Johannes de Ketham was probably a Viennese physician whose real name was Johannes Kircheim. The treatise's sober line drawings are noted for recalling those of the great artist Andrea Mantegna.

were a hundred times worse off than before he came. He would certainly have been murdered if he had stopped."

The changes that anatomy was undergoing had more to do with the body's representation than with its care, more to do with external surface than with internal humours: no anatomist was in a position to help patients better than any doctor had before. But the innovations were significant. Eventually they would lead to the dismantling of humoural theory. The shifts did not happen overnight, and not without resistance, but the culture was changing whether one liked it or not. Vesalius's lecture-demonstrations in Padua were immensely popular. Duke Cosimo I de' Medici even helped provide corpses for Vesalius's dissections and was keen to follow the show.[52] It was in fact a relatively smooth process, especially in Italy, to obtain for dissection the corpses of condemned convicts, though these had to be from a town other than that in which they were to be flayed, or the families might have protested. Since there were few available corpses, however, anatomists supplemented these by robbing the occasional grave.

The governing structures that defined the medical establishment were changing too. In England, the Physicians Act of 1540 united the Guild of barber-surgeons and the Fellowship of surgeons within one Company of Barbers and Surgeons, allowing their respective practices to be better regulated: surgeons could now no longer perform the duties of barbers, while barbers had to be content to act mainly as dentists. This new Company was now alone authorized to teach anatomy and surgery, and was entitled annually to receive a few bodies for dissection (of condemned criminals, of course). Two years later, the 1511 Surgeons and Physicians Act was amended to allow anyone with a knowledge of herbalism to provide treatments for everything from wounds to kidney stones: the often lethal monopoly of one Guild over medical knowledge had officially ended.

9. Beauty Beneath the Skin

THE INFORMATION ANATOMISTS gathered did not remain in the hands of a few specialists: a new genre of anatomical representation began to be produced as "fugitive" sheets, multilayered, popular anatomical engravings, some of better quality than others but all addressed, for the first time, to a lay audience. They had a wide circulation—especially throughout northern Europe—and delivered the latest available information about the body's makeup even to those who could not read.[53] These popular sheets absorbed the revisions that anatomists, especially Vesalius, were making thanks to their dissections of human corpses. It was not only that a field that had not evolved much in over 1,000 years was effectively being transformed: now, heretofore specialized information was circulating even among the unlettered.

Fugitive sheets were one outcome of a new marriage between art and science: the revolution in anatomy was an aspect of a broader revolution in visual culture that was affecting artistic as well as scientific representation. Now that a novel attention to apparent form was slowly replacing idealized, Platonizing representations of the physical world, the artistic representation of the human figure was changing. This had begun earlier—Leonardo da Vinci's extraordinary anatomical drawings are of unsurpassed quality. But others were beginning not only to advocate seeing over reading, but actually to take a good look at what stood or lay before them and to convey the human form on paper, or on canvas. Anatomy was serving artists, some of whom attended dissections in order to carefully study the flayed body; and artists, in turn, were beginning to serve anatomists by ably executing the illustrations that, from now on, would be central to anatomical treatises.

Artistry is just what one gets in *De dissectione partium corporis humani,* by the Parisian Charles Estienne, another pupil of Dubois. It actually appeared before the *Fabrica* came out, although it was published two years later, in 1545, complete with sophisticated engravings of the human figure. Those representing the female body, including internal reproductive organs, were largely recycled from a series of erotic prints on the theme of the loves of the gods.[54] Estienne's *De dissectione* offered a long, impressively detailed, informed text on anatomy that also made occasional use of small sketches set in the margins of the text and illustrating it. But in fact, the few, main engravings seemed designed instead to attract and provoke. The figures were depicted in landscapes of various sorts, as if they were alive, full participants in the physical world. The skeletons apart, they were all attractive, flawless nudes in contemplative, often sexually suggestive positions, where one part of the body (especially the abdomen, though in one case the cranium) had been "opened" up to reveal just a little of the internal organs.

Anatomical illustration in Estienne's manual was aestheticizing more than informative; it was also eroticized, conjoining knowledge of the body with its desirability. This new anatomical art helped to overcome the earlier resistance to the full pictorial depiction of nudity. Painters like Michelangelo were now shamelessly representing unclothed bodies. The human animal was arising from the dissecting table; death served to reevaluate the living body, to take a better look at the beauty of the soul's material envelope.

In this way, the sphere of the visible was expanding. The three-dimensional body was also becoming increasingly measurable, and its representation by anatomically informed artists was increasingly embedded within that of nature, rather than depicted for the sake of religious edification. In 1538 Vesalius had published, in Venice, his *Tabulae anatomicae sex,* six anatomical tables for use mostly by medical students. They were clear, didactic descriptions of the Galenic organism, heavily annotated and practical rather than provocative or pleasing, and so popular that they were often plagiarized in those years. With the *Fabrica,* however, there was a leap in quality.[55] The images now carried an enormous amount of information, as well as corrections of Galenic errors. They showed, for example, how untrue was the old notion that men had one less pair of ribs than women. This was a momentous revision: Eve had not emerged out of Adam's rib.

These images were more accurate and innovative than those of Vesalius's predecessors and contemporaries. They were also of high artistic quality. Who exactly executed them remains a mystery. Giorgio Vasari, in his *Lives,* had attributed them to a Flemish artist and pupil of Titian in Venice, Jan Stephan van Calcar. Calcar definitely did create the engravings of the *Tabulae,* but probably not those of the *Fabrica.* Whoever did, at any rate, reached in sensitivity and precision a level second only to that achieved by Leonardo in his anatomical drawings—which were still unknown in the mid-sixteenth century.

These striking engravings were also eminently usable renditions of the human anatomy that it had been possible to investigate, given good dissection and drawing tools and acute observational skills. The *Fabrica*—along with its abbreviated popular version, the *Epitome,* published by Vesalius in the same year—was revolutionary in its clarity and impact. It turned anatomy into a dignified field, worthy of study by the best minds. Vesalius was adamant that doctors should engage in the very activities

Fig. 19. Charles Estienne, print from *De dissectione partium corporis humani* (Paris, 1545), p. 295.

they had begun to despise after the advent of what he called, in his preface to the *Fabrica,* the "Gothic devastation." From then on, he wrote, they had "quickly degenerated from the standard of ancient physicians, abandoning the technique of cooking and food preparation to those attending the sick, the composition of medicine to druggists, and surgery to barbers. And so in the passage of time the science of healing has been so sadly torn asunder that certain doctors, peddling themselves as physicians, have taken for themselves exclusively the prescription of medication and diet for hidden conditions; other branches of medicine they relegated to those whom they call surgeons and scarcely respect as servants, disgracefully spurning what is a principal and most ancient branch of medicine, one preeminently reliant (if anything is) on the investigation of nature."[56]

In the same year of 1543, there appeared another aesthetically impressive work of natural science, which promoted the investigation of nature: Leonhart Fuchs's *New Kreüterbuch,* or the *New Herbal.*[57] Fuchs was a successful physician, a professor of medicine at the university of Tübingen, and a prolific writer in his field, who criticized Mondino, promoted dissection, and introduced the use of Vesalius's *Fabrica* in his teaching, writing a two-volume commentary on it.[58] His monumental catalogue of plants contained over 500 color plates, all carefully depicting a wide range of botanical species. Fuchs relied here on Theophrastus, Dioscorides, and Pliny, as had his predecessors, but he also studied the plants with an eye that was attentive to botanical details. He found new species, and he was in many cases the first to illustrate and describe plants in terms that are scientific even by modern standards. He noted for each plant its structure, annual cycle, favored climactic and light conditions, and so on. He also listed the plants' medicinal qualities, in humoural terms, in the belief that all doctors, and not just apothecaries, should know their *materia medica.*

Vesalius, along with most other physicians trained as humanists, would have agreed with this, since he too believed that doctors should be their own apothecaries and their own surgeons. Doctors, he wrote, had "lost the knowledge of simple medicines that was absolutely essential to them; it is their fault that the workshops of apothecaries teem with barbaric words and indeed false drugs, that we lack so many of the best compounds of the ancients, and that many are still not even known to us."[59] But neither in his view nor in that of Fuchs did the new focus on technique and

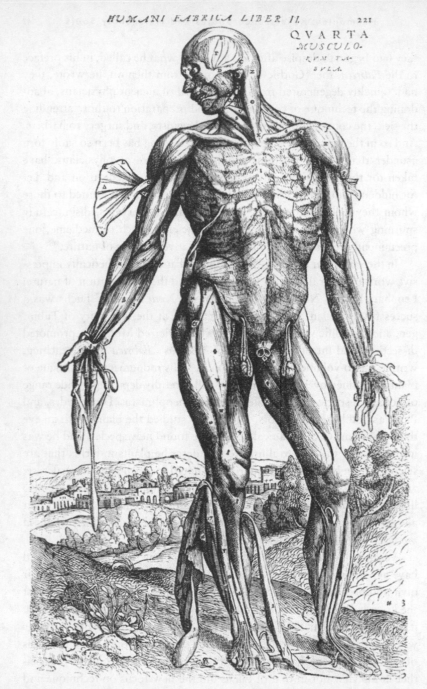

Fig. 20. Vesalius, print from *De humani corporis fabrica* (Basel, J. Oporinus, 1555).

apparent structure, along with the concurrent concern for the art of depicting the visible, physical world, entail the demise of humours. Vesalius did not want to destroy the past. He was merely adamant about the need to acknowledge and correct the errors that Galen had made, since the latter had "himself never dissected a human body, but in fact was deceived by his monkeys (granted a couple of dried-up human cadavers came his way) and often wrongly disputed ancient doctors who had trained themselves in human dissections," like Herophilus and Marinus. As a result, Galen had "departed much more than two hundred times from a true description of the harmony, use, and function of the human parts."[60]

So it was not enough to read better editions of the great medical authority: one had to use a critical eye, bloody one's hands, and dare to contradict this authority. Until now, dissections had been conducted from on high—the physician would read in Latin from Galen while the lowly surgeon was to cut into the body. If the surgeon's Latin was insufficient or nonexistent, this would happen under the guidance of a Latinate assistant. But it was time, said Vesalius, for the educated, Galenist physicians to literally step down from their "lofty professorial chairs," to abandon "their egregious conceit," and to wield the knife themselves. It was by doing so, at any rate, that the intrepid young man had been able to deepen knowledge of human anatomy. The errors he showed up in the *Fabrica* were no mere details. They would eventually lead to the necessary reorganization of the human body: here, there was no trace of the Galenic *rete mirabile* (still depicted in the *Tabulae*, however), in conformity with what Berengario had suggested earlier. The *vena cava* that departed from the heart was not in fact connected to the liver, as Galen had supposed. The septum between the two sides of the heart was not quite as porous as Galen had believed.

However commonsensical Vesalius's approach, though, it was all a bit too much for a number of his colleagues, notably for his old Parisian teacher Sylvius, who had himself encouraged his students to conduct dissections firsthand but who had probably not braced himself for the actual consequences of this practice. He had wanted solid, novel Galenic scholarship, certainly, but not revolution. He vehemently attacked his former pupil, and tried to reason his way out of the younger man's novel conclusions. Rather than accept that Galen had made serious mistakes, Sylvius supposed, for instance, that the *rete mirabile*, clearly present in beasts, had probably been present in humans at the time of Galen: our anatomy had simply changed. In the revised, 1555 edition of the book, Vesalius inserted in his preface a sharp, barely coded reference to "the attacks of those who have not, as we have done in the Italian schools, seriously applied themselves to anatomy, and now, old men wasting away with envy of the true discoveries of the young, will be ashamed, like other followers of Galen, to have been blind until now and not to have noticed the things we now set forth, even though they claim a great name for themselves in this art."[61]

Vesalius was right, of course. And he was not alone. His followers in the sixteenth century included anatomists such as Juan Valverde de Amusco and Bartolomeo Eustachi, who, taking their cue from him, also produced highly impressive, detailed, finely observed renderings of the human body, collated from the evidence that they gleaned from dissections.[62]

And there were those, like Paracelsus, who practiced before bothering to preach or draw, notably a groundbreaking surgeon, the Frenchman Ambroise Paré. He, too, believed in the necessity of sticking to observation rather than to texts or to blind "followers of Galen." He started out as a lowly barber-surgeon, knew little or no Latin, let alone Greek, practiced dissections as an apprentice at the Hôtel-Dieu hospital in Paris, became a military surgeon, and, though he insisted on working also for the poor, would eventually work for three successive kings—Henri II, then Charles IX, and finally Henri III. Paré revolutionized the technique of treating wounds by purely empirical means. It had been usual to pour hot oil over wounds—a dangerously and painfully ineffective method. One day, having run out of oil in Turin during the Italian campaign in 1537, when he was serving as a military surgeon with François I's troops, he devised a concoction of egg yolk, rose oil, and Venice turpentine, along with diachylum (a mixture of plant juices) and oxycrate (a mixture of vinegar and water). He realized that the wounds healed rather well, and faster, with these compresses; inflammations and fevers were fewer. It was none other than Sylvius who encouraged Paré to publish his account of wound treatment: it came out in 1545 as *La méthode de traicter les playes faictes par hacquesbutes*. A few years later, during the siege of Danvilliers on the eastern front during a Franco-Austrian war, Paré had to amputate the leg of a soldier of the count of Rohan. He decided not to cauterize the wound with a hot iron, as had been the norm until then, but rather to ligate the arteries at the base of the leg. The Hippocratics had used the technique;

but Paré was the first to resort to it again. Hemorrhage stopped swiftly. Recovery ensued. The ligation of arteries became an established practice, to the benefit of all.

Paré, incidentally, did meet Vesalius: the two were called to the bedside of Henri II as the king lay dying in July 1559, of wounds he received during a tournament.

10. New Bodies, Old Science

BUT THE "TRUE DISCOVERIES" of this new generation concerned the visible realm of anatomy, not the invisible realm of physiology. Humours were the only certain key to the causal operations of the body, which otherwise remained for the most part enclosed within a set of secrets; and humours were still difficult, if not impossible to represent. They remained tenacious despite their invisibility, even as anatomy proceeded without them. In his *Journeys to Divers Places* (his account of his decades on the road as a military surgeon), Paré himself told of an adventure he had as a prisoner in 1553, when he took care of a young gentleman's eventually fatal wound. He recounted how, after applying his compresses, he "drew five basons of blood from his right arm, considering his youth and his sanguine temperament."

In other words, the larger, humoural picture continued to prevail. Despite the demise of the *rete mirabile,* furthermore, animal spirits continued to turn into vital spirits as they entered the brain with the blood, and to account for cognition and emotion. Certainly, by observing fresh brains Vesalius had been able to disparage the ventricular picture of brain function that had been the established theory for a good 1,500 years. But nothing clear replaced it.

Meanwhile, although the medieval correspondences between the heavens and the body were slowly unraveling, aspects of the culture that had sustained them remained alive: astrology continued to make sense, just as humours continued to circulate, unseen but manifest. Agostino Chigi was not the last plutocrat to play with planetary spheres. Medical astrology was a fashionable activity in the sixteenth century, at a time when medicine could only do so much for the seriously ill. Despite the promise of renewed knowledge in a world that had grown geographically larger, death stood at every corner of every walk of life, and physicians had to be used to failing in order to survive in their profession.

It is indicative of just how much the old and the new overlapped that the very years in the sixteenth century during which anatomy was maturing also saw the span of the career of Girolamo Cardano. He was both a doctor and an astrologer whose original and profound engagement with each discipline brought him wide fame, in Italy and abroad. Cardano was born in Pavia, near Milan, as the illegitimate son of a lawyer whose talent in mathematics was such that he was sought out for advice by Leonardo da Vinci. He studied medicine in his native town, and then at Padua. When he became rector of the university of Padua, he was already wasting time and money on an addiction to gambling, which nearly ruined him and the family he had just started. Repeatedly rejected by the College of Physicians in Milan—in part because of his illegitimate status—he set up a small medical practice in the little town of Saccolongo, near Padua. (He later claimed to have spent the six happiest years of his life there.) Eventually, he acquired a lectureship in mathematics in Milan, while continuing to see patients.

Though it did eventually recede, his love of gambling would never die, even after he had found success through his astrological charts of the famous and through his writings, which encompassed an impressively wide range of subjects. There were encyclopaedic works[63] and some 200 volumes on mathematics, mechanics, geology, cryptography, philosophy, medicine, natural philosophy, metaphysics, music, alchemy, astrology, religion, and much more. He was a multitalented man who eventually became renowned as one of the greatest physicians of his day, was named a professor of medicine at Pavia in 1543, and would be called on by rulers from France, Germany, Switzerland, England, and Scotland.

Despite his self-professed arrogance and what one scholar has called his "in-your-face frankness and penchant for risky claims,"[64] his star shone. His connections were powerful enough that he survived the stifling and paranoid atmosphere of the Counter-Reformation, some embarrassingly wrongheaded horoscopes, and nasty attacks from his peers. He even weathered the dishonor he suffered when his eldest son was put to death for the murder of his wife, or when his daughter, a prostitute, caught syphilis (he did write a treatise on the "French malady"); or when, in 1570, he briefly ended up in jail, on an accusation brought by his own younger son, for casting the horoscope of Jesus Christ. The charge of heresy did cost him the professorship he then had, in Bologna, but he moved to Rome and practiced medicine there too, on a pension from the pope, until he died a few years later.

That his reputation survived these rather extraordinary dramas testifies to the regard in which his work was held. He was a great theorist of probability who exerted this skill in his games of chess, dice, and cards, and who by the same token understood deeply the power of astrology. He obsessively drew up genitures (birth horoscopes) for the great and the good as well as for himself. He prided himself to no end on his medical successes, which he attributed not only to his own talents but also to his capacity for a foreknowledge that fed what, in his original and resolutely upbeat autobiography, *The Book of My Life* (first published in 1575), he boasted were his unfailingly accurate diagnoses of patients' ills. He experienced spiritual "visitations," while claiming that his insights had everything to do with a keen understanding of nature rather than with any familiarity with the supernatural. But he described these visitations ambivalently, as if to show that they marked him out as having a privileged access to the body's secrets.

He had no such privileged access, of course. Still, he vaunted to posterity that he was good at prognosis, because he had read Hippocrates carefully on the matter, learning from him (especially from the *Epidemics*) how to use case histories. Such histories had their roots in the medieval *consiliae,* close in kind to Galen's own accounts of individual cases; and they were not unlike horoscopes.[65] All along, Cardano was a rationalist in his practice. He understood, as he put it in his autobiography, "how from a comparison of the cures of one member and another, some understanding of the causes of the disease and method of treatment may be deduced."[66] If astrology continued to coexist with rationalist medicine at these highest levels of clinical practice and theory, it was perhaps because empirical common sense was quite compatible with the calculation of chance. In any case, the combination of the two seems to have enabled Cardano to make accurate prognoses. For instance, he cured the archbishop of Scotland of asthma by correctly attributing it to an intolerance to the feathers in his pillows.

For Cardano, as for other practitioners of astrological medicine, such as his contemporary Andrew Dygges in England, astrology continued to function as a medical tool precisely insofar as the body was humoural. Dygges adopted the medieval zodiac man in a tract he published in 1555, noting the correlations between body parts, planets, and zodiac signs that could be of use to doctors. Aries governed head and face; Taurus, the neck; Gemini, arms, hands, and shoulders; Cancer, breast, stomach, and ribs; Leo, heart and back; Virgo, bowels and belly; Libra, kidneys, navel, and buttocks; Scorpio, "secret membres"; Sagittarius, thighs; Capricorn, knees; Aquarius, shins and legs; and Pisces, feet. To each area of the body corresponded the organs, bodily functions, and diseases that were affected by its constellation, and to each constellation corresponded a dominant planet. Out of this set of functions emerged a temperamental picture of the individual. The formula was clear, simple, and easy to use.

Horoscopes today still refer to each sign's respective propensity to certain illnesses—so Virgos have sensitive digestive systems, Libras tend to have kidney problems, and Aries are prone to headaches and fevers. But horoscopes, of the sorts one reads in magazines or on Web sites, are mostly sold on the promise that they will tell us who we "really" are, as if an external, impersonal, God's-eye description of our character had the capacity to reveal us to ourselves in a true light. They describe what the day will look like, how compatible we are with our partner, what emotional characteristics will dominate this week or the year. They rely on a simplistic determinism that calls for the abdication of further analysis.

The horoscopes that Cardano spent his life calculating, or that someone such as Chigi consulted, were not so simplistic. It is true that they helped people have a sense of control over their destinies just as they do today; and that they were only of use insofar as one could read a relevant meaning into them. In the sixteenth century, too, they could seem absurd enough. Shakespeare has Edmund, in *King Lear*, poke fun at the abuse of astrology for the sake of a reassuring determinism or the strengthening of prejudice—against bastards, in his case. Edmund says it is

> the excellent foppery of the world, that, when we are sick in fortune,—often the surfeit of our own behaviour,—we make guilty of our disasters the sun, the moon, and the stars; as if we were villains by necessity, fools by heavenly compulsion, knaves, thieves, and treachers by spherical predominance, drunkards, liars, and adulterers by an enforced obedience of planetary influence; and all that we are evil in, by a divine thrusting on: an admirable evasion of whoremaster man, to lay his goatish disposition to the charge of a star! My father compounded with my mother under the dragon's

tail, and my nativity was under *ursa major;* so that it follows I am rough and lecherous. 'Sfoot! I should have been that I am, had the maidenliest star in the firmament twinkled on my bastardizing.[67]

Cardano could have said 'sfoot too, especially when the College of Physicians first refused to take him in. But he certainly did not believe that all astrology was foppery. It relied, as did any divination practice, on a sophisticated calculus—one that Cardano mastered fully—and so the context within which horoscopes signified anything at all in the Renaissance called for reflection, not abdication of responsibility. It was the same context in which humoural explanations made sense of physical and emotional states, and it pervaded the whole culture.

11. Diagnosing Melancholy

A ND SO IT was quite literally under the heavens that we suffered our humourally provoked passions and lived out our never fully determined fates. Our humours obeyed universal laws and yet, by knowing their course, or for that matter the course of our relevant stars, we could at least hope to act upon these external forces and exert reason in order then to act wisely. Passions, after all, were disturbances inflicted by the body on the occasionally passive soul housed in the brain, and they called for action. But although passions were clearly humoural, no surgeon, anatomist, or physician had discovered how they really functioned. Localized ailments like asthma, or phthisis—a condition that Cardano had professed to be able to cure with great facility—were quite different from more generalized ailments, and from the soul's spells of unease.

No one, for that matter, was ever able to cure the illness of Isabella de' Medici, daughter of Cosimo I de' Medici—the Duke who was so enthusiastic about Vesalius. Her tragically early death was tinged with scandal when rumors—probably unfounded—arose that her husband had killed her in a fit of jealous rage, ensuring her a dubious posthumous fame. Married to Paolo Giordano Orsini in 1553 at the tender age of fourteen, in an arrangement that suited both powerful families for political reasons, she found herself on her own in Florence from 1566, when Pope Pius V recalled her husband into service. The two, it seems from her letters, were fond of each other—and it also emerges that the long-distance relationship did not suit her in the least.[68]

In that year Isabella suffered intense headaches that she herself attributed to her "great loneliness and affliction." By springtime, she was also experiencing fevers, cold sweats, vertigo, and loss of appetite. She sought isolation in order to *umorezzare*—indulge in her humours—as she put it in response to a letter from Paolo, who told her to remember one who "adored" her and entreated her to remain "joyful" in order to "chase away the humour." The headaches were growing so painful that she could barely write, and the doctors forbade her to do so. But she wrote all the same, one letter after another, to her husband who remained on duty in Rome, enmeshed in politics, while in Florence, she was surrounded by various members of the Medici clan. Doctors prescribed bleedings, but she refused them. They suggested the usual dietary prescriptions, and they too recommended the pursuit of joy. She wondered, in one of her letters, how anyone who had "no cause" for joy could be asked to be joyous. There were rumors that, in her desperation, she was unfaithful (with her secretary).

Paolo did eventually reappear in Florence. Isabella became pregnant a number of times and had a number of miscarriages. The first treatises on obstetrics were appearing; more medical attention was given to childbirth and less power to uneducated midwives.[69] But Isabella's miscarriages might have had less to do with their mistakes than with her weak constitution. At last, however, two children were born, a girl first and then a boy, to the great delight of both lineage-obsessed parents. Still, Isabella remained ridden with anxieties. Cosimo died in 1574, but Isabella's ambitious brother Francesco de' Medici refused to part with the inheritance that she had expected to receive. Paolo was in debt, and away on increasingly fraught business most of the time. Isabella now complained of swollen legs and lost looks. As she was preparing to join Paolo in Rome, she caught smallpox, recovered, and prepared again to leave. But her brother would still not fund her; in the end she stayed put. In March 1575 she was diagnosed with tertian fever. Paolo visited for a few months. Soon she was unable to walk, and then unable to write. By now her husband's career in Rome was floundering, and he finally returned to stay by her side. Soon after, while they were traveling together, she died. She was only thirty-three years old.

What was the illness that struck her in 1566? Why did she need to *umorezzare?* One can infer from the tone of her letters that she was predisposed to anxiety and melancholy; and, given the torments she had to endure, one can imagine that, from a humoural point of view, her melancholy had turned into an acute acedia. This was an ailment quite unlike that suffered by Francesco Sforza over 100 years before, though it would have been due at first to the same black humour. Galen had written that "most of the ancient physicians give the name *black humour* and not *black bile* to the normal portion of this humour, which is discharged from the bowel and which also frequently rises to the top" of the stomach contents.[70] Black bile, as opposed to black humour, was thus "that part which, through a kind of combustion and putrefaction, has had its quality changed to acid."

In Isabella's case, her whole organism was eventually debilitated, weakened all the more by the stress and frustration caused by her circumstances. But her ailment was also the outcome of her temperament. We do not have her astrological chart, but it might have contained indications that Saturn would be sitting in both her sun and moon throughout her life. At any rate, when doctors suggested that she seek *allegria,* they were speaking in the same terms as those used earlier by Ficino; music, for that matter, might have been suggested too. But Isabella's circumstances were such that she was seemingly beyond help.

In 1586, a decade after Isabella's death, Timothy Bright published a *Treatise of Melancholy* in which we may find a few clues to help us understand what Isabella herself might have thought was wrong with her. Bright was a clergyman and an established doctor in London whose treatise would become well known—it probably exerted an influence on Shakespeare, informing the character of Hamlet to some extent. It perpetuated some aspects of the Aristotelian account of melancholy as a potentially creative force, and began by carefully making a distinction between the humour itself and the emotional state the word also denoted. Although they shared the same name, the humour and the emotion were not identical, thought Bright. Melancholy had to be understood under two aspects: one clear, physical, visceral, and the other correlated with the first but broader, and describable in poetic terms rather better than in scientific ones.

Not long after, a poet called Nicholas Breton was indulging in such poetic descriptions of the melancholy soul in a collection of poems, *Melancholicke Humours, in Verses of Diverse Natures,* published in London in 1600 and described as "the fruits of a fewe melancholike humours:

which chiefely he commendeth to spirits of his own nature, full of melancholy, and as neere Bedlam, as Mooregate" (that is, as near to Bedlam, the hospital for the mad, as is Moorgate, located right near it, in what is today London's East End).[71] The poems, close in kind to the lyrics used by his contemporary, the composer John Dowland, were unremittingly bleak, depressed, and despairing of love and fortune. In *A Solemne Sonnet*, Breton wrote: "Fortune hath writ characters on my heart, / As full of crosses, as the skinne can holde: / Which tell of torments, tearing every part, / While death and sorrowe doe my fate unfolde." The spirit of these complaints was not unfashionable in Elizabethan times, and indeed it was self-consciously aestheticizing, almost joyous in its warm embrace of absolute misery.

The humoural status of black bile was ambiguous. It was of course an aspect of the humourally defined organism—but as the product of a burned distillation of yellow bile, or in some cases, explained Bright, of the denser parts of blood or of salty phlegm: it emerged out of other humours, and so was in a sense a meta-state, rather than an identifiable type or *krasis*. Whereas it was a given that changes in the proportions of humours present in a person produced illnesses characteristic of certain temperaments, determined by their cardinal humour, black bile alone pointed to a wide range of disturbances. One could not suffer an onset of phlegma, or blood, in the sense that the dyscrasia would not affect one's mind-set or worldview—it had localized, recognizable pathogenic effects. But one could suffer an onset of melancholy, that is, an attack of black bile whose effects could be radically mind-altering, not necessarily pathogenic, and could vary tremendously from person to person and from case to case.

So, although Francesco Sforza had seemed to suffer from a purely humoural problem, it could in many other cases be tricky to diagnose black bile as a cause of ailment, when so many ailments actually fell under the category of melancholy. But Bright did explain that "natural"—healthy—melancholy could overload the blood if it became too warm, producing vapors that caused cognitive perturbations once they entered the brain. Inversely, "unnatural" melancholy occurred when residues from an excess in the grossest part of the blood—called the "excrement"—failed to be expelled from the organism. The corruption and transformation of blood or yellow bile into black bile overstimulated the passions, which in turn might overwhelm reason, disturbing ordinary cognitive functions and triggering extreme states of fear, or despair. In both cases, changes in temperature and in the quantity of melancholic humour normally present in the body produced the disturbances and "alterations" identified as melancholy. It would seem that Isabella was afflicted with "unnatural" black bile, concocted out of a large dose of the "natural" kind.

12. Love-Maladies

THE ANATOMY OF MELANCHOLY is one of the most extraordinary books ever written in the English language. It is by an Oxford clergyman and scholar, Robert Burton, who spent decades expanding on the volume that he first published in 1621. By the time it appeared in its final, posthumous edition in 1660, it had reached nearly 1,500 pages. Magisterial and immediately successful, it is replete with quotations from countless scholarly sources, classical and contemporary, that touch in some way or other on the subject of melancholy.

Burton's tome could also be quite easily used as a guide with which to further diagnose Isabella. If one looks for an account of melancholy in women, one finds that it could occur in widows "with much care and sorrow, as frequently it doth, by reason of a sudden alteration of their accustomed course of life," as well as to "such as lie in child-bed" and to "nuns and more ancient maids, and some barren women."[72] That was because "the midriff or *diaphragma*, heart and brain," he wrote, citing the fifteenth-century Spanish doctor and gynecologist Luis de Mercado, were "offended with those vicious vapours which come from menstruous blood." Isabella, though, was neither a widow nor "in child-bed." The cases did not stop here, however. According to another source, there might be in melancholic women

> an inflammation of the back, which with the rest is offended by that fuliginous exhalation of corrupt seed, troubling the brain, heart and mind; the brain, I say, not in essence, but in consent; *universa enim hujus affectus causa ab utero pendet, et a sanguinis menstrui malitia,* for, in a word, the whole malady proceeds from that inflammation, putridity, black smoky vapours, etc.; from thence comes care, sorrow, and anxiety, obfuscation of spirits, agony, desperation, and the like, which are intended or remitted, *si*

amatorius accesserit ardor [should the amatory passion be aroused], or any other violent object or perturbation of mind.

In Isabella's case, sorrow itself would have been at fault. Burton explained that sorrow generally "hinders concoction, refrigerates the heart, takes away stomach, color, and sleep; thickens the blood;[73] contaminates the spirits (Piso); overthrows the natural heat, perverts the good estate of body and mind, and makes them weary of their lives, cry out, howl and roar for the very anguish of their souls."[74] All this could conceivably have driven Isabella to the infidelities her husband was said to have suspected, leading her, in the gossipy "crime of passion" scenario that has survived in the historiography even to this day, to her death at his hands.

Isabella's sorrow was not unrelated to another species of melancholy, known as love-melancholy, the erotic malady, or erotomania. It had an ambivalent reputation—bad if one leaned toward Stoicism or aspired to chastity, but good if one believed that a capacity for strong passions was the mark of a fine soul. In any case it was an illness, whose consequences could be devastating and which called for medical treatment. Love-melancholy was what had afflicted Antiochus (among the many characters enumerated by Burton) and what Erasistratus, the illustrious Alexandrian anatomist, had diagnosed. It was typically the product of painfully unrequited love, of desire unfulfilled, of extreme bittersweet romantic longing, of a desire that could turn metaphysical, of an obsession that could cause insanity.

It was the source of sonnets and the foundation of chivalry, the name of blind lust and the dark underside of passion, the cause of extreme joys and fears—of "fire" and "ice," in Burton's words. It had been the reason for Dido's suicide, and, as Burton reports, the fount of rash, mad behaviour, such as Medea's, in whom "Reason pulls one way, burning lust another, / She sees and knows what's good, but she doth neither."[75] Most lovers, said Burton,

are carried headlong like so many brute beasts; reason counsels one way, thy friends, fortunes, shame, disgrace, danger, and an ocean of cares that will certainly follow; yet this furious lust precipitates, counterpoiseth, weighs down on the other; though it be their utter undoing, perpetual infamy, loss, yet they will do it, and become at last *insensati*, void of sense; degenerate into dogs, hogs, asses, brutes; as Jupiter into a bull, Apuleius an ass, Lycaeon a wolf, Tereus a lapwing, Callisto a bear, Elpenor and Gryllus into swine by Circe.

This was a state in which humoural flows ruled the brain, and whose "symptoms are either of body or mind; of body, paleness, leanness, dryness, etc." Lovers "pine away, and look ill with walking, cares, and sighs," Burton here quoted the Dutch physician Jason Pratensis, arguably the first to focus on diseases of the brain (long before they were called neurological): "Because of the distraction of the spirits the liver doth not perform his part, nor turns the aliment into blood as it ought, and for that cause the members are weak for want of sustenance, they are lean and pine, as the herbs of my garden do this month of May, for want of rain."[76]

On the other hand, Burton observed, "there be some good and graceful qualities in lovers, which this affection causeth." Cardano himself had stated, citing Plutarch, "As it makes wise men fools, so many times it makes fools become wise; it makes base fellows become generous, cowards courageous." Pullulating examples of this positive aspect of the condition included the same story of Acis and Galatea that Raphael had represented in Chigi's villa. Burton reports what happened "when the hirsute cyclopical Polyphemus courted Galatea" by quoting Ovid: "And then he did begin to prank himself, / To plait and comb his head, and beard to shave, / And look his face i' th' water as a glass, / And to compose himself for to be brave."[77]

This was more than a treatise about the black bile that caused or was correlated with melancholy; Burton had written a paean to the literary, mythological, historical, and philosophical richness of the states that melancholy named. This gathering of scholarship was itself a melancholy enterprise, a testimony to the ultimate vanity of human knowledge and endeavor. But it remains a veritable encyclopedia of everything there was to know, at the time, that could conceivably have anything to do with the melancholic condition—or indeed with the human condition. As Burton put it, "[T]he tower of Babel never yielded such confusion of tongues, as the chaos of melancholy doth variety of symptoms."[78] But Burton knew the physiological basis of this variety of symptoms, recounting it prosaically: "The gall, placed in the concave of the liver, extracts choler to it: the spleen, melancholy; which is situate on the left side, over against the liver, a spongy matter, that draws this black choler to it by a secret virtue, and feeds upon it, conveying the rest to the bottom of the stomach, to stir up appetite, or else to the guts as excrement."[79]

It is intriguing to think of humoural secretions as the source of all human culture—of the highest and lowest aspects of human life. Yet, that is exactly what the concept of melancholy made possible for the literary imagination to accomplish. Earlier, Thomas Walkington had composed a much less illustrious treatise, entitled *Optick Glasse of Humors* (now nearly forgotten), in which he distinguished the beneficial sort of melancholy from the detrimental one. The first was a "precious balme of witte and policy: the enthusiasticall breath of poetry, the foyson of our phantasies, the sweete sleepe of the senses, the fountaine of sage advise and good purveiance," whereas the second "causeth men to bee aliened from the nature of man, and wholly to discarde themselves from all societie, but rather heremits and olde anchorets to live in grots, caves, and other hidden celles of the earth."[80]

Isabella, for her part, had neither written poetry nor hidden away from society. Doctors tried to treat her throughout her short life for a variety of diseases, but her first symptoms were clearly those usually ascribed to melancholics. These were hard to account for because, to the minds of creative melancholics such as Burton himself, there were as many causes of melancholy as there were human stories. It was clear that Isabella had pined for her husband and had suffered from longing, sorrow, and financial insecurity—all causes of what today one calls stress—and a fragile disposition. A diagnosis of melancholy was, at any rate, not unrespectable; and it was appropriate that she should have longed for her husband, insofar as it was well known that abstinence was not recommended for married women: it could give rise to the "vicious vapours which come from menstruous blood."

13. Uterine Fury and Satyriasis

BEHIND THE NOTION that women's lust was a potential threat to their own health if it was mismanaged, lay a prejudice whose origin lay in the Aristotle that Galen had absorbed and recycled. In 1623, two years after the first edition of Burton's book, there appeared in France a treatise on love-melancholy by a physician called Jacques Ferrand: *De la*

Fig. 21. *The melancholic,* woodcut from Cesare Ripa, *Nova Iconologia* (Padua, 1618: 5th edition).

maladie d'amour ou mélancholie érotique.[81] There, in a chapter entitled "Whether Love in Women Is Greater and Therefore Worse Than in Men," he explained that, although women had a naturally cold and humid temperament, lacking in heat, while the erotic impulse was stronger in hot and dry temperaments, women in fact tended to be more passionate, "witless, maniacal, and frantic from love" than men, simply because, as Galen had written,[82] they did "not have the rational powers for resisting such strong passions."

Ferrand also took seriously the idea that women's natural lasciviousness was due to nature's having "placed the spermatic vessels in the woman very close together, joining the horns of the uterus," while in men these were pushed "a fair distance outside the abdomen, for fear that the principal faculties of the soul, the imagination, the memory, and the judgment, would be too inconvenienced by the sympathy and proximity of these genital parts." As a result—and this was an idea also found in the Hippocratic *On the Diseases of Young Women*—"woman experiences more violently this brutal desire, and not unreasonably so, since nature owes her some compensating pleasures for the suffering she endures during pregnancy and childbirth."

Despite his citation of the widespread Aristotelian belief that women were imperfect men, Ferrand himself does not come across as especially misogynistic, however misogynistic was the old idea that women were less rational than men, and that those who were not sexually active might be in danger of going mad. He would have understood and probably commended Isabella's healthy need to sleep with her husband in the same terms as a man's need to sleep with his wife. But her kind of melancholy did not entirely correspond to Galen's definition of the condition. In Ferrand's words, from a chapter entitled "Of Melancholy and Its Several Varieties," it was "a form of dotage without fever accompanied by fear and sorrow"; and although Isabella did suffer fear and sorrow, the diagnosis of love-melancholy does not exactly apply to her.

Indeed, all melancholics had an excessively dry and cold brain (they could even become epileptic if the humour spread to the brain's ventricles, according to the Hippocratic *Epidemics*). There were three broad types of melancholy, Ferrand explained: one arose "from black choler engendered in the brain"; another was "produced when the humour is spread over the entire body through the veins"; the third was "the flatulent or hypochondriacal, because the disease is situated in the hypochondries, containing the liver, the spleen, the mesentery, the intestines, the pylorus, the vein of the womb, and other adjoining parts of the body." Erotic melancholy belonged to the latter sort, "because it is the liver and the surrounding parts that are principally affected and because the essential faculties of the brain are corrupted by the blackish vapors that rise from the hypochondries to the citadel of Pallas, that is to say the brain."

Although Francesco Sforza's actual brain and reasoning faculties were not affected by the "blackish vapors" of melancholy, he was clearly affected with a type of hypochondriac melancholy, which was, said Ferrand, "hot and dry through the adustion of yellow bile, blood, or natural melancholy," and was "the principal cause of erotic melancholy or erotic mania, for which reason the Aristotelian writer in his *Problemata* says that 'melancholics are subject to incessant sexual desire' "—as indeed Francesco was.[83] The Duke would have produced "a variety of flatulent vapors" that tickled him, driving him "to extremes of lasciviousness," compounded, furthermore, by his sanguine temperament. Indeed, semen, as Galen had explained, was concocted out of blood, and sanguine types tended to have a large sexual appetite.

Love-melancholy, as opposed to erotic mania, also belonged to this third type of melancholy. It was especially dramatic. The most powerful attacks of melancholy tended to be caused not by any "external occasion" but by an "internal illusion," as Timothy Bright had put it in his *Treatise of Melancholy*—and this was definitely the case in love-melancholy, where an object of love affected the spirits by its absence, rather than by its intensely desired presence. Bright believed that those who benefited from a generally balanced temperament and healthy heart were able, regardless of the dominant humour in their complexion, to respond with the proper emotion to all external objects. Others could fall victim to the hold of the imagined object and to a longing intensified by its own prolongation. These victims were wan and hollow-eyed, devoid of appetite, distracted, exhausted by sleeplessness, and obsessed with the object of desire. Obsessive love, wrote Ferrand, could destroy the lover, cause derangements in reason, a depraved imagination and incoherent speech: "You will see him crying, sobbing, and sighing, gasp upon gasp, and in a state of perpetual inquietude, fleeing all company, preferring solitude and his own thoughts."[84] Given that the noxious humour could corrupt the mental faculties of reason and imagination, doctors had to take the condition seriously, and endeavor to treat those who had it by counteracting the excessive dryness of the brain.

Women afflicted with an extreme form of the pathology, "uterine fury"—to which Ferrand devoted a whole chapter—experienced "a raging or madness that comes from an excessive burning desire in the womb, or from a hot temperature communicated to the brain and to the rest of the body through the channels in the spine, or from the biting vapors arising from the corrupted seed lying stagnant around the uterus." If they were unaware of their pain, then that must be because, as one Hippocratic aphorism had it, they "must also suffer from an intellectual disorder," due probably to "a surfeit of acrid seed and flatulency," suffered especially by "young girls, widows, or women of a warm temperature who delight in dishonest pastimes and pleasures, who dine on rich foods, socialize, and think of nothing but satisfying their sensuality."[85] Uterine fury, like satyriasis in men, affected the brain and therefore the capacity to exert some self-control over the overwhelming desire to have intercourse in whatever way was possible. The best cure for either of these two conditions was marriage, or at any rate, as Avicenna himself had recommended in *De anima*, legitimate coitus. Failing that, the doctor could avail himself of this duty, manually triggering the woman's "purge." For, left uncured, these conditions could lead to uncontrollable mania and, indeed, to madness.

But not all love-melancholy was madness, of course. Ferrand made extensive use in his guide of the Hippocratic-Galenic physiology, referring to Plato, Plutarch, Pindar, Aristotle, Avicenna, Arnau de Villanova's late-thirteenth-century *Tractatus de amore heroico*, Ficino, and his own contemporaries. Unhappy, tormented, unfulfilled love was known as *amor hereos*—"heroic," or "lordly"—as Ferrand took the concept from Arnau, either "because the ancient heroes or demi-gods were often afflicted with this ill according to the mythical recitations of the poets, or because the great lords and ladies were more inclined to this malady than the common people, or finally because love rules and dominates the hearts of lovers." In any case, *amor hereos* was exceptionally well explained by humoural theory and scholastic psychology. The medical literature on the phenomenon—on the banality of spectacularly obsessive love, so to speak—was nearly as substantial as are the myths, epics, tragedies, and verses inspired by it, and Ferrand did not need to imagine much on his own.

Amor hereos was initially caused by overheating of the vital spirits in the heart, causing a perturbation throughout the body; and once they reached the brain and became animal spirits, the latter overheated the brain, too, which dried up. Arnau had explained that since the retention of perceptions in the imaginative faculty required dryness, that part of the brain became all the more dehydrated if the cogitative faculty was overactive, thus drawing all the humidity to itself—as happened in the case of *amor hereos*, where the lover ruminated constantly on the image of the beloved. As a consequence, this was the only image that eventually remained imprinted in the now arid imagination, to the exclusion of all other internal or worldly preoccupations. The humours were in disarray; the body's main organs—liver, heart, brain—were dysfunctional. It was a sad, dangerous fate.

14. Anti-Melancholy Antidotes

BUT THE AVAILABLE CURES for love-melancholy were plenty, and some were straightforward enough. If, in the case of men, either marriage or union with the beloved was impossible, one could always resort to courtesans, though this was not highly recommended, since sexual activity could also aggravate the condition. Ferrand seems to have believed that fornication actually enhanced the predilection for "lust and wantonness."[86] Distance from the beloved was another cure. So was cold water. Bleeding the "liver vein of the right arm" could work if one wanted to resort to surgery,[87] and satyriasis could be cured if one also bled a little the ham vein at the back of the knee; cupping "on the thighs near the genitals" after opening the arm might do the trick as well. Pharmaceutical remedies, on the other hand, could take the form of "an enema containing cooling and moistening ingredients," to which one should add "hemp seed, agnus castus [known in English as monk's pepper], and the like."[88]

This should be followed the next day by "a bolus of cassia, catholicon, diaprun, or tryphera persica with a few grains of agnus castus, or else quite a mild purgative"—Avicenna had warned against the use of "violent cathartics and strong laxatives" and was one of those who had suggested "a decoction of beets, mallows, and [the herb] mercury." Hemlock might work too: it had been used effectively by the Athenians; but of course, as it was the official poison (the condemned Socrates had, famously, taken it as his last drink), it risked killing the person along with the desire. Against all too potent philters,[89] the stuff of legend, myth, and *Romeo and Juliet*, the superstitious among Christians could use "charms, amulets, or medications hung on the body," although these were pagan cures. But the best remedies, said Ferrand, were really "prayer, the reading of good books, and other serious activities."

Unless one had been the victim of a philter, it was possible in any case to prevent the condition through a simple regimen, also outlined in Ferrand's complete guide.[90] So "a cool, moist environment" was good. Avicenna had recommended "warm air for men and cold for women." Rhazes had warned against "clothes lined with fur, ermine skins, or velvet because they heat the blood"; and it was best to avoid "highly scented musk perfumes, Cyprian powder, civet, amber, French moschatel and moschatel ointment." It was worth exercising enough, refraining from sleeping on the back (that heated the kidney area), and avoiding dances or comedies. One should drink water, not wine; eat lightly ("love never lodges in an empty belly"), especially if one had "a sanguine, well-tempered, or choleric complexion, in which case all his meats should be lightly nourishing but refrigerative."

Dietary recommendations thus included salads and broths of "purslane, garden sorrel, endives, chicory, and lettuce, which is so effective for this disease that Venus, wanting to forget her illicit loves, had her beloved Adonis burried under a bed of lettuce." Fruits such as "melons, fresh grapes, cherries, plums, apples, pears" and the like were also effective in calming lust, as were "breads made of rye, barley, millet, or spelt." Sauces should be composed of "vinegar, lemon juice, citrons, sorrel, verjuice, and similar liquids"; and "spicy, salty, and fricasseed foods" should be avoided "because salt induces sensual desire by virtue of its heat and acrimony if one uses too much of it." Ferrand also advised the potential or stricken lover to stay away from "meats that are highly nourishing, hot, flatulent, and melancholy, such as soft eggs, partridges, pigeon, sparrow, quail, young hare, and especially goose" (hot, strong, airy meats multiplied the sperm, one of the first causes of love, said Ferrand); from "pine nuts, pistachios, hazelnuts, chives, artichokes, cabbage, turnips, carrots, parsnips, ginger conserves, eringoes, satyrion, onions, truffles, and rocket"; and from "oysters, chestnuts, chickpeas," and the like. Clearly Francesco Sforza had been doing everything wrong and might even have been afflicted with satyriasis.

Burton was rather more literal and less medical. Against melancholy in general, he did recommend "clarified whey, with borage, bugloss, endive, succory, etc., of goat's milk especially, some indefinitely at all times, some thirty days together in the spring, every morning fasting, a good draught." But he otherwise delighted in telling stories about the cure of love-melancholy in particular by the other, more humanly dramatic means, such as enforced distance from the beloved. He recounted a tale of Petrarch's, for instance, about "a young gallant, that loved a wench with one eye, and for that cause by his parents was sent to travel into far countries; 'after some years he returned, and meeting the maid for whose sake he was sent abroad, asked her how and by what chance she lost her eye? No, said she, I have lost none, but you have found yours,' " signifying thereby that all lovers were blind. If, on the other hand, this simple device did not work, "then other remedies are to be annexed, fair and foul means, as to persuade, promise, threaten, terrify, or to divert by some contrary passion, rumour, tale, news, or some witty invention to alter his affection."[91] There were ways of thawing the worst obsessions, by describing the beloved in monstrous terms, stimulating the imagination, allowing the mind to reflect on other objects, letting the world matter to it again. A balance could be found: reason could be reactivated even if, as Shakespeare had Theseus exclaim in *A Midsummer Night's Dream,* "Lovers and madmen have such seething brains, / Such shaping fantasies, that apprehend / More than cool reason can comprehend."[92] However powerful was physiology, the brain's seething humours were not all.

For there was and is more to melancholy than physical causes, more to anguish about life and death than black bile, indeed more to the psyche than the sum of its humoural parts. Certainly, it was, and is, useful to be able to name a common condition that, rather than an actual illness, was a "disposition," as Burton acknowledged. It was, he wrote, a transitory state "which goes & comes upon every small occasion of sorrow, need, sickness, trouble, feare, griefe, passion, or perturbation of the Minde"—indeed, that which is "the Character of Mortalitie" itself.[93] But if the awareness of mortality characterized our conscious lives, then conscious life itself was the cause of melancholy, and its manifestation would be the excessive state of consciousness, the excessive presence of one's mind to oneself. Described in this way, the condition would indeed need to encompass every work ever created, every love story, every lifetime—and end up larger still than Burton's book, bursting out of its spine as this one already was.

Passions and perturbations of the mind would continue to fascinate beyond Burton, of course. During the seventeenth century, numerous treatises on the passions appeared that provided commonsensical guides to behavior. These works of psychology tended to be much more usable—and all were shorter—than *The Anatomy of Melancholy.* Burton's encyclopedic tome would be the last of its kind. The classical learning it gathered would still be in use over the centuries to come, but never quite as actively as it had been in the Renaissance. New ways of describing the world—from the cosmos to geography, from the human body to the world of animals and plants—were supplanting the classical models. Experiment was supplanting written tradition. Galen was increasingly questioned; new physiological explanations of the organism were becoming available, and the humours were losing their theoretical ground. The scientific revolution was now in full swing, and it is usually at this juncture that one dates the beginning of modern science. But the ancient and powerful humours did not actually disappear: instead, a new phase in their history began.

V

Nature: Of Blood, Airs, and Reasons

(Scientific Revolution: Seventeenth Century)

I am telling you that I now abandon you to your poor constitution, to the intemperance of your bowels, to the corruption of your blood, to the bitterness of your bile and to the starchiness of your humours.

MOLIÈRE[1]

Men that look no further than their outsides, think health an appertenance unto life, and quarrel with their constitutions for being sick; but I that have examined the parts of man, and know upon what tender filaments that fabrick hangs, do wonder that we are not always so; and considering the thousand doors that lead to death, do thank my God that we can die but once.

THOMAS BROWNE[2]

1. New Science, Old Bodies

WE ALL HAVE some sort of belief about how our body works, what is good for it and bad for it, what diseases are and how they can be stopped; but few of us are experts about our physical selves. Even a doctor's knowledge is limited, although we need to believe that someone knows more about the processes of life and death than we do. No society exists that does not allow for the possibility of entrusting our body to an authority, a tradition, a theory that would explain or at least account in some way for its most obscure workings. Unmediated self-knowledge, whether of the body or of the psyche, is probably impossible to obtain: we require words, mirrors, maps in order to make sense of ourselves. And because it pertained to mind as well as to body, humoural theory had offered a perfect mediation for self-knowledge—and not only for the doctor's knowledge of a patient.

But it had also been a part of speech, integrated within the wider culture, requiring, in order to function, a full belief in its credibility: it had presented itself as a truthful account of the body rather than just as a creased, overused map of its byways. During its reign, if one believed that our humours determined us, then to know one's humoural self was, in effect, to know oneself.

Humoural physiology came under assault during the seventeenth century, and once its unraveling began, it took little time for it to be discarded as an outdated map. But the map would serve again: a functional map will always remain valid, unless geography itself changes. The geography of the human body had not changed, but the guidance offered by the humoural map was ambiguous, for it also was a reflection of our mapmaking minds.

Certainly there were powerful forces uprooting the classical foundations on which humours had been based. This was a time when natural philosophers all over Europe were eager to shed their dependence upon Aristotelian physics. They were experimenting, measuring, calculating from scratch, examining animals and plant samples from the New World that had never been seen before. Scholars and collectors were gathering in their cabinets of curiosities bones, dried body parts, fragments of ruins, fossils, metals, powders, mummy fragments, dried plants, rare glass, oddly shaped seashells, archaeological remains, coins, anatomical drawings, and scientific instruments. The learned were discussing discoveries, forming scientific societies, testing the world's surface and structure, endeavoring to open the book of nature for themselves and to find the codes that would unlock its secrets.

Already in the 1500s, these curious minds had been cataloguing physiognomies or novel species, and discussing monstrous births and singularly shaped adult humans or animals.[3] Monsters were a popular topic. It was believed, as it had been in the Middle Ages, that perception affected reproduction: a woman who thought about her lover during intercourse with her husband might give birth to a child that looked like the lover; if a pregnant woman was frightened by a frog, she might give birth to a froglike creature.

But monsters were now natural kinds; they could be studied and classified along with everything else. By the early decades of the 1600s, the entire natural world seemed open to classification, waiting to be ordered into taxonomies. Wondrous novelty, errors of nature, the search for regularities via irregularities, for laws via exceptions, for God via the work of demons or occult forces—all these became fascinating as religions clashed, as the world ceased to be a closed system and new theories emerged about its origins, its age, and its status within the universe. Fossils seemed to be telling the planet's history at least as loudly as the Bible. Other worlds were now imaginable. The moon might have inhabitants—and no one could know what minds or bodies lunar species might have.

But the new did not simply displace the old. The willful shedding of increasingly useless thought structures was all work in progress, and the human body did not yield its secrets and laws so easily. Faced with mortal lives, distressing illnesses, and repeated epidemics of plagues and other diseases, physicians continued to resort to the old, tried and tested treatments, bleeding their patients and prescribing enemas and remedies based on classical pharmacopoeia. Empirical, "folk" knowledge continued to serve, as did the newly edited corpus of Galenic-Hippocratic writings. The cutting edge of innovative, imaginative theoretical work was not happening at the patient's bedside or in the doctor's cabinet. Patients suffered from this, particularly the most skeptical among them, such as Montaigne:

"Even the very promises of physic are incredible in themselves," he had written in his *Essays,* in the late 1500s; "for, having to provide against divers and contrary accidents that often afflict us at one and the same time, and that have almost a necessary relation, as the heat of the liver, and the coldness of the stomach, they will needs persuade us, that of their ingredients one will heat the stomach, and the other will cool the liver."[4]

But doctors had little else to provide, and the gap between theory and practice, never absent before, was growing wider. Old ideas and new insights overlapped confusingly. The "scientific revolution" that historians would identify after the event was not a clearly defined episode, and a significant number of those at its forefront continued to hold on to astrological, magical, supernatural beliefs that they sometimes managed to square with the new, empirical science they were eager to practice.

2. Campanella's Heavens and Galileo's Revolutions

MEDICINE AND ASTROLOGY could still be entwined, at least in theory. They certainly were entwined in practice for the Calabrian Tommaso Campanella, originally a Dominican friar who lived in the sixteenth and seventeenth centuries. Following the sixteenth-century philosopher Bernardino Telesio, he espoused a naturalism with regard to the world that, early on, got him into trouble with ecclesiastical authorities, just as it had led to the burning of the philosopher Giordano Bruno on the Campo dei Fiori in Rome, in 1600. (Bruno's statue is still standing there, in the midst of the market, nowadays a lively meeting place rather than a severe memorial.) A provocateur and an anti-Aristotelian, Campanella also became close to Giacomo della Porta and was deeply involved in natural magic and astrology. He is especially remembered today for *The City of the Sun,*[5] a utopian fantasy in which he outlined the plan of an ideal state, modeled to a large extent on Plato's *Republic,* and conceived in exact counterpoint to the religious and political chaos that prevailed throughout Europe during this period of the Counter-Reformation.

In this fantasy, everything was to be ordered and controlled, including the body. Reproduction was planned according to optimal astrological conditions. Priests (the "officials") were discouraged from reproducing because "those who are much given to speculation tend to be deficient in animal spirits and fail to bestow their intellectual powers upon their progeny because they are always thinking of other matters."[6] They were of the melancholic type like that depicted by Dürer or embodied by Hamlet, consumed in metaphysical preoccupations, far away from the land of sense, and burdened with abstruse knowledge.

Campanella wrote this text in jail: the Inquisition had arrested him in 1599 for his heresies and for his plan to overthrow the Spaniards (then ruling Naples and Sicily) in order to set up his utopian society. This did not stop him from continuing to argue for the reality of astral influences, or from writing dozens of books on astrology, magic, medicine, and metaphysics. Freed only in 1626, thanks to the intercession of Pope Urban VIII with the king of Spain, he was jailed in Rome by the Holy Office but released again in 1629. He now became astrological adviser to the pope, who, worried by his bleak astrological charts, needed the friar's expertise. Campanella tried to provide him with more favorable charts and to reassure him, resorting to natural magic in order to counter the ominous predictions.

He described his magical techniques in a treatise called De siderali fato vitando. These techniques included the use of white garments and decorations against a predicted eclipse, the "sprinkling of essences and distilled waters, the playing of soothing music," and "the lighting of seven torches representing the sky and seven planets" to replace "the darkened, menacing" sky. As it happens, these were also remedies against melancholy. The Pope was afflicted with ill health, and there were various plots to unseat him during those years. The bad omens and intense political intrigue around him exacerbated what was treated as melancholic anxiety. But De siderali fato vitando had appeared (in 1629) without papal approval—and without Campanella's consent. This unauthorized, very public account of the Pope's peculiar sessions with Campanella led to the latter's disgrace. This seems to have been deliberately arranged to wrong Campanella, and—successfully, as it turned out—to jeopardize the plan, in discussion then, of making him a member of the Holy Office. Campanella was charged with heresy and branded as superstitious.[7]

He defended himself against the charge by writing an Apologeticus. By now Campanella had lost favor with the Pope, who was waging a campaign against astrology in spite of his own extensive reliance on it in private (he was infuriated by its abuse at the hands of those who were plotting his death). In the apology, Campanella appealed to past authorities from Scripture, the Church Fathers, Augustine, Albertus Magnus, Aquinas, and Ficino, in order to show that his magic was neither heretical nor superstitious nor idolatrous. It was not witchcraft either, he wrote; nor was it based on a pact with the devil. He was dealing with fully "natural" causes within a perfectly Christian framework. His effort to affect the stars was not against faith—on the contrary, it was "advantageous to faith. In fact, by asserting that astral fate cannot be avoided, one subjects free will to the stars." There was nothing occult about his methods, many of which were based on medical knowledge in any case: "all doctors have shown that vinegar, scented distilled water, aromatic perfumes and flowers purify noxious air and dissolve or keep away the seeds of pestilence and bad influences—especially Marsilio Ficino, Florentine canon, great theologian and philosopher, in his short book *De peste*, where he lists all the physicians who agree on these points." (This was true, and indeed when the plague struck Tuscany again, in 1630, these old techniques were standard—cloths boiled in vinegar did seem to counter miasmas and stop the spread of infection.)

Campanella in this way turned the accusation on its head, using elaborate logical arguments to transform what certainly looked like occult practices into perfectly reasonable, Christian methods of treatment. He was determined to win the day. But in 1631 Urban VIII reaffirmed an earlier, sixteenth-century papal bull that punished harshly anyone who practiced divination and judicial astrology—that is, the casting of horoscopes. (Pope Urban was still alive after the advent of the solar eclipse that conspirators had conveniently predicted would signal his death.) This did not include the use of astrology to determine such things as crop yield, weather, or, indeed, health. It was only supposed to undermine practitioners of predictive arts, not to deny that astronomical objects had an influence on terrestrial life. Campanella's *Apologeticus* fell on deaf ears. But soon after, he found refuge at the French court, where he would die, a protégé of Cardinal Richelieu.

In 1622, Campanella had been an extraordinarily outspoken defender of another unorthodox thinker keen on looking at the stars, Galileo Galilei. Galileo's friend and patron Cardinal Maffeo Barberini would become Urban VIII the next year.[8] In 1616, Galileo had been reprimanded by Cardinal Bellarmine, in the name of the Church then led by Pope Paul V, for daring to suggest that Copernican heliocentrism was a fact borne out by observation rather than a mere hypothesis, and that the earth really moved, both around its own axis and around the sun. From his prison, Campanella composed an *Apologia pro Galileo mathematico*, in which he sought to show, just as he would in the *Apologeticus*, that what looked like heresy in fact was not—that the search for the truths of nature did not contradict the truths of Scripture, and that both led to knowledge of God. Galileo had been the first to protest his profound Catholicism, and he genuinely believed that there was room within the faith for an interpretation of Scripture that allowed for the earth not to be at the center of the universe, without its contradicting the word of God.

Conservative theologians at the service of an embattled Church thought otherwise, however. The particulars of the case were as politicized as those in the case of Campanella, and they continue to be rehearsed and analyzed over and over again. But Galileo was no politician, and much less of a rhetorician than his fellow freethinker—he cared little for persuasion, being convinced that the facts spoke for themselves; and indeed Campanella was critical of Galileo's failure to engage the orthodox on their own ground by acknowledging their concerns. It was not enough to tell them simply to take the viewing glass and watch for themselves the phases of Venus or the mobile moons of Jupiter; nor did it suffice simply to claim, as Galileo did, that the Church's rejection of heliocentrism would damage its reputation in the long run.

With the publication in 1632 of his *Dialogue on the Two Chief World Systems*, Galileo incensed the Pope because he seemed there to have disobeyed the injunction brought on him sixteen years before, according to which he would be left alone so long as he would never teach, "hold or defend" the Copernican hypothesis. This major book, produced when Galileo was old and ill, staged a dialogue between a Venetian upholder of heliocentrism called Giovanni Sagredo—whom Galileo had known—and a fictional Aristotelian whom he called Simplicio. Although he tried to argue at his trial that he was not defending any one position in the book and that he had taken care to present Copernicanism in the form of a hypothesis (and although there was in fact confusion with regard to the legality of the original injunction of 1616), the Inquisition could not afford to let him go and Urban VIII was not in a position to protect him.

The *Dialogue* joined Copernicus—as well as the important works of Johannes Kepler, the brilliant German astronomer who was among the first to uphold Copernican heliocentrism—on the Index of books forbidden by the Church, where it would remain until 1835. By 1633, Galileo was allowed to return to his villa in Arcetri, in the hills outside Florence, but was condemned to house arrest for the rest of his life. The next year, Campanella was off to France and a better life.

The story of Galileo's accusation is complex, given his professed and sincere Catholicism, and his belief that his work was not at all heretical. He cared for the Church and thought that churchmen would realize how badly its reputation would suffer if it remained blind to the truth. (That turned out to be wishful thinking.) He was a skeptic who did not believe in systems, whose only certainty was that we must trust observation, but who knew that we had a limited capacity to understand the world we observed. This is partly why he is considered today one of the fathers of modern science.

3. Atoms and Humours

THERE WERE OTHERS like Galileo, of course, at this time of extraordinary ferment but also of tremendous confusion. Questions were teeming about the nature of knowledge and belief, about the status of the Church and Christian dogma, and about the place of man in nature. Nature itself was beginning to look puzzling to those who studied it, and the investigation of its laws was fraught with the dangerous possibility of taking away from God his own creation. Atomism was gaining ground, on the basis of the fifth-century BC Leucippus, of the slightly later Greek Epicurus, and of the first-century BC Roman Lucretius—who, in the poem *De rerum natura,* had described the universe as consisting entirely of particles.

Atomism was now central to the thought of Pierre Gassendi, a seventeenth-century French priest who was also an anti-Aristotelian professor of philosophy. He developed a model of nature that relied largely on Epicurus, whom he studied at length, brilliantly managing to accommodate Epicurus's heathen materialism to Christianity. Gassendi depicted the universe in terms of particles in motion that constituted everything from stones and plants to the sensitive soul. These particles flowed through empty space but had been created by a Christian God at the dawn of time. The question would next arise whether God had created these moving particles only once, at the beginning of time, leaving the running of things to natural processes, or whether a divine force was continuously behind all motion.

Gassendi lived relatively peacefully until his death in 1655, close to the great minds of the period, including the philosopher René Descartes; Descartes's early protector and friend, the friar Marin Mersenne, chief among the early exponents of the notion that the world should be understood as a mechanism, reducible to mathematical relations; and Nicolas-Claude Fabri de Peiresc, the best-connected and most prolific correspondent of Europe, an erudite scholar, antiquarian, and collector, of whom Gassendi wrote a biography. Peiresc was also a friend of Campanella and Galileo, and of Cardinal Francesco Barberini—the uncle and guardian of Maffeo, the future Pope Urban VIII. The scholarly world was densely interconnected and politically relevant; scientific communication was commensurately fast and efficient. New alternatives to scholasticism circulated rapidly—and they mattered.

Innovation mattered in art, too. Further north, in Mantua and then in Venice, the composer Claudio Monteverdi was revolutionizing music, almost entirely abandoning the contrapuntal polyphony that had prevailed until then in favor of predominantly melodic compositions that expressed and transmitted the precise emotions of individual voices. In what is arguably the first "modern" opera, *Orfeo*, the meanings of words resonated with melodic mood, and characters emerged out of story lines in a way no one had ever heard before. (Its first performance was in 1607, at the Gonzaga court in Mantua.) Monteverdi also wrote powerful religious music and, throughout his life, a large number of enchanting secular madrigals, or songs. In the 1620s, established in his position as *maestro di cappella* at Saint Mark's Cathedral in Venice, he was at work on a new set of songs. This was the eighth such book, containing madrigals of war and of love.

It is clear that he conceived their effects in terms that can only be called humoural: the songs of love were designed to appease listeners, while the songs of war were supposed to stimulate them. Robert Burton would have approved of this; he had written in his *Anatomy of Melancholy* that music was "a roaring-meg against melancholy, to rear and revive the languishing soul," citing the sixteenth-century Flemish physician and canon Levinus Lemnius to the effect that, "affecting not only the ears, but the very arteries, the vital and animal spirits, it [music] erects the mind, and makes it nimble."[9] Monteverdi's songs of war might indeed have accelerated the course of the animal spirits, stimulated the pulse, perhaps increased the secretion of blood in phlegmatic individuals; the songs of love, by contrast, would have reduced the secretion of choler, slowed down the pulse, regulated the flow of animal spirits, and reduced melancholic anxiety, turning it into a deeper, calmer contemplation.

But only seven years separate the publication of the first edition of Burton's *Anatomy of Melancholy* from the appearance of a book that, had it dealt with music, might have accounted for its impact on the body in a radically different way: *An Anatomical Essay Concerning the Movement of the Heart and the Blood in Animals,* by William Harvey, first published in Latin in Frankfurt, in 1628. Burton had paid his respects to Galileo, quite content with the idea that the earth "moved about the sun" along with other planets that were equally inhabited.[10] He was familiar with the new ideas around him, but it is fair to say that he was not so much interested in their innovative brilliance as sensitive to their literary, poetic appeal. It made perfect sense to him to write that people afflicted with melancholy or mania manifested symptoms "according to their temperament or crasis, which they had from the stars and those celestial influences, variety of wits and dispositions"[11]; and that these symptoms proceeded "from the temperature itself and the organical parts, as head, liver, spleen, meseraic veins, heart, womb, stomach, etc., and especially from distemperature of spirits (which, as Hercules de Saxonia contends, are wholly immaterial), or from the four humours in those seats, whether they be hot or cold, natural, unnatural, innate or adventitious, intended or remitted, simple or mixed, their diverse mixtures and several adustions, combinations."

To an extent, Harvey's essay spelled the undoing of the humours: this was the account of blood circulation, based on lectures he had given in 1616—the same year that marked the start of Galileo's official troubles—to the College of Physicians, of which he was a member and, later, president. He had been a student first at Gonville and Caius College in Cambridge—headed at the time by John Caius, a pro-Galenic student of Vesalius—and then at Padua's medical school, under Fabricius of Acquapendente, who was an illustrious heir of Vesalius and an advocate of comparative anatomy. Harvey became a physician at Saint Bartholomew's Hospital in London in 1609, and its Lumleian lecturer in 1615. Soon he was royal physician. Especially close to Charles I, he effectively found himself on the embattled Royalist side during the bloody Civil War. He was awarded a doctorate in medicine in 1642, the year of its outbreak, but was ejected from Saint Bartholomew's the next year on orders from the House of Commons; he was also made warden of Merton College in 1645, but for only a brief stint, since Oxford fell to Cromwell's troops the next year. Harvey's wife died then, preceded by his two brothers. At that point, he retired from public office, but he continued to work in London.

It is often said by scholars of this chaotic, brutal, distressed period of English history that Harvey's thought mirrored his times—the status of the center of the so-called body politic was in crisis just as he was redefining the function of the human heart. He owed his insights in part to his Paduan training with Fabricius of Acquapendente. But Harvey also made judicious use of Aristotle's extensive analyses of his own comparative dissections.[12] Rather than dismiss his predecessors, Harvey looked anew, and found that the past could be used. After all, it was Aristotle himself who had said, "Faith is to be given to reason if the things which are being demonstrated agree with those which are perceived by sense: when they have become adequately known, then sense should be trusted more than reason."[13] Harvey cited this approvingly in his essay *Movement of the Heart and Blood in Animals,* which was published in 1627.

Because comparative dissections had been impossible to practice for over a millennium, the available conclusions about the body's structure had been stuck, wrote Harvey, in the form of "a universal syllogism on the basis of a particular proposition." This "universal syllogism" is in part what accounts for the success of the Galenic system. No revision was possible without further facts, of the sort that Vesalius and his successors had begun to garner. It was important to break out of the teleological form of reasoning at which Galen had excelled, and which had allowed him to be convincing about anything, since he sought to explain the body's organs and functions in terms of a given "reason": nature had foresight, and Galen was, in a sense, its fastidious, confident messenger.

Galen had written in *On the Usefulness of Parts,* "Nature established the heart in the very centre of the cavity of the thorax because she found this place to be most suitable for protection and for uniform refrigeration from the whole body of the lung. It is the common opinion that the heart is not situated exactly in the centre but more to the left, but people are deceived by the pulsation apparent in the left breast, where that ventricle lies which is the source of all the arteries."[14] If a reader today ignorant of all anatomy were to read this account, there would be little reason not to believe it. Galen worked around the evidence he had, without questioning the validity of this evidence or of his interpretations, in thrall to his own brilliance and deductive powers. It was a hefty heritage that Harvey respected—he had no problem referring to Galen as "immortal." But what he found out about the functions of the heart and the motion of the blood was as revolutionary as what Galileo had found out through his telescope about the motion of the earth.

4. From Spirits to Circulation

THE HUMAN BODY was closer to hand than Venus or the moon, but proving that the blood circulated via the heart was no simple task. It had been a given until then that the septum in the heart, which divided it into its two ventricles, was porous, thus enabling the passage of blood from one side to the other. The lungs acted as the refrigerator of the heart, which was, as Aristotle had believed, innately hot; this explained, evidently, why the two sets of organs were so near to each other. This picture of the organism had been developed in Alexandria during the third century BC and later. Galen had summed it up, explaining that "the encephalon [brain] and the spinal medulla are the source of all the nerves (the encephalon being in its turn the source of the spinal medulla itself), that the heart is the source of all the arteries and the liver of the veins, and that the nerves receive the psychic faculty from the encephalon, the arteries the faculty of pulsation from the heart, and the veins the natural faculty from the liver. The usefulness of the nerves, then, would lie in conveying the faculty of sensation and motion from its source to the several parts; that of the arteries is to maintain the natural heat and nourish the psychic *pneuma;* and the veins were formed to produce the blood and also to convey it to all the parts."[15]

Within veins and arteries circulated not only blood but humours and *pneuma,* the air or spirit without which there could be no life. The veins thus originated in the liver, and within them circulated natural *pneuma,* the instrument of appetitive functions—principally nutrition, growth, and reproduction. The arteries, on the other hand, originated in the heart. The heart transformed the natural *pneuma* into the more refined, vital *pneuma* through the heart's diastole—the dilating phase in the heartbeat, followed by the systole, during which Galen believed the heart rested. This vital *pneuma* was the instrument of the sensitive functions, principally motion, perception, and sensation. The psychic *pneuma*, refined in the brain, was the instrument of reason as well as emotion. Natural, animal, and vital *pneumata* or "spirits" flowed within their respective *pneuma* and literally embodied the connection between sensation and cognition, emotion and movement.

The heart itself had little to do with the actual movement of the blood, which Galen and Galenists had assumed moved around the body through the independent pulsation of arteries. *Pneuma* traveled from the lung via the pulmonary vein to the heart's left ventricle, where it produced the arterial blood. The heart, in turn, supplied "the lung with nutriment from the blood as a sort of exchange and to do the lung this service in return for the air which the heart receives from it."[16] Respiration was "useful to animals for the sake of the heart," rather than the other way around; and the heart in turn required "the substance of the air" in order "to be cooled because of its burning heat." The cooling of the heart was effected by the inspiration of air into the lung. During expiration, remainders of the air that had been processed—"sooty vapors"—were expelled, helping in this way to cool the heart.

The anatomical revolution had begun—somewhat—to dent the authority of this enduring story. Vesalius had already noted, *contra* Galen, that the vena cava did not originate in the liver and that the septum between the ventricles was not so porous. And decades before Harvey, the blood's circulation had been conceived by a Spanish contemporary of Vesalius, Miguel Serveto, or Michael Servetus, who knew his Galen well, and who had also worked with Gunther von Andernach in Paris, practicing comparative dissections there. Serveto infuriated both Catholics and Protestants with his anti-Trinitarian convictions and would be burned at the stake by the Inquisition, in Geneva and with the approval of John Calvin.

Fittingly enough, Serveto's description of blood circulation was embedded within a theological treatise, *Christianismi restitutio,* published in 1553. This was his last book—the one that would get him into trouble with Calvin and ended up on the 1559 Index of books prohibited by the Church. It was based on the biblical notion that "God breathed the divine spirit into

Adam's nostrils together with a breath of air" and "maintains the breath of life for us by his spirit." Serveto thought that the heart was "the first living thing, the source of heat in the middle of the body. From the liver it takes the liquid of life, a kind of material, and in return vivifies it, just as the liquid water furnishes material for higher substances and by them, with the addition of light, is vivified so that [in turn] it may invigorate." The "divine spirit"—the soul, in other words—was not, as Galenists believed, "in the walls of the heart, or in the body of the brain or of the liver, but in the blood, as by God himself in Gen. 9, Levit. 7 and Deut. 12."

The vital spirit, wrote Serveto, originated in the heart's left ventricle, assisted by the lungs for its generation. It was a "rarefied spirit, elaborated by the force of heat," as he put it, a "clear vapor from very pure blood, containing in itself the substance of water, air and fire. It is generated in the lungs from a mixture of inspired air with elaborated, subtle blood which the right ventricle of the heart communicates to the left. However, this communication is made not through the middle wall of the heart [the porous septum], as is commonly believed, but by a very ingenious arrangement the subtle blood is urged forward by a long course through the lungs," where it "becomes reddish-yellow and is poured from the pulmonary artery into the pulmonary vein. Then in the pulmonary vein it is mixed with inspired air and through expiration it is cleansed of its sooty vapors. Thus finally the whole mixture, suitably prepared for the production of the vital spirit, is drawn onward from the left ventricle of the heart by diastole."[17] The demonstration continued. One of its conclusions was that the rational soul was seated not in the brain's mass but rather in the vessels that girded it.

Others after Serveto, and before Harvey, would also mention circulation. There were the sixteenth-century physician Andrea Cesalpino, who agreed with Serveto's picture of lung function, the anatomists Juan Valverde de Amusco and Realdo Colombo, and Fabricius of Acquapendente, Harvey's mentor at Padua, who identified venous valves. And Serveto was not even the first to have had the idea of pulmonary transit: it seems to have been mooted in the Arab world, in the thirteenth century, by a doctor and professor of medicine in Damascus and Cairo, 'Ala'-ad-Din 'Ali ibn Ali-l-Hazm al Qurashi, known as Ibn an-Nafis. He was the author of a commentary on the *Canon* in which, noting the absence of passages in the septum between the heart's ventricles, he concluded that the blood cir-

Fig. 22. Print from Realdo Colombo, *De re anatomica* (Venice, Bevilacqua, 1559).
Taken from Meyer-Steineg and Südhoff, *Geschichte der Medizin im Uberblick mit
Abbildungen* (Jena: Fischer, 1921), p. 232.

culated. His insight had no immediate repercussion within the Islamic world, but it might possibly have had an impact on Serveto.[18]

5. Harvey's Blood

H ARVEY'S IDEA, then, was not entirely novel. But its time had come. Moreover, he was the one who did the systematic job of testing, dissecting humans and animals, ligating the arteries of vivisected frogs, and observing the beating hearts of chicks—demonstrating, noting, and recounting the results in treatises devoted to the subject, rather than to theology. In order to arrive at his conclusions, Harvey had relied on observation, not on the Bible. In the essay *Movement of the Heart and the Blood in Animals,* he described the heart as a pumplike muscle (acknowledging the author of the Hippocratic *De corde,* who had also called it a muscle). Its "action and office," he observed, were to "contract and to move something, in this case the blood contained within it." Diastole and systole were in fact echoed in the pulse—and this vindicated what Erasistratus had suggested over 2,000 years before. Moreover, since "the beat of the heart is continuously driving through that organ more blood than the ingested food can supply, or all the veins together at any given time contain,"[19] it was impossible for the liver to be producing the blood. Using such arguments, and careful descriptions of experiments he had carried out, Harvey thus reconstructed the empirical process that led him to his brilliant conclusion: in order for the venous blood to change into arterial blood, it necessarily had to circulate throughout the body, via the lungs, to and from the beating heart, which contracted at systole (rather than resting, as Galen had believed), and expelled the blood at diastole.

The heart, therefore, had a newly central function. The liver was no longer the place where the body's life-giving forces were concocted. The whole Galenic edifice was in danger, and conservatives knew that. One of Harvey's opponents was Jean Riolan, against whom he composed his later essay *The Circulation of the Blood,* published in 1649, the year when his protector, King Charles I, was executed. There, he wrote that the heart was "the starting point of life and the sun of our microcosm just as much as the sun deserves to be styled the heart of the world." It was, he had discovered, present in the chick embryo before either liver or brain (though

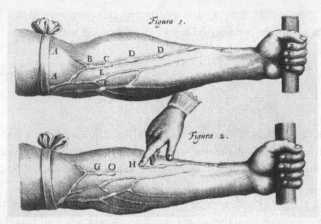

Fig. 23. William Harvey, depiction of two out of four stages in an experiment conducted and described in *Exercitatio anatomica de motu cordis et sanguinis in animalibus (On the Motion of the Heart and Blood in Animals)*, first published in 1628. One of Harvey's most famous experiments, it demonstrated that the blood in veins flows in only one direction, towards the heart.

probably not before blood) and so, "together with all the veins and arteries and the contained blood," it was "to be regarded as the beginning of the things that are in the body, the creator, fount and spring, and the prime cause of life."[20] It was "the tutelary deity of the body, the basis of life, the source of all things, carrying out its function of nourishing, warming, and activating the body as a whole."

Harvey's findings were revolutionary; but he himself was not, at least not consciously. The Aristotelian order beloved by Galen did not cease to find favor with him. In 1649 he still used a language imbued with scholastic reflexes, talking of the "native heat, or innate warmth" of the blood itself as "the common instrument of all operations, and also the primary efficient cause of the pulse."[21] He even dismissed Francis Bacon, the anti-Aristotelian natural philosopher who had been the first to call for the creation of a community of like-minded gentlemen engaged in the pursuit of experiments for the sake of understanding the mechanisms of nature and improving the human lot.[22] The community Bacon had dreamed of would eventually be created after his death: his vehemently anti-scholastic program was enshrined as the mission of the Royal Society, fully established when it received its charter from Charles II in 1662.[23]

Harvey's work would actually give birth to much of the research conducted under the aegis of the Royal Society. In fact, despite his professed disdain of Bacon, he had worked rather intensely with the group of natural philosophers in Oxford who would later form the core of the Royal Society. The motto of this group, *Nullius in verba*—"Nothing by word alone"—signified that the modern scientific investigation these young researchers and scholars were creating was based on hypothesis, experiment, and proof, not on verbal testimony. (Many of them would themselves begin experimenting with the blood in the 1650s, trying out the first transfusions, for instance.) Harvey was an experimentalist of the first order. And while he was indeed wary of giving up the past, he did not entirely take on board the spirits that putatively flowed within the blood. He noted that "there are many and opposing views" regarding their nature, "what is their state in the body, and their consistence, and whether they are separate and distinct from blood and the solid parts, or mixed with these. So it is not surprising," he went on, "that these spirits, with their nature thus left in doubt, serve as a common subterfuge of ignorance." Whatever spirits existed were "no more separate from the blood than is a flame from its inflammable vapour."

Yet here too he preferred to refrain from driving too far ahead. His rather conciliatory view was that "in their different ways blood and spirit, like a generous wine and its bouquet, mean one and the same thing. For, as wine with all its bouquet gone is no longer wine but a flat vinegary fluid, so also is blood without spirit no longer blood but the equivocal gore." In this sense, spirits were in fact quite real. One could deem them constitutive of the blood; arterial blood was simply "more spiritous and more heavily endowed in the former vital force."[24] And so in the end, spirits, in Harvey's mind, continued to make all the difference between life and death, between what should be called blood and what gore. The notion that it was something in the air that made the difference between blood and gore, indeed between life and death, was not yet mooted. It would emerge later on, as did the valves, unknown to Harvey, that allowed for the passage of the blood between arteries and veins—discovered later in the seventeenth century by the Italian physician Marcello Malpighi.

The doctrine of blood circulation as it now stood provided an explanation that had not existed before for the sorts of physiological and psychological phenomena Galenic physicians and philosophers had described, and for the workings of humoural remedies, whose effects on the organism were not really understood in any case. Harvey believed that the centrality of the heart to the body explained contagion, the effects on the whole body of a "poisoned wound," of venereal disease, or of the bite of a snake or mad dog: the poison traveled from the affected part to the heart, in the "returning blood," and from there contaminated the whole body.[25] If remedies such as colocynth and aloe were laxative and cantharides was diuretic, if garlic on the soles of the feet promoted expectoration and cordials promoted strength, as he wrote, that was in all likelihood because "the veins through their openings absorb a fraction from the substances placed on the outside, and carry it inside together with the blood, just as those in the mesentery suck chyle out of the intestines and carry it together with the blood to the liver." The blood's circulation also accounted for the effectiveness of bloodletting and cupping: such practices relieved "all pain as if by charm" because of the "compression and artificial constriction of the artery taking the blood flow to a part."[26]

And crucially, circulation explained why emotions had visible, physical manifestations. It was possible now to suppose that emotions did not correspond to individual humours, and that, instead, one could see the mark of the blood itself, and of its spirits, in "the way in which our body reacts differently in every affection, appetite, hope, or fear." It was the blood's flow that explained why "in bashfulness, the cheeks are lavish with blushes; in fear, disgrace, and shame, the face is pale but the ears are red as if about to hear ill: in adolescents touched with desire, how quickly is the penis filled with blood, erected and extended?" Indeed: the blood was the engine of life—and in ways that Harvey and his successors could not have imagined. Blood had now replaced individual humours.

But this replacement did not change the nature of the passions. To some extent, blood served the same explanatory functions as those fulfilled by humours. It could very well accommodate the notion that some emotions and dispositions were warm and others cold, and that people could be identified according to their temperament. Fear, desire, repulsion, lust, pride, shame continued to have physiological manifestations. Burton's *Anatomy of Melancholy* was the anatomy of something real. His critical distance from his lavish accounts of the beliefs transmitted by elevated humanism belies the absence of an alternative story to the humoural one. Nothing else seemed able to match ordinary common sense in explanatory power. "Folk" beliefs that had always informed humoural theory survived this onslaught from on high, simply because no better explanation for our passions was available, and because humours were surely being secreted somehow, somewhere. Perhaps they did not take the pathways that had been imagined until now, but there had to be more to the story than blood. Its contents were still mysterious, and Harvey himself did not know what its "spirits" really were.

6. Cartesian Souls

NO ONE KNEW, really, what these "spirits" were. One could only speculate as to the seat of the soul. In 1650 a fifty-four-year-old Frenchman who had claimed that the soul was entirely separate from the body died during a rough Stockholm winter. He had claimed that the body was an automaton of sorts that functioned according to the laws of mechanism, and that the soul was completely immaterial, in contact with the body only via the pineal gland in the brain. He was René Descartes, a philosopher who did all his most important work on the road mostly outside France and in the Netherlands, a refugee there from the political and religious intolerance at home.

Descartes's ambition had been to replace the all-embracing Aristotelian system, and its world of abstruse, unverifiable, teleological claims about the order of things, with his own, equally all-encompassing, universal system. To an extent, he succeeded: he is still thought of as the father of modern philosophy. He was also a natural philosopher, however, who invented analytic geometry and for whom truths about physics, or nature, had to take the form of demonstrably "clear and distinct" ideas, as sharp as the truths of mathematics. The world of physical things was ultimately reducible, in his mind, to the world of mathematical relations. And so he defined the realm of thinking apart from that of the body, on the basis of what has become known as the *cogito*, shorthand for his famous formula in Latin, which in his *Discourse on Method* (written as a preface to scientific works) appeared as *je pense donc je suis*—"I think, therefore I am," *cogito ergo sum* in the later Latin version. There is nothing humoural about the *cogito*. It is supposed to define what we are in terms of our self-reflexiveness, our self-aware rationality, our capacity to stand apart from our own thinking. It is supposed to celebrate, in a way, the ineffability of the soul. With this formula, humours would seem to have been made truly and finally redundant.

Later, the demonstration was tightly and beautifully developed within Descartes's six *Metaphysical Meditations,* first published in Latin in 1641 and dedicated to the deans and doctors of the Faculty of Theology in Paris, clearly to avoid political condemnation. Very schematically, it proceeded from the initial premise that one's senses on their own could be deceiving; they could even distort the world, in the same way that yellow spectacles, say, will make everything look yellow. And so Descartes had to find the basis on which to ground his belief in the reality of the world, which included mathematical truths. He therefore endeavored to doubt everything—imagining that an evil genius had been telling him all along that he and the world existed—in order then to find out what could be left of the realities and truths he took for granted, including his own existence. But the very fact that he doubted entailed that he thought, so at least his mind existed. And since he could have a clear and distinct idea of God, then it followed that there must be a cause of that thought, which must be God; hence God existed. And since God was infinitely good, He would not allow an evil genius to deceive Descartes. Hence, Descartes's very capacity to think entailed the reality of the physical world. Only reason—the reason that enabled him to conceive of God—could yield what Descartes called "clear and distinct ideas," which were necessarily true and existed beyond the world of sense, such as the truths of mathematics. True knowledge, then, depended on the mind, not the senses; and since Descartes conceived of body and soul as distinct from each other, each one really must have been of a distinct nature: the body mortal and divisible, the soul immortal and indivisible. So the mind, the *res cogitans* or thinking thing, was necessarily separate from the senses and from the body, the *res extensa* or extended thing.

The *cogito,* and everything that it encompasses, is at once a puzzling instrument, an open-ended thought, and an argument full of fallacies. It happens also to be a child of the self-reflectiveness that Augustine had profoundly analyzed in his *Confessions:* Descartes, who had gone to a Jesuit school, had remained imbued with the lessons of Augustine and of early Christian theology. But Descartes called the idea "clear and distinct" because in his view reason was a clear and distinct idea, as simply defined as the neat, perfectly designed realm of the passionless body. Despite the scholastic origins of the *cogito,* there was something new about it: it was an extreme form of mind-body dualism which did not coexist with the tripartite soul Augustine himself had held on to. Although body and soul were intimately connected—via the pineal gland—the soul, on principle, did not need the body. Some extreme Cartesians even believed that the body was an impediment to the perfect communication between minds, and that language itself was an approximation we had to make do with because we were fallen creatures, condemned to lumber around these bodies of ours.[27]

But not everyone was convinced by this novel dualism. Descartes himself chose to have a series of impressive, albeit highly respectful objections appear in the first edition of the *Meditations,* along with his own answers. One set of objections was from Marin Mersenne, who asked, rather daringly, why it was impossible for this thinking thing whose existence was guaranteed by the *cogito* to be a body, rather than a disembodied soul: perhaps it was precisely by virtue of the body's movements that we had thoughts—perhaps thoughts were themselves corporeal movements.[28] Questions of this sort and others multiplied. The debates were intense, complex, and convoluted. Many people wondered how, for instance, an entirely immaterial soul could logically have any impact on the body. Descartes had relegated all the functions of what had been the Scholastics' sensitive soul to the rational one, turning all thoughts and all emotions into the attributes of human consciousness. As for the pineal gland, it was perhaps a convenient refuge for the immaterial soul, but this could not work conceptually; nor did it make any sense anatomically.

Because nature was revealing itself to the eye, passionate questions surrounding the status of human nature itself were multiplying by now, especially after Descartes's death. One significant participant in these discussions was the Danish physician Niels Steensen, or Nicolaus Steno. He had done important work on muscles and glands, and is best known today for his research on fossils. He would give up science after converting to Catholicism in Italy and becoming a bishop. But before then, one day in 1664 or 1665, he undertook a brilliant demonstration of the errors wrought by anatomists of the brain. Speaking at the Parisian academy of an intrepid traveler called Melchidésec Thevenot (the French Académie Royale des Sciences would grow out of this academy), he succeeded, within an hour or so, in discarding Descartes's idea that the soul was seated in the pineal gland. It is notable that he began his lecture-demonstration before the natural philosophers, amateur experimenters, anatomists, and men of letters gathered that day with an appropriate avowal of ignorance: "Gentlemen, instead of promising to satisfy your curiosity regarding the anatomy of the brain, I here confess, sincerely and publicly, that I know nothing about it." The lecture, however, was insightful. Steno was probably the first to notice that the brain's white matter was replete with microstructures.

A decade later, the physician Daniel Duncan would talk of the "pineal gland, rotten and as big as a nut, that, in Montpellier, I saw being taken out of the head of a woman who had reasoned perfectly well until her death"—proof enough, in his view, that the pineal gland had nothing at all to do with reason. The inference actually seems rather paradoxical: if the pineal gland had nothing to do with reason, then its physical state would not reveal anything about the state of a person's rational soul in the first place. In fact, it was impossible to define the functions of the rational soul by pointing to its physical location; and this difficulty arguably remains today.

But anatomical and conceptual paradoxes were only part of the problem with Descartes. There were ethical and metaphysical issues, too. The integrated, three-tiered organism of Aristotle and Galen was now imploded, split in half. Descartes had done this for the sake of his system's internal coherence, but the problem was that it also followed, logically, that he had to regard all nonhuman creatures as mere automata. Animals were now doomed to live only with an appetitive function, soulless, devoid of any capacity to experience pain, joy, or any emotions. This was a devastating result, and debates erupted in France and England, continuing well into the eighteenth century, over what sorts of minds animals might have within this new world. In some ways, the debate still goes on.

It seemed to some opponents of this brutal mechanism that if Descartes was right, then humans were also potential automata. On the other hand, to a Cartesian like the Protestant Jean-Marie Darmanson, for example, who wrote a tract in 1684 defending the beast-machine thesis, if animals were "accorded the slightest degree of knowledge, joy, sadness, pleasure, pain, hate and love," then one had to assume they had a soul like ours, "entirely separate from the body." But if that soul were immortal, then nothing would separate humans from beasts. So it had to be mortal, and so, by equivalence, our soul would have to be mortal too. Yet in this case as well, nothing would separate us from beasts. Hence, if one wanted to hold on to the exceptional status of human beings and the immortality of the human soul—as one did if one wasn't a *libertin*—one had to accept that animals had no sensations.

And so the arguments about animal minds went on, often technical, sometimes absurd, at times brilliant, in many guises and versions. They were catalogued and analyzed at length by Pierre Bayle in his colossal *Dictionnaire historique et critique*, arguably one of the first encyclopedic dictionaries, initially published in 1697. (Bayle was a Protestant also living in exile in Holland, an erudite scholar and original thinker, and one of the great skeptics of the seventeenth century.)[29] Older questions now resurfaced that showed the difficulty of holding on to human exceptionalism in a world where Galenic final causes were no longer functioning, where one could no longer answer "how" with "because," and where the form of bodies could not be explained in terms of their function.

What were humans made of? What sorts of human beings were those who were born deaf and mute? Were feral children human? Were babies born with two heads, six toes, or one lung fully human? What were these monsters and marvels that, during the 1600s, had become so fashionable a topic of conversation, so serious a program of inquiry, and so controversial an issue? Just as mechanism became fully entrenched in habits of thought during the eighteenth century, so anything that seemed to escape its laws, or, inversely, demonstrate its riches, attracted intense attention. One woman in Surrey named Mary Toft became a famous case in 1726 when she claimed to have given birth to rabbits—seventeen of them in all. She eventually confessed that the claim was a sham. But while it was happening, the event had been considered credible enough.[30]

7. Anti-Humours

B Y THE MID-1600S, humans no longer seemed to have a secure place in the natural world. Nor was the political, urban world very safe. Absolute monarchies were doing little to alleviate economic distress among populations that experienced nature as a fierce enemy that had to be domesticated. Political and religious grievances continued to burden England. The wars of religion had destabilized the whole of Europe, and new religious sects were thriving in the Reformed lands. In the south, the mood of the Counter-Reformation was still sustaining the Inquisition, and censorship was as sharp as the thoughts that were finding expression in spite of it. In the world of learning, Scholasticism was now in crisis; and the ancients, although still deeply a part of lettered culture, were no longer considered useful for scientific work. Anti-Galenic currents of thoughts that had emerged from northern Europe in the sixteenth century were gaining momentum. Galen was no longer an absolute authority in the world of scientific societies in London, Paris, or Florence. Aristotle was frequently derided, though he was still taught. Matter was being redefined and taken apart. Just as reason was taking off in its Cartesian, disembodied universe, new layers of matter were appearing, disrupting ancient physiologies. New notions were emerging about the nature of the blood.

The elaborate thinking of Joan Baptista Van Helmont had given birth to the most effective anti-Galenic ideas since Paracelsus; they would resonate especially in England. Van Helmont was a Flemish scholar and physician who trained at Leuwen, and his profound knowledge of the classical authorities led him, not unlike Paracelsus, to pursue an innovative set of theories about body, world, and cosmos. He described these most influentially in his *Ortus medicinae,* published posthumously in 1648, four years after his death. It combined Neoplatonism, alchemy, Paracelsianism, keen observation, and a dislike of any orthodoxy. His vision was highly spiritual and definitely vitalistic. He posited an *archaeus* as a vital principle, as did Paracelsus (of whom he was nevertheless critical), and "signatures" or correspondences between all the matter in the world.

Van Helmont is mostly remembered today as the man who coined the term *gas* (probably from the Greek *khaos)* to designate the essence of matter. He described it as the volatile, universal state in which everything finds itself and from which, he thought, everything was born. All matter was essentially made of water and air, but without gas, no particular objects would emerge out of this primal matter; and when a fruit, for instance, was disintegrating, it was reverting to its gaseous state. Matter, in other words, conserved itself. Van Helmont was an impressively imaginative and visionary chemist. He described about a dozen gases, including CO_2—and narrowly escaped death by gas poisoning. That gas could kill was no surprise to him: even the *pneuma* in the blood was gas, he thought, and illness could be due to an overload of gases that were foreign to the organism.[31]

This was a man who dismissed traditional humours as nonsense. He was able to identify the acidity of gastric juices, partly by observing that a piece of glass swallowed by a hen would be smooth by the time it was eliminated. There was no such thing as an "excremental" black bile; nor did the spleen store it. Rather, bile had a positive function, connected to that of the stomach, which was central to the organism. (Paracelsus also had thought of bile as a crucial substance that was not a "humour.") Digestion, then, was due primarily to acidity, not necessarily to heat; excessive acidity in the stomach, not black bile, was the cause of digestive problems, and it could be corrected with an alkaline solution. Van Helmont even re-created gastric acid, by combining an aqueous solution of saltpeter, vitriol, and alum with sal ammoniac. He also analyzed the urine, and saw that it had nothing to do with gastric bile. From this he concluded that the humoural habit of using urine as one of the main diagnostic tools was pointless.

Bloodletting was nonsensical too, as were purgatives and laxatives. Herbs and mercury-based medications that increased sweating were much more effective, he thought, in the case of fevers, which were the expression of the "fury" of the *archaeus*. (That mercury was a poison and a potential killer did not seem to matter.) He was not alone in thinking so: Paracelsus had thought a nitrous chemical was a cause of fever; and Campanella, who was Van Helmont's contemporary, had written in a treatise of 1620 that "fever is not a disease but a war against the disease."[32] As for the notion of a phlegmatic disease—of the negative effects of catarrh descending from the head—Van Helmont thought that was wrong too. Conditions that the Galenists had attributed to its excess, from apoplexy, dropsy, and pleurisy to asthma, phthisis, and coughs, in fact each had their own particular causes. Van Helmont had no affinity with the holistic thinking at the core of humoural theory. Disease was not a global condition, and it had nothing to do with a dyscrasia. Each disease was a local event, the outcome of specific "irritants" acting on specific organs, which reacted according to their own vital force, as if they each had an autonomous agency and were endowed with what he called "tissue-intellect."

In place of the Scholastics' hierarchy of souls or appetites, Van Helmont saw a hierarchy of life forces, all of which were interconnected, in sympathy or antipathy with each other. It was this interconnection that explained why one could cure a wound at a distance, for instance, with a weapon salve—by imbibing the injuring weapon with the blood from the wound and with an ointment salve. In the words of the German physician Daniel Sennert, a professor at Wittenberg—who had argued that the processes at work in the healing of wounds were natural, not magical—the recipe consisted of: "Scull-Mosse [moss from a human skull], two Ounces, Mummy, halfe an ounce, Mans fat, two ounces, Mans blood, half an ounce, Linseed Oyle, two Drams, Oyle of Roses, and Bole Amoniack, of each one ounce. Mixe them together and make an Oyntment: into the which hee puts a Stick, dipp'd in the Blood of the wounded person, and dryed, and bindeth up the wound with a rowler dipt every day in the hot Urine of the wounded person. For the annointing of the Weapon hee addes moreover; Honey, one ounce, Bulls Fat, one dram."[33]

This was just one of many versions of the recipe, popular already in the sixteenth century. It was wrongly attributed to Paracelsus himself and was enthusiastically taken up by Paracelsians, in the early 1600s. It relied on a belief that action at a distance could work; and indeed, in a world where astrology and incantations were considered potent, there was no reason for this not to be an available cure as well, just as it was possible to cause illness at a distance too. Some people, in particular a Flemish Jesuit, Jean Roberti, worried that dark, demonic forces might be involved in action at a distance, and the young Van Helmont joined a debate that was pitting this Jesuit against a German Protestant, Rudolph Goclenius. Van Helmont argued in his pamphlet on the matter that the principles of sympathetic attraction and antipathetic repulsion were not supernatural and that to appeal to them was not to engage in magic, either natural or demonic: the phenomenon could simply be explained by magnetic attraction between the blood on the weapon and the blood in the wound.[34] What seemed magical was in fact natural. Van Helmont got into trouble for mocking the Jesuits. The Spanish Inquisition initiated a long-term trial against him, and, like Galileo, he was eventually condemned to house arrest.

The weapon-salve was also the subject of a famous, more innocuous tract by the colorful English character Kenelm Digby, a diplomat, soldier, traveler, writer, philosopher, natural philosopher, turncoat, chancellor to Queen Henrietta Maria, and founding member of the Royal Society. During decades he spent exiled in Paris, he was a member of Mersenne's circle and met everyone from Descartes to Hobbes, from Ben Jonson to the Oxford group of natural philosophers. He agreed with the beast-machine thesis, dabbled in alchemy, created remedies, and even wrote a cookbook. The tract, published in 1658 as *A Discourse Made in a Solemne Assembly of Noble and Learned Men at Montpellier in France, by Sir Kenelme*

Digby, Knight &c., Touching the cure of wounds by the powder of sympathy, argued for the efficacy of this remedy. This came from a man who was at the forefront not only of political and religious activity in his time, but also of intellectual and scientific innovation.[35]

Not everything was known about matter: and so anything could be legitimately imagined. What was unknown about the physical world was no longer considered to be metaphysically unknowable. Even physical processes that seemed incomprehensible and therefore looked like magic could in fact be deemed to pertain to nature. This was still the case during the 1700s—and someone like the rabbit-bearing Mary Toft had, astutely or rashly, decided to exploit to her advantage the culture in which she lived. The weapon-salve recipes were still circulating in the eighteenth century: there was always room for a little magic.

8. Empiricism

THERE WAS, then, no one orthodoxy, and no certainty, about the physical order of things. This is partly why, despite Harvey and the alternative offered by Van Helmont, humours were not fully dismantled; nor did they become a mere antiquarian curiosity. New ideas did not necessarily cohere with the culture in which they emerged, and they did not necessarily result in innovative medical practices. Harvey himself had bled his patients, and "empirics" were taking clinical care into their own hands, away from experts and Galenic rationalists.

There were of course established physicians who doubted the practical usefulness of Galenic medicine. One of them was Thomas Sydenham, a member of the Royal College of Physicians who was among London's most respected doctors in the 1660s and 1670s. He was trained mostly at Oxford, and he would be remembered posthumously as the "English Hippocrates." Indeed he did advocate a return to a Hippocratic rather than Galenic, nondogmatic approach to medical care. He closely observed the brutal plague that struck London in 1665 and 1666; researched and treated epidemics such as scarlet fever and smallpox; established the classification of conditions such as chorea (a disease of the nervous system); and discovered the use of cinchona (or Peruvian) bark, whose active ingredient is quinine, for the cure of malaria. He was a keen, innovative observer. He had no use for what he called the "speculative theorems," the—presumably Galenic—"principles" and "maximes" about nature that had sustained medical practice for so long. To him, experience and reason were quite sufficient.

There was a philosophical basis for this rejection of theory. He was convinced that the main duty of the physician "was simply to confine his observations to the 'outer husk of things,' God having so shaped his faculties as to perceive only the superficies of bodies, and not the minute processes of Nature's 'abyss of cause.' "[36] Sydenham had faith in observation and in the capacity of humans to unravel the causal order in the world and in the body. But he also was convinced that there was a limit to this capacity, that God had foreknowledge of this causal order, and that only by knowing God could one have any control over events. We shared with animals a lowly soul, but we also had a higher, rational soul with which we had to worship God and God's natural realm; and it was also up to us to use the rational soul in order to live beyond the passions that might lead us to self-destruction. Aristotelian forms and teleology, he thought, did nothing to explain the body, its anatomy, physiology, and ailments, yet we had to understand our nature in terms of our ultimate, final function, which was the contemplation of God. The order of things was in fact teleological, just as it had been for Aristotle. Religion and science were intimately conjoined, and the one justified the pursuit of the other.

Sydenham was friendly with the Oxford natural philosopher Robert Boyle, a visionary chemist who shared this highly religious view of his role. Boyle wrote in the aptly titled *The Sceptical Chymist,* published in 1661, about the moral, not just methodological importance of preserving skepticism with regard to one's research and one's capacity to know God's world. Boyle was a central figure in the Oxford group of "virtuosi" who were conducting experiments and redefining matter, and who were to form the core of the Royal Society. Some of them, like Robert Hooke, were engaged in revealing heretofore unseen dimensions through the microscope. The use of the lens to see the very small was at first the result of a reversed telescope, as Galileo had realized in 1624.[37] The first exact scientific illustrations began then, as did taxonomies. Microscopes became increasingly sophisticated in the next decades. In 1665, Hooke published the results of his research in the great *Micrographia, or Some Physiological Descriptions of Minute Bodies Made by Magnifying Glasses,* where he depicted the magnified world of, for instance, poppy seeds and the feet of flies. Antoni van Leeuwenhoek discovered living "animalcules" or "globules" floating around and seemingly reproducing in rainwater, and described them in a letter he sent in 1681 to the *Philosophical Transactions* of the Royal Society. He had seen for the first time what later would be called bacteria.

The philosopher John Locke would be fascinated with these hidden worlds. He was another friend of Sydenham, whom he met through Boyle as a medical student at Oxford. Locke knew that it was the mission of the natural sciences to uncover such worlds, though he did not believe that we were in a position to understand fully what we did not see directly. As he put it in his magnum opus of 1690, which would establish the foundations of modern empiricism, the *Essay Concerning Human Understanding*, once our senses were altered with contrivances such as microscope lenses, "the appearance and outward Scheme of things would have quite another Face to us; and I am apt to think, would be inconsistent with our Being" because they would "produce quite different Ideas in us."[38] To imagine physiology was one thing—it was, one might say, consistent with our being—but to see fragments of hidden structures was somewhat unsettling.

Hidden structures were becoming increasingly visible. But, as had been the case 300 years before when Berengario da Carpi began to look inside the body, nothing could be seen without a theoretical underpinning. The notion that the world was made of invisible microscopic particles had been gaining ground steadily since Gassendi had revived ancient atomism, but it supposed a disjunction between our experience of matter and the actual makeup of the physical world much more radical than the disjunction between humours and skin. Earlier, in a treatise of 1623 called *The Assayer*, Galileo had described the physical world as separate from our sensory experience of its qualities. Descartes had done the same. Locke had thought deeply about this disjunction, concluding that we had experience only of the secondary qualities of matter.

But once this gap was established, there was room for particles, however unsettling these might be. Boyle integrated them into what he called his "corpuscular philosophy," which combined his chemical and alchemical research with aspects of Cartesian mechanism. Descartes himself had imagined that animal and vital spirits were such particles, though since he followed Aristotle's notion that "nature abhors a vacuum," he also believed that particles had to be finite in order to move within the plenum that was the universe, and that their motion created whirlpools. But regardless of their shape, size, and ultimate nature, particles constituted all matter, including our bodies, which, it seemed to all the innovators, were a sort of machine designed by the divine Architect. To study the body now came to be seen as a way of honoring this Architect.

9. Medical Secrets and Popular Healers

CONCEIVING of the body as a machine did not make it any less transparent than it had been before these innovations. The medical scene during the early decades of the seventeenth century was understandably confused, and by the 1660s the profusion of available cures had eroded the authority of most of them.

Already during the Renaissance, techniques of diagnosis and cure were less homogeneous than they had been in the medieval period. The highly successful French physician Jean Fernel had introduced the notion of "hidden causes" by the mid-sixteenth century. He too was interested in astrology and alchemy, and he was sympathetic to Neoplatonism. He was also the first to develop the notion, in his *Medicina* (published in 1554), that one could divide medicine into physiology, pathology (from *pathos,* suffering or disease in Greek), and therapy. This was a variation on the Galenic division of medicine into diagnosis, cure, and prognosis. He also distinguished local ailments caused by a humoural dyscrasia from those that affected the entire organism, which he thought were due to "an affection of the body contrary to nature," and not to humoural problems.[39] The causes of disease in the body's "whole substance," as Fernel had put it, were not necessarily knowable, because they were the work of God.

Humoural explanation could only go so far, then; it did not tell us everything there was to know. By the mid-1600s there was increasing room for alternatives to it, and especially for the chemical, anti-Galenic medicine derived from Van Helmont, also known as iatrochemistry, or medical chemistry.[40] Iatrochemistry acquired a certain standing among natural philosophers; in 1665, they even came close to founding a Society of Chemical Physicians as a rival to the Royal College of Physicians over which Harvey himself had presided. The middle classes were growing, and therefore so was medical demand. In response, clinical activity expanded beyond its previous, official boundaries.

The apothecary, herbalist, astrologer, and popular physician Nicholas Culpeper contributed to this expansion when he recycled, to great effect, the data gathered in the old *materia medica*. A radical and a Parliamentarian during the Civil War, he distrusted the medical profession ever since Hippocrates, disparaged bloodletting and laxatives, and made some use of Paracelsianism. He denounced the abuse of expensive remedies and the profit sought by those who sold and fabricated them out of spices and herbs from India, in favor of the "simples" that one could make easily and cheaply at home, given a modicum of knowledge. In the spirit of making medicine available to all, he opened a successful practice in the heart of London, where he saw patients from all walks of life, often for no fee. In 1649 he went so far as to contravene the rules of the Royal College of Physicians by translating from the Latin and by publishing, without its authorization, the *Pharmacopoeia Londonensis* in order to make the remedies of *materia medica* available to a wider public. Two years later, he did something rather unusual for a man by publishing a *Directory for Midwives*. It must be said that the beliefs he revealed there regarding monstrous births, the genetic determinism of gender, or the processes at work in breast-feeding were not particularly novel—they depended on the notion of humoural, overly secreting female bodies. But he was intent on bringing what he considered sound medical knowledge to the public, regardless of its origin. He even put out a translation into English of the old Galenic compilation, *Ars parva,* in 1652.[41]

That same year, the book by which he is still known appeared, *The English Physitian: Or An Astrologo-Physical Discourse of the Vulgar Herbs of This Nation.* It listed what he thought were the most useful of the hundreds of remedies available in England along the lines of John Gerard's earlier herbal but with an added astrological dimension. It is significant that, for all his disparaging of traditional medicine, he used humoural categories to explain the properties of herbal remedies, adopting, as the title makes clear, the categories passed on within astrological medicine. He noted about hemp (cannabis), for instance, that it "consumes wind, and by too much use thereof disperses it so much that it dries up the natural seed for procreation; yet, being boiled in milk and taken, helps such as have a hot dry cough" while "the Dutch make an emulsion out of the seed, and give it with good success to those that have the jaundice, especially in the beginning of the disease, if there be no ague accompanying it, for it opens obstructions of the gall, and causes digestion of choler." Dioscorides was his explicit source with regard to pennyroyal (a type of mint), which "makes thin tough phlegm, warms the coldness of any part whereto it is applied, and digests raw or corrupt matter. Being boiled and drank, it provokes women's courses, and expels the dead child and afterbirth, and stays the disposition to vomit, being taken in water and vinegar mingled together. And being mingled with honey and salt, it voids phlegm out of the lungs, and purges melancholy by the stool." A white beet "doth much loosen the belly, and is of a cleansing and digesting quality, and provoketh urine: the juice of it openeth obstructions both of the liver and spleen, and is good for the headaches and swimmings therein, and turnings of the brain."

As for the vervain, it was as potent here as it had been in the *Circa instans:* "This is an herb of Venus," he wrote,

> and excellent for the womb to strengthen and remedy all the cold griefs of it, as Plantain doth the hot. Vervain is hot and dry, opening obstructions, cleansing and healing. It helps the yellow jaundice, the dropsy and the gout; it kills and expels worms in the belly, and causes a good color in the face and body, strengthens as well as corrects the diseases of the stomach, liver, and spleen; helps the cough, wheezings, and shortness of breath, and all the defects of the veins and bladder, expelling the gravel and stone. It is held to be good against the biting of serpents, and other venomous beasts, against the plague, and both tertian and quartan agues. It consolidates and heals also all wounds, both inward and outward, stays bleedings, and used with some honey, heals all old ulcers and fistulas in the legs or other parts of the body; as also those ulcers that happen in the mouth; or used with hog's grease, it helps the swellings and pains in the secret parts in man or woman, also for the piles or haemorrhoids; applied with some oil of roses and vinegar unto the forehead and temples, it eases the inveterate pains and ache of the head, and is good for those that are frantic.

Culpeper had an inkling that his book would meet with success. It did, and modern editions still sell internationally. In his lifetime, however, he was rather unfairly branded a quack, or charlatan, on a par with the herb-

alists and such who had been officially sanctioned by the 1542 amend-
ment to the Physicians and Surgeons Act. As William Clowes, a surgeon
at the service of Queen Elizabeth, had exclaimed back then: "Nowadays
it is apparent to see how tinkers, tooth-drawers, pedlars, ostlers, carters,
porters, horse-gelders and horse-leeches, idiots, apple-squires, broom-
men, bawds, witches, conjurers, sooth-sayers and sow-gelders, rogues,
rat-catchers, runagates, and proctors of spital-houses, with such rotten
and stinking weeds, which do in town and country, without order, hon-
esty and skill daily abuse both Physic and Chirurgery, having no more
perseverance, reason or knowledge in this art than hath a goose."[42]

By the 1600s, a quack was still most likely considered to belong to
these lowest orders of society. But the charge of quackish incompetence
had already begun shifting terrain, and ironically, by the eighteenth century the term was applied to those who abused the language and methods
of humoural medicine, especially if they did so without a license, and primarily for financial gain. Many of these practitioners met with success
and enriched themselves. Yet the term remained misleading. In 1719, the
surgeon general to the army wrote a fast-selling popular treatise promoting the use of mistletoe as a cure for epilepsy; but not long before, Boyle
himself had also advocated mistletoe.[43] The beliefs of scholars did not always differ from those of quacks. There was but a fuzzy line between folk
belief in traditional medicines, professional use or support of *materia
medica,* and abuse by unlicensed medical practitioners of the general population's need to believe in a miracle drug.

This abuse was noteworthy. In 1722, Daniel Defoe, the prolific jour-
nalist and author of *Robinson Crusoe,* retrospectively described in his
Journal of the Plague Year the frenzy that had descended upon London
with the plague in 1665 and 1666. People, he wrote, "were more addicted
to prophecies and astrological conjurations, dreams, and old wives' tales
than ever they were before or since. Whether this unhappy temper was
originally raised by the follies of some people who got money by it—that
is to say, by printing predictions and prognostications—I know not; but
certain it is, books frighted them terribly, such as Lilly's Almanack, Gad-
bury's Astrological Predictions, Poor Robin's Almanack, and the like; also
several pretended religious books, one entitled, Come out of her, my People, lest you be Partaker of her Plagues; another called, Fair Warning; another, Britain's Remembrancer; and many such, all, or most part of which,
foretold, directly or covertly, the ruin of the city."[44]

People ran "to fortune-tellers, cunning-men, and astrologers to know their fortune, or, as it is vulgarly expressed, to have their fortunes told them, their nativities calculated, and the like; and this folly presently made the town swarm with a wicked generation of pretenders to magic, to the black art, as they called it, and I know not what; nay, to a thousand worse dealings with the devil than they were really guilty of. And this trade grew so open and so generally practiced that it became common to have signs and inscriptions set up at doors: 'Here lives a fortune-teller,' 'Here lives an astrologer,' 'Here you may have your nativity calculated,' and the like." People were also

running after quacks and mountebanks, and every practicing old woman, for medicines and remedies; storing themselves with such multitudes of pills, potions, and preservatives, as they were called, that they not only spent their money but even poisoned themselves beforehand for fear of the poison of the infection; and prepared their bodies for the plague, instead of preserving them against it. On the other hand it is incredible and scarce to be imagined, how the posts of houses and corners of streets were plastered over with doctors' bills and papers of ignorant fellows, quacking and tampering in physic, and inviting the people to come to them for remedies, which was generally set off with such flourishes as these, viz.: "Infallible preventive pills against the plague." "Neverfailing preservatives against the infection." "Sovereign cordials against the corruption of the air." "Exact regulations for the conduct of the body in case of an infection." "Anti-pestilential pills." "Incomparable drink against the plague, never found out before." "An universal remedy for the plague." "The only true plague water." "The royal antidote against all kinds of infection";—and such a number more that I cannot reckon up; and if I could, would fill a book of themselves to set them down.

There did exist a demarcation beyond which a practitioner was a quack rather than a legitimate doctor: the quack sold magic, prognostication, the promise of guaranteed cure—the sort of broadly advertised, all-purpose cordial that one can see in Westerns, on placards in American nineteenth-century frontier towns. In a sense, guaranteed success of a cure or medicine was as good an advertising tool as any for a physician and especially an apothecary in search of financial gain keen to sell his drugs. A good Hippocratic doctor would always have held on to skepticism, but in the 1700s the College of Physicians in London gave up legislating medical practice. It is no surprise that mountebanks and quacks of the sort described by Defoe proliferated after then.

10. Transfusions and Confusions

N THE LATE 1600s, legitimate medical practice nevertheless was still humoural, based on the usual bleedings and purges. But, unlike quackery, it no longer advertised itself. Moreover, now that questions regarding the failures of clinical care could be asked, its official legitimacy was itself no longer absolute. The foundations on which it had rested were increasingly shaky. Among natural philosophers and medical practitioners, mechanism was now an established system, mostly in its corpuscularian form, but there were also numerous Gassendists, Helmontists, and iatro-chemists in England as well as France. For the Baconian philosophers, however, none of these theories preceded experiment. Observation alone mattered, along with the provision of adequate testimony. Robert Boyle was one of those who believed that all of nature was chiefly made of corpuscular matter in motion, but along with the other "virtuosi" and founders of the Royal Society, he was mostly intent on furthering the heritage of Harvey.

This meant picking up where Harvey had left off, and experimenting as much as possible. Soon these men ventured into elaborate explorations of the blood and lungs. Blood transfusion in particular was an exciting prospect, and thought to possibly offer a cure for madness: a change of blood would perhaps change the person. (Thomas Allen, the physician who then headed Bethlem, or Bedlam, the hospital for the mad, wanted to protect his patients and was rather opposed to the idea; but to no effect.) The first animal-to-animal transfusions were tried in 1665; a number of dogs died in the process. A transfusion on a human being took place in 1667: Arthur Coga, "the subject of a harmless form of insanity," in the words of one of the masterminds of the project, volunteered to be the recipient of the blood of a sheep.[45] "They purpose to let in about twelve ounces," Samuel Pepys recorded in his diary before the event. Coga was just "a little frantic," and was "poor and debauched." Pepys reported that "Some think it may have a good effect upon him as a frantic man by cooling his blood, others that it will not have any effect at all." At any rate, Coga lived, and Pepys noted that, although the man was "cracked a little in his head," he nevertheless spoke "very reasonably and very well."[46]

Across the Channel, the Académie Montmort (made up of those who later gathered at the home of Thévenot and formed the core of the Académie Royale des Sciences) sponsored the physician Jean-Baptiste Denis's attempts to transfuse the blood of a lamb into an ailing young man. The volunteer not only survived but actually took a turn for the better. An attempt on another man also succeeded; a third attempt resulted in death. The fourth became an infamous case: it involved a man deemed manic, named Antoine Mauroy, who received the blood of a calf, given to him in the hope that it would calm him down. The operation seemed a success, until, not long after, the man died. His wife apparently had poisoned him, so Denis escaped prosecution; but the case was never quite resolved. Soon after, the transfusion craze abated.

But experiments continued. Harvey's heirs were making one discovery after another,[47] and now they were using Van Helmont's chemical ideas to further their understanding of respiration. No one yet understood what was the substance in the blood that was necessary to life. It was clear by this point that blood circulated in the fashion described by Harvey, but its nature remained mysterious; the mechanism by which it conferred life on the organism was quite unknown. If the soul was in the blood, then one had to discover what this soul was.

A key contributor to these investigations was the natural philosopher Thomas Willis, whose research combined alchemical principles with the new corpuscular theories. He worked closely with other "virtuosi" at Oxford, including Richard Lower and Robert Boyle. Boyle created an air pump, which demonstrated that there was such a thing as a vacuum in nature, despite what Aristotle and Descartes had thought.[48] The group notably experimented with saltpeter, otherwise known as niter—a combination of what today is called potassium nitrate with sodium nitrate, long recognized to have both heating and cooling properties. It had been used as gunpowder for centuries, worked well in meat curing, and in the 1600s began to be used also as a fertilizer. Paracelsians had worked with this substance, and the Oxford Harveians pursued the research. Niter seemed increasingly to be an essential ingredient in air.

Willis's long-standing study of fevers culminated in the notion that they were a fermentation of the particles in the blood—a matter of chemistry, not of humours. In 1665, Richard Lower published his account of the impressively skilled, if gory, experiments he had conducted with Willis on vivisected dogs. He proposed that "aerial niter," first described by another collaborator, Ralph Bathurst, as the substance that nourished the spirits in the blood, was absorbed by the blood in the lungs, causing it to ferment and turn bright red. In 1669, Lower recounted how arterial blood emerging out of the lungs was brighter than venous blood because it had been transformed and fermented by a substance—the "nitrous spirit" in the air that entered the lungs.[49]

This was not quite right, of course. Our blood undergoes oxidation, not fermentation. Oxygen, though, had not been found. But these early chemical experiments, many originating in the sometimes far-fetched inquiries of the alchemists, were by now transforming the stuff of which physiology was made, turning our humours into precise substances. The map of the inner body was slowly filling in; the questions that needed answering were beginning to emerge. The Royal Society played an important role in publicizing findings and establishing exchanges between natural philosophers throughout Europe.

In Italy, the Bolognese Marcello Malpighi, a well-known physician and professor of theoretical medicine, was making important contributions to anatomy. He helped to consolidate the understanding of the circulation of the blood; conducted microscopic studies of the lung (especially in frogs) and saw its capillaries; was the first to understand how arterial blood passed into the veins, also through tiny capillaries; described red blood cells, as well as taste buds; and studied plants and their capillaries. Malpighi was a progressive: as a professor at Pisa, he frequented Galileans, and he fled for a while to the Sicilian city of Messina to avoid angry Galenists. When he returned to Bologna, he became a correspondent for the Royal Society. Eventually he was chief physician to Pope Innocent XII, as well as a professor of medicine in Rome. His correspondence is extant, and testifies to the radical changes that medical diagnosis at a high level underwent now that new anatomical structures and physiological operations had been found.

Malpighi was one of those whose intense investigations of the body, now envisioned as a part of the natural world and on a par with the bodies of animals, yielded insights that were central to the development of modern medicine. But new concepts and discoveries remained embedded within an old vocabulary, especially when it came to the ordinary, practical care of patients. Even for the most progressive physicians, there were few alternatives to remedies that had been in use for so long.

In the 1690s, a Bolognese gentleman and amateur physician, Giovanni Antonio Volpi, who mostly treated the poor—for charity—wrote to Malpighi regarding the case of a good friend of his, a man aged about fifty. The man was "of a robust complexion, plump, and sanguine, used to riding and prone to sweating continuously," and "began, some fifty five days ago, to suffer from an acute pain in the thighs and legs, especially in the evening." The pain then went to the "left shoulder and the ribs, with fever, difficulty breathing, sleeplessness, dryness, most acidic urine." Volpi's diagnosis was "acrid, serous putrefaction of the humoural mass," which, in this liquefied state, entered the nerves, "descending into the legs and feet," and causing the troubles elsewhere. Volpi had ordered some bloodletting from the arm, fearing that the lung might collapse; this relieved the patient, but only for a while; another phlebotomy followed (six ounces of blood were taken), and this time pain in the superior part of the body ceased. One foot was out of commission, but swelling in the face and feet went down once he was "regular" again; and he was sleeping better. But to stabilize the patient and encourage weight loss, Volpi was not sure whether or not, on top of the venison gelatin he was already prescribing, he should prescribe milk with a decoction of rose roots and quince, and whether it should be cow's or goat's milk; hence his letter.

Malpighi's response was that, as far as he could tell from this difficult case, the "fixation of irritating serum and salts" in the upper body hindered the motion of the nervous juices and therefore encouraged the tensing of the muscles in the lower body, all of which triggered "exhaustion and weight gain." The patient needed remedies that promoted motion and perspiration or excretion, unless there were signs of the "French malady" here; and also remedies that had oily particles able to nourish each part of the body. He would need to take the gelatin of horn of venison with food; and in the morning, a broth with ground pine, chicory, cowslip, cinquefoil, couchgrass and other ingredients, mixed with a few drops of sal ammoniac or terebinth if urine was excessive. Once the weight had stabilized, and with the better season, he could take donkey milk for a month, then a little cow's milk with some barley water. Hibiscus syrup

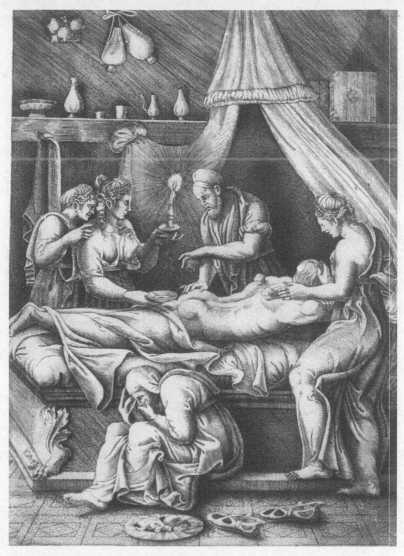

Fig. 24. Cupping scene: *Allegory of sickness,* engraving by Giorgio Ghisi, after Giulio Romano, 16th century (Bartsch catalogue: vol. 31, p. 410, print n. 63).

with sulphur balsam would help too, and he could soak the affected parts in camphor water. Culpeper himself would have recognized these treatments and might have approved of Malpighi's advice.[50]

Here again, high and "folk" concepts were intimately combined, although humours themselves were now background metaphors rather than explanations. More specific structures and functions were at the foreground of Malpighi's medical imagination, while Volpi still spoke of the "humoural mass." But even the top doctors like Malpighi had to make do with the remedies they knew, and, in illness as in health, the humoural pathways of the body seemed to remain, filled now with secretions, blood, inflammation, and irritation, rather than with the more general hot, cold, dry, wet qualities embodied by the humours.

This was also the case with the soul. As the centuries passed, it ceased to be the exclusive property of morals or churches. Just like matter, it was becoming the object of a new science and entering the brain, inhabited by spirits and humours old and new.

VI

Brain: Passions and Nerves

(The Making of Modernity: Seventeenth to Nineteenth Centuries)

—you have all heard, I dare say, of the animal spirits, as how they are transfused from father to son, &c. &c.—and a great deal to that purpose:—Well, you may take my word, that nine parts in ten of a man's sense or his nonsense, his successes and miscarriages in this world depend upon their motions and activity, and the different tracks and trains you put them into.

LAURENCE STERNE[1]

We believe the brain to be an organ whose internal equilibrium is always in a state of change—the change affecting every part.

WILLIAM JAMES[2]

I. Cartesian Humours

QUEEN CHRISTINA of Sweden, who had invited Descartes to Stockholm, liked to have her philosophy for breakfast. Swedish winters are cold, especially in the long dark mornings; and the winter of 1650 was one of the coldest ever. Descartes had been pleased by the royal invitation, but these strenuous summonses hardly suited a man who had never gotten up very early. He eventually succumbed to pneumonia—any humoural doctor could have predicted the risk. At his death, he was still firm in his belief that it would be possible one day for medicine to ensure long lives. The world that he left behind in 1650 had radically changed from the world he had been born into in 1596, and he had played no small part in changing it. He had split human beings into two; he had called bodies automata and destroyed scholastic psychology.

But not entirely: the old *pneuma* and spirits that had been part of the body's humoural network did not simply vanish. Nor would they disappear soon. Where once humours had flowed, particles now circulated. Mainstream physicians, meanwhile, did not change their practices. Those, like the Helmontians, who were opposed to humoural conceits and preferred homeopathic to allopathic treatments, were by and large the minority. In his treatise on human physiology, *Of Man,* published in 1644, Descartes had actually recycled the Galenic model into his mechanistic framework.[3] His method certainly was innovative: he resorted to a thought experiment, entreating his readers to imagine a "statue or earthen machine," which he set out to describe before moving on to the soul, and then to the matter of the connection between the machine and the soul. This was how he would explain how the soulless body worked. No forces were needed for its functioning, nor was willpower needed; mechanism alone sufficed.

Philosophers in our day have been known to resort to a similar thought experiment, imagining what they have called zombies, creatures like us in all ways but devoid of consciousness. Descartes created his own experimental zombie. But the details of its functioning were rather derivative.

The animal spirits, as he saw it, ensured the body's movements, traveling from the arteries to the nerves and muscles via the "pores" of the brain, causing us to perceive, feel, move, and remember, just as they had with the Galenics. As Descartes had also explained in the *Discourse on Method,* which appeared in 1637, the spirits were propelled by the heart's heat and changed according to the blood's composition, determining all passions and bodily states. The movement of the blood was a product of heat and fermentation and was the cause of the heartbeat, itself caused by the heart's heat, which was "a fire without light," as he put it, rather than Harvey's pumplike muscle. Had it been a muscle, Descartes explained, the animal spirits it produced would have had to travel to it in order for it to function; and this clearly was impossible. Once in the brain, the animal spirits passed through the ventricles and entered the pineal gland, or *co-narium,* from where they kicked into action. Via the nerves, they triggered the muscles' flexion and distension, causing the body's movements; and via the same nerves, they affected the brain according to the part of the body they were coming from, causing in this way all sense perception and all sensation. This was how one was aware that a feather was tickling one's cheek, for instance, or that one's left foot hurt.

This was old hat, not revolutionary: Descartes's physiology differed little from what had been common currency in the Renaissance. This was also the case with extreme Cartesians, radical dualists like the priest Nicolas Malebranche, who could write in his magnum opus, *The Search after Truth,* about "the natural and mutual correspondence of the soul's thoughts with the traces in the brain, and of the soul's emotions with the movements of animal spirits."[4] What was new in Descartes was that he compared the body to a machine, such as a clock, and represented the various cords that connected its parts to the brain in terms of a perfectly conceived mechanism—a notion that was quite fashionable at a time when automata of all sorts were being created and exhibited at courts and in public places. In Descartes's "earthen machine," the agitation of a cord in the foot, say, triggered an agitation at the end of the cord, as occurs when one rings a bell. Awareness that all this was happening was due merely to the presence of the rational soul in the pineal gland. But the physiological operations that took place inside the body that Descartes imagined (rather than observed) did not otherwise differ from what others had imagined before him.

His dualism even ceased to be of any use to him when he finally ventured into the realm of psychology and wrote about the passions. Although he produced comparatively little on the subject, it was one of concern to him: he saw passions as inalienably human characteristics. By knowing them, one sharpened one's moral conscience. Rational self-knowledge, and in particular the awareness of one's emotions, served the cause of virtue. Descartes developed such ideas in the epistolary exchange he had with the Bohemian Princess Elizabeth: highly intelligent and fretting in her political exile in The Hague, she looked to the best minds in search of some sort of moral clarity. She arguably was the one who entreated Descartes to think explicitly about the emotions and the humoural flows underlying states of mind—for the purpose of living a morally better life. The treatise, *Passions of the Soul,* which he did end up writing, and which was published the year before his death, in 1649, is really an exercise in moral psychology, surprisingly similar in style to the sorts of treatises on the passions that had been already in vogue for about a century.

2. The Physiological Self

SUCH TREATISES on the passions were guides to self-analysis, self-help books of sorts; and they became increasingly popular during the sixteenth and seventeenth centuries. They traced the genealogy of emotions and indicated how to use reason to curb their excesses—in line with an even older tradition of moral psychology based on a combination of Aristotelian ethics and Stoicism, one that Galen had picked up in his own treatise on the passions. The role of these books was to examine what exactly happened—what sort of experience took place—when one felt upset, joyous, or angry. The point was to help mitigate the effects of passion and to encourage the good use of reason for the sake of self-improvement. Definitions of passions and prescriptions for their control helped readers identify, and, if need be, undermine their effects. The contents of such a work would therefore typically contain chapters on the body's physiology; our impulses, sufferings and pleasures; the nature, effects, and causes of love, hate, desire, anger; and the ways in which passions distorted reason, and reason modified passions.

An emotional state was not only a pathology, curable through recourse to medicine and pharmacopoeia. But it was that, too. Ever since Hippocrates, and especially Galen, there existed remedies to counteract the excesses of some passions, especially those attributable to melancholy. The humoural underpinning of our emotional life was always there. The experiences of joy, pain, anguish, and fear each had their temperature, their match in some sort of stuff in the body whose motion modulated the emotion. Scholastic psychology had been a solid framework that allowed for the integration into an organic whole of bodily functions, passions, sensations, cognition, reason, and volition.

In a treatise published during the Renaissance, in 1538, *De anima et vita beata,* by the sixteenth-century Spanish philosopher and humanist Juan Luis Vives, there was a description, for instance, of how

> a mere commotion of our fantasy bearing some resemblance to an opinion or judgment that a given object is good or bad, is enough to disturb our soul with all emotions: we fear, rejoice, cry, feel sad. This is also why our emotions seem to converge toward that part of the body where the fantasy prevails, and also why we will actually attribute bodily qualities to emotions and call them warm, cold, dry, or a mixture of those. Internal and external causes can exacerbate and repress the influence of our bodily temperament. Among the internal causes we find the emotions themselves: sadness makes us cold and dry, joy makes us warm and wet. Emotions both reflect and contribute to the temperament of the body.[5]

Some fifty years later, Timothy Bright described melancholy from without and from within—from the medical perspective, in terms similar to those of Vives, and from that of the suffering subject, in terms proper to those of a religious man. And indeed, from the outside in, from the perspective of the clinician looking for symptoms of humoural disturbance, it looked like a pathology; whereas from the inside out, from the perspective of psychological observation or self-analysis, it was a mode of apprehending and inhabiting the world, an aspect of our cognitive makeup. Emotions, then, were ways of knowing the world.

But they were also a way of distorting it. Strong emotive states could even be the work of the devil. Symptoms that might resemble those of melancholy (including lycanthropy, the belief that one was a werewolf) could be signs of possession: the wildly popular *Malleus maleficarum,* which had appeared in 1485 with the blessing of Pope Innocent VIII, had argued just that. It described madness or delirium in terms of sinful demonic possession, and for centuries it would serve as a justification for countless forms of Inquisitorial and popular witch-hunting. In 1562, the physician Johann Weyer had argued in his *Of Deceiving Demons* that possessions were often cases of melancholy. But the debates over sorcery went on into the seventeenth and eighteenth centuries. Even then, they would often turn on whether the accused should be considered stricken with the pathology of religious melancholy or possessed by supernatural forces. In 1633–1634, there would be just such a debate—and a trial— concerning the seemingly "possessed" Ursuline nuns at Loudun in France, in the wake of a plague epidemic there.[6]

The drive to persecute is itself a passion, and someone like Timothy Bright would have considered the obsession with demons a case of religious enthusiasm. He was religious, not dogmatic. His guide to the passions had profoundly theological inflections, insofar as he was concerned with the acedic, sinful aspect of melancholy, and he had sober, humane reasons for arguing that a balance between states was optimal not only medically, but also morally.

The difficulty of finding such a balance had been recognized by all those who, melancholically perhaps, were driven to ponder the matter of the relation between emotion and reason. An example in point is Thomas Wright, an English Jesuit who was periodically under some sort of house arrest during the reign of Elizabeth I and was banished from England when James I came to the throne. He had occasion enough to think about the passions, and he suggested in his definitive *The Passions of the Mind in General* (1601), that

> we may compare the Soul without Passions to a calm Sea; with sweet, pleasant, and crispling streams; but the Passionate, to the raging Gulf swelling with waves, surging by tempests, menacing the stony rocks, and endeavouring to overthrow Mountains; even so Passions make the Soul to swell with pride and pleasure; they threaten wounds, death and destruction by audacious boldness and ire; they undermine the mountains of Virtue with hope and fear, and, in sum, never let the Soul be in quietness, but ever either flowing with Pleasure or ebbing with Pain.[7]

Wright thought passions were "perturbations of the mind," which we shared with beasts, "bordering upon reason and sense" but emanating from our sensitive soul; that they were called passions "because when these affections are stirring in our minds they alter the humours of our bodies, causing some passion or alteration in them"; and that they were perturbations because "they trouble wonderfully the soul, corrupting the judgment and seducing the will, inducing, for the most part, the vice, and commonly withdrawing from virtue; and therefore some call them maladies or sores of the soul."[8] He was interested in their processes, and wrote, rather lucidly, that while passions engendered humours, humours also bred passions. He thought that melancholy, for instance, was first a passion that engendered the humour, and that anger was first a humour that triggered a passion, as occurred when one was exposed to the color red, believed to inflame the blood.[9]

Yet reason was as close to passion as were the senses, so much so that it could end up laboring in the service of sensuality: witness the arts of cooking, tailoring, building, wrote Wright. (And indeed, cookery, fashion, and design are still the triad at the heart of newspaper style sections.) Passions, which were seated in the heart, could also "seduce the will," bribing it with pleasure by means of the wit. Moderate pleasure brought health, "because the purer spirits retire unto the heart," which then engendered more numerous, purified spirits. An overwhelming pleasure, on the other hand, could cause "great infirmity, for the heart, being continually environed with great abundance of spirit, becomes too hot and inflamed," and so produced "choleric and burned blood."[10] Other passions, such as sadness and fear, could also have negative consequences in their excessive states.

3. The Sensitive Soul

THE SENSITIVE SOUL was complex, precisely because imagination, sense, and common sense were intertwined within it. Reason was potentially powerful enough to steer the passions, given good discipline and a thoroughly religious disposition; but it was not exactly in opposition to them. This was the sort of scholastic psychology that really did offer an account of why we were aware of our emotions and affected by them, why the body's motions affected the soul, and why the soul's motions affected the body. It had always been a truism, among humouralists and anti-humouralists alike, that the mind had an impact on the body, that passions and perturbations could cause bodily distress and serious disease, that sadness and indeed melancholy could precipitate even death, and that the very fear of illness—even of the plague—could cause its symptoms.[11] We were fundamentally psychosomatic creatures.

So when Descartes got rid of the middling, corporeal soul, this picture could have changed radically. But it did not. The scholastic picture was a more realistic one than Cartesian dualism, which had no noticeable impact on either psychology or medical practice. Six years after Descartes's death, during the outbreak of plague in Rome in 1656, Matteo Naldi, a physician to the Pope, could thus argue—wrongly but suggestively, and in line with mainstream beliefs—that "the perturbations of the soul and of the blood, either because of their affection for those who die first, or because of the fear of having received the contagion from the latter, or because of some other trauma; these perturbations usually give rise to the progression of the disease, as a kind of final trigger, like a little spark from a stone landing in kindling that is all ready to catch fire."[12]

Descartes himself would not have disagreed with this view that illnesses of the body were connected to movements of the soul. He also knew that it was because of the union of mind and body that guides to moral life were needed, and his treatise on the passions was as devoted to the cause of ethics and virtue as its predecessors had been. Like Vives and like Wright, he readily acknowledged the role of emotions in a conscious, ethical life. He even stated in the *Principles of Philosophy* that he regarded ethics and medicine—along with mechanics—as the topmost branches of the tree of philosophy, whose roots were metaphysics and whose trunk was physics. And he made explicit room within his *Passions of the Soul* for an intense interaction between emotion and reason, between bodily processes and consciousness, all in the belief that only by knowing one's passions could one steer them properly and prevent reason from being enslaved to them.

Here too he recycled the physiology that had sustained scholastic theories of psychology. While the ethics came from Augustine, the psychology came from Thomas Aquinas. Throughout, Descartes held on to humoural conceits, and, just as in the *Treatise on Man,* to the animal and vital spirits that accompanied them. Of anger, for instance, he wrote that it was an "agitation of the blood" that arose out of the combination of desire and self-love and was caused by courage; in the case of hate, the agitation was transmitted by "bilious blood from the bile and the little veins in the liver" which "entered the heart, where, given the blood's abundance and the nature of the bile it was mixed into, it excited an ardent heat stronger still than the heat provoked by love or joy." Emotions—all of which he thought were derived from six "primitive" ones, admiration, love, hate, desire, joy, and sadness—were a matter for moral conscience, but they were also at their most "clear and distinct" in their humoural guise.

Of course Descartes did want to show that all this activity was "just" mechanical, that consciousness was a function of the immaterial soul, that without the disembodied *res cogitans*, we were just well designed organisms, zombies, unaware, incapable of experiencing, conceiving, remembering, or emoting meaningfully. But since we are conscious of ourselves and are not zombies, this sort of mechanism is bound to remain a mere thought experiment, and could never correspond to reality anywhere but in science fiction fantasies. Descartes had wanted to turn emotional cognition into a disembodied activity, but whenever he broached ethical discourse, he actually threw it back, complete with its animal spirits, into the old fold bequeathed by Galen. In spite of himself, he was unable to divorce ethics from medicine. Reason and physiologically mediated emotion remained intimately, humourally connected. That was why we were moral, creative creatures at all.

Dualism was a philosophical mistake—one that Plato had committed too, and that Christianity had inherited from Augustine onward. But in the end, even for Descartes, the notion of a man-machine would have been abhorrent, as "inconsistent with our being" as magnified worlds later seemed to John Locke. It was a very successful mistake, though; it tapped into our need to represent the body to ourselves as a machine in order to understand it. Certainly, it was a way of mediating self-knowledge, building on our tendency to conceive of our bodies in the possessive: we talk about "my hand," "my back," "my eyes," as if, like a handbag or a pair of shoes, they merely belonged to the separate entity that was the self, rather than constituting the self. Descartes presented a recognizable map to the mind, flattering our deepest intuition as thinking and speaking creatures that our thoughts are disembodied, separate from the humoural beings that we are. He also played an important role in recycling the Galenic physiology for service within a dualist, rationalist, mechanistic world. But our emoting and even our moral selves are physically embodied; and whether he liked it or not, Descartes himself knew that.

4. Introducing "Neurologie"

WHILE DESCARTES, DEFTLY enough, endeavored to square the circle that was his overly systematic program, other physicians and natural philosophers embraced alternatives to his harsh dualism in their quest to understand the nature of the soul. One system that seemed to make much more sense of the soul was Gassendism. Gassendi had made it possible to reconcile naturalism with Christianity: the soul, or *anima*, did not have to be a disembodied entity. It was material; it corresponded to the operations of the mind that the Scholastics had called sensitive and appetitive (or corporeal and vegetative) and that it was the mission of natural philosophers to study. And there was still room in this system for the existence of the *animus*, the immortal rational soul which partook of divinity, and whose study was the safe ambit of theologians.

Thomas Willis was one of the natural philosophers who followed Gassendi in defining the soul as an "igneous fire" in the actual blood. He was also the man who eventually coined the term "neurologie" to designate what he called the "doctrine of the nerves."[13] In the late 1650s, in the course of his research on fevers, Willis, who was a trained, practicing physician in Oxford, had already noticed that a fever could cause a patient to lose the capacity to speak rationally or coherently. He surmised that the brain might be a chemical (or alchemical) instrument, a "marvelous alembic" that distilled the spirits out of the blood. And he realized that only by looking closely at the brains of dead patients might he begin to understand their minds.[14]

The brains Willis dissected turned out to differ so widely from Vesalius's descriptions that he decided to undertake a complete, new anatomy of the brain and nerves. The ambitious project took shape over a few years. Thanks to a chemical invention of Boyle's, brains could now be preserved: it had become possible to study the brain without it decomposing before one's eyes; and it was possible to actually see, for instance, that ventricles were empty spaces, not filled with the faculties imagined by the

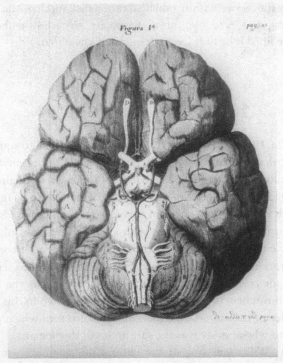

Figura 1ᵃ

Fig. 25. Christopher Wren, an illustration for Willis's *Cerebri anatome* (1664): arteries.

Scholastics. Willis found that the blood and its spirits circulated within the microstructures that Steno had identified in the white matter of the brain, entering through the arterial anastomosis at its base—what Galen and Galenists had confused with the *rete mirabile* and is now called the circle of Willis. He distinguished the brain's gray from the white substance, and was the first to correctly identify cranial nerves. He recorded all this and more in *Cerebri anatome,* which appeared in 1664 and had striking illustrations by the brilliantly versatile Christopher Wren. The book was a resounding success, and after its publication the now celebrated physician moved from Oxford to London, where he counted the great and the good among his patients.

Willis avowedly dismissed humours. He did not believe, for instance, that a "Melancholick humour" had any role to play in melancholy, which was "a complicated Distemper of the Brain and Heart." Instead, melancholy could either proceed "from the vice or fault of the Brain, and the inordination of the Animal Spirits dwelling in it"; or, when the afflicted became "very sad and fearful," it should be "attributed to the Passion of the heart." It could be either a cognitive malfunction or a matter concerning the relation of reason to passion.[15]

But despite this rejection of old beliefs, the nervous system he depicted did remain full of concoctions. Out of the arterial blood, the "animal spirits" entered the cortex in both cerebrum and cerebellum, which he found were connected with each other. Perception and sensation traveled through the cranial nerves and the corpus striatum, he thought. Actions and sensations depended on the spirits' flow through the nerves, which he still conceived to be hollow; and emotional intensity was a function of the amount of blood that the arteries let in, regulated as they were by the nerves. Willis understood, correctly, that the cerebellum was responsible for involuntary motion such as the heartbeat, but also that higher functions such as imagination and language were processed in the cerebrum; and he housed the rational soul in the corpus callosum, where, he wrote, it responded to the spirits transmitted by the sensitive soul to its perceptions and sensations.

These spirits in turn could change thought processes, moods, and emotions. In cases of delusions, for instance, distorted spirits drew distorted paths in the cerebrum. The psychosomatic organism Willis described was not so different from the one described by humouralists: he was recycling the vocabulary of spirits just as Descartes had done. Willis too matched mental functions with movements of substances within the brain, and imagined that our mental capacities were due to complex concoctions. Of course liver and heart were no longer places of humoural secretion. But when it came to practice, Willis bled, cauterized, and purged his patients as his predecessors had. His medical care was admittedly, and famously, gentler than theirs; but it was, by and large, just as humoural.[16]

The gap between his research and his clinical activities did not seem to bother him much, however. The old and the new coexisted, and it was still in this spirit of synthesis that he went on to develop his exploration of the sensitive soul in a book that appeared in English in 1683, *The Soul of Brutes, Which is that of the Vital and Sensitive of Man*. He knew, as Steno did, that the human soul "understands all things but her Self," as he wrote. But he nevertheless wondered, "in this Age, most fruitful of Inventions . . . why may we not also hope, that there may be yet another disquisition concerning the Soul, and with better luck than hitherto? Therefore, however the thing may be performed, I shall attempt to Philosophise concerning that Soul at least, which is Common to Brute Animals with Man." Beasts had souls just as they had (or were) bodies; moreover, it was only by practicing "Psychology," by studying the corporeal soul common to animal and man, dependent on the body and essential to its life and functions, that one could differentiate it from the "Rational Soul, Superior and Immaterial."[17]

Willis needed to divide the soul into the corporeal one he was studying and the rational one that lay beyond the realm of scientific investigation. This was why Gassendi's view that the sensitive soul and the rational soul were continuous with each other was much more useful to him than Descartes's dualism could ever be. It was also why he followed Gassendi in the belief that the intercostal nerve connected brain and heart, innervating the organs of the hypochondria, the diaphragm, and the face: this nerve, Willis thought, accounted for the uniquely human ability to control passions with reason.[18]

Steno was one of the many who had admired Willis's *Cerebri anatome;* but he did find mistakes in it, and he was rather more skeptical about the possibility of localizing functions in the brain. He even thought that Descartes had been right to resort to his "earthen machine" to explore the organism, since it was a way of discussing cognitive functions without presuming any knowledge of the body. Willis, on the other hand, was fairly optimistic about his science and believed that by stretching the old tradition of comparative anatomy to the realm of the brain, one would eventually be able to correlate mental functions with cerebral parts. What is significant is that these functions still included Galenic animal spirits, which he thought were a "fiery substance," produced "in the Cortical or Barky substances of the Brain and Cerebel" before they descended "by and by into the middle or marrowy parts, and there are kept in great plenty, for the business of the Superiour Soul."

This "Superior Soul"—higher reason—remained beyond the realm of study. But it is actually difficult to understand how separate Willis thought he could keep it from the functions pertaining to the old sensitive soul and its spirits, which he studied so assiduously. He was quite sure that the cerebral pathways he was able to trace accounted primarily for the sensitive, corporeal soul, and that his research did not downgrade the higher, immaterial, contemplative immortal soul so necessary to religious sensibilities—the conscious mind that, still today, may seem to elude its own grasp. He did avoid being branded an atheist for wanting to study the material nature of our mind. Indeed he never got into trouble; he was embraced by both Church and state, and would be buried with pomp at Westminster Abbey, no less. But in the end, Willis was all the same engaged in naturalizing the mind. Slowly but surely, the foundations for believing in a higher, rational soul were crumbling.

5. Enlightened Thinkers, Old-Fashioned Doctors, and the Embodied Mind

WILLIS WAS NOT ALONE in turning the study of the mind into the study of the brain. Materialism and even atheism were on the horizon now, often still in the form of Gassendi's atomist philosophy. In France, Claude Perrault (brother of the more famous Charles, author of children's fairy tales), an established architect, theoretician, natural philosopher, natural historian, and physician, was asking questions similar to those of Willis. So was the anti-establishment materialist physician and natural philosopher Guillaume Lamy, a vociferous participant in the debates surrounding the Antoine Mauroy case of blood transfusion and an irate *libertin* whom even Pierre Bayle called an "extreme Epicurean." Lamy saw no point in looking for divine intention in nature. As he wrote in his *Discours anatomiques,* which appeared in Paris in 1675, matter was endowed with motion by nature, not God, and the parts of bodies "were formed by the blind necessity of matter's motion," constrained to be what they were from the particle level upward, "just as the sum total of three dice must figure between three and eighteen."

And so it was, he wrote, echoing the Roman poet, Epicurian, and atomist Lucretius[19] and turning Galenism on its head, that "one should not say that eyes were made for seeing but that we see because we have eyes." We were evolved creatures, just as those animals that were "equipped with feet walked, those with wings flew, those with neither feet nor wings swam in the sea or crawled on the ground, those with teeth chewed, the strongest or most agile became masters of the others, in such a way that there is no need to look for ends in those kinds of principles." As for humans, they were too imperfect to be what Galenists had supposed—the culmination of God's work. Such a God, furthermore, would be inappropriate, Lamy added ironically, for it was unlikely that he should have exhausted all his resources in the creation of man. The notion of an "intelligent design" was nonsensical. The "Author of nature" had labored entirely for his own pleasure, creating matter and particles, whose motions produced an infinitely varied array of shapes, including man. And all the natural philosopher could study were the particles that composed matter.

Matter included the soul, which was housed in the brain, "where it exerts its most noble functions." Brain injury, for example, usually went along with disorders in mental functions; and it took just a little opium, or a few glasses of wine, for the soul to follow in the body's weakness. Since mind and body interacted and affected each other, the soul must be material in some way. Lamy explained this further in another book he published in 1677:[20] what we perceived with our senses was communicated to the brain by these flowing bits of soul. The functions of the *sensus communis* were identical to the various impressions "caused in the soul by the action of objects" on the animal spirits in the nerves. Different sorts of matter corresponded to different bits of soul.

Like Willis, Lamy claimed to be interested merely in the old sensitive soul, replete with its animal spirits. But he went farther than Willis, because he knew that he was in fact talking about all our cognitive faculties— about what allowed us to be aware of our feelings, memories, and dreams, of the difference between dreams and wakefulness, of the sense of time and so on. Lamy and his fellow *libertins* were then still somewhat marginal, but their presence was symptomatic of the revolutions that the corpuscularian and mechanistic philosophies were breeding, potentially at least.

At the same time, the debates over the minds of animals were more active than ever, and the problems triggered by the experiments conducted on the brain by Willis and others were beginning to appear in a guise that might seem familiar to us: if one could explain the soul mechanistically, what were humans exactly? How did they differ from animals? In a way, humours had been inoffensive: they had accounted for our psychology, nothing more. Humouralists had not meddled with immortality. But now psychology itself was expanding, in part via the new "science of nerves," which Willis had called "neurologie"; and its territory was beginning to encompass what had been the domain of morals and theology before.

Jonathan Swift absorbed these lessons and used them for his own satirical purposes. In 1704, he published *A Discourse Concerning the Mechanical Operations of the Spirit,* an essay in which he sharply mocked religious enthusiasts. He reduced their mystical experiences to biology, to the result of the "perpetual motion of *see-saw*" combined with half-closed eyes, collective incantations, and tobacco that tricked them into a trance. It was still a little *risqué* in 1704 to describe religious revelation in such terms. Given the extreme unorthodoxy of this text, Swift had it appear in the guise of an anonymous letter (he was a consummate user of authorial masks) in which he went on to write how "you may observe their eyes turned up in the posture of one who endeavors to keep himself awake; by which . . . it manifestly appears that the reasoning faculties are all suspended and superseded, that imagination hath usurped the seat, scattering a thousand deliriums over the brain."[21]

It was clear to Swift that "the seed or principle which has ever put men upon visions in things *invisible,* is of a corporeal nature; for the profounder chemists inform us that the strongest *spirits* may be extracted from *human flesh.*" And since the spinal marrow was "a continuation of the brain," it created "a very free communication between the superior faculties and those below."[22] Physicians were right to say that "nothing affects the head so much as a tentiginous humour, repelled and elated to the upper region, found by daily practice to run frequently up into madness." Quakers had seen many female patients suffering from uterine fury. One should not be fooled: "However spiritual intrigues begin, they generally conclude like all others; they may branch upwards toward heaven but the root is in the earth."[23] The business of flesh and blood was not contemplation but "matter"—and not only for women. Willis had explained hysteria as a convulsion of the brain and nerves, freeing it from its old etymological and physiological association with the uterus. Although Swift did not refer to this, it would have made sense now to understand religious melancholy as merely an instance of hysteria.

As more and more details about the body's physiology and anatomy emerged, all-embracing theories that would account for everything about it, from illness to madness, were proliferating, taking the place of the integral, humoural organism. A "student of physick" named Thomas Tryon, in *A Treatise of Dreams and Visions,* published in 1689, called Galen's humours "Forms and Words, rather than Realities." He proposed that mental diseases arose "either from Irregular passions of the Mind, or poysonous ferments, occasioned by ill Dyet, or improper Physick in the Body." Physiology held all the answers.[24]

But Tryon himself was merely replacing words with other words. Humoural conceptions did not have to be discarded. They were simply being replaced by their physiological equivalents. Where once there had been a general structure in which the whole organism fit, now there would be even more talk of nerves, fibre, tissue tone, irritability, electricity, and soon cells, which were increasingly studied from the eighteenth century on, and whose ubiquity as the constituent elements of all living beings would be understood by 1838.[25] It was becoming possible, in *metaphysical* terms, to forgo talk of immortal souls. Humours, though, were still useful *psychological* models, and there was conceptual room for them in this world where mind was turning into matter. The problem was that they were also still sustaining what already were theoretically outdated medical practices. The best physicians—of which Willis was one—bled, cauterized, purged, prescribed brutal enemas, dangerous arsenic, or, as ever, the deadly hellebore that the Greeks had used as a supposed cure for madness. Urine, stool, and pulse were still the central tools of diagnosis, as if time had stopped after the medieval Galenic treatises.

No wonder physicians had a bad reputation; and in this atmosphere of confusing medical pluralism, no wonder members of the establishment were on the lookout for the quacks and charlatans who practiced humoural treatments, now considered too ineffective for their unpleasantness to be tolerated. It was increasingly easy to caricature measures whose potentially quackish "folksiness" had been disregarded until just a few decades before. In many ways, though, the distinction between what was established and what was "folk," or suspect, lay in the eye of the beholder. Molière, the star comic dramatist of seventeenth-century France, constantly poked fun at the doctors of the day, as in the early play *Le médecin volant,* for instance: "Hippocrates says, and Galen with much good reason convinces that a person who is ill is unwell. You are right to pin your hopes on me; for I am the greatest, most effective, most knowledgeable doctor there could be in the vegetal, sensitive, and mineral faculty."[26] Nearly a century later, in 1736, William Hogarth was satirizing physicians in his etching *The Company of Undertakers,* which depicts them closely examining urine or sniffing the tops of their canes, with, at the foot of the image, the mordant, rather cynical motto *Et plurima mortis imago* ("And many an image of death").

These caricatures corresponded to the sorts of doctors who literally bled George Washington to death and who practiced phlebotomies on women about to give birth. Opposition to change and resistance to discoveries were real. Quite a few mainstream doctors, for instance, were wary of the wildly innovative, highly effective practice of inoculation against smallpox. Lady Mary Wortley Montagu—traveler, woman of letters, and wonderful correspondent—had observed such inoculations in Constantinople in 1716–1718, while her husband was ambassador to Turkey; and she had successfully promoted the practice back home. Her own daughter was the first person in Britain to benefit from it, in the midst of a smallpox epidemic in 1721. Various aristocrats followed suit, as well as King George I himself, who inoculated his son, the prince of Wales. Inoculation became a fairly fashionable procedure, and inoculation centers were set up. But it was costly and impractical, requiring that subjects undergo long periods of preparation and recovery. It was also only partially successful. The Calvinist theologian and philosopher Jonathan Edwards, for instance, died of it in 1758 shortly after becoming president of what has since become Princeton University.

The rapidly accumulating discoveries in anatomy and physiology were clearly not easy to apply to the care of patients; in all ways, the apparatus for the leap from theory to practice was lacking. But these discoveries were progressively leading to new conceptions of the body and of the psyche. In time, they would also lead to new medical treatments of the physically and mentally ill.

6. Of Mechanism and Vitalism

I T IS COMMON to believe that the Enlightenment gave birth to an alarmingly mechanistic body from which all life had been drained out, that it spawned a disembodied reason presiding over a puppetlike Cartesian organism. It is true that mechanism reigned in physics, and that great leaps were made in part thanks to the calculations mechanism afforded. But this was not the only novelty produced by the intellectual ebullience of the early Enlightenment. In the realm of biology, mechanism was much less in contradiction with the old traditions than one might suppose. There is a tendency today to think of herbalism, for instance, as a return to pre-mechanistic sources, to old, "popular" knowledge. But medical traditions have never been fully unified at any one time. Official, "higher" medicine was never formally incompatible with bodies of empirical knowledge like herbalism.

And now that physiology and anatomy could no longer rely on the officially discarded humouralism, there was room for a multiplication of theories. What complicates matters is that humours persisted *de facto* among practitioners—along with the Lucretian-Gassendist version of the old idea of a soul within the blood. There also appeared an alternative to biological mechanism, a doctrine known as vitalism, which reconstituted anew the psychosomatic organism previously populated by old humoural substances. It was not quite so new when it emerged as such, since it grew out of the iatrochemistry of sixteenth- and seventeenth-century northern Europe—actually out of the anti-humouralism of Paracelsus and van Helmont. But over the course of the seventeenth century and beyond, it did develop as a tradition that allowed for a novel conception of the living organism, now seen as endowed with innate forces in the blood and nerves.

Biological mechanism certainly had its exponents. The seventeenth-century mathematician, astronomer, natural philosopher, and physician Giovanni Alfonso Borelli had been a student of Benedetto Castelli, an associate and defender of Galileo. And Borelli is remembered for the ingenious way he applied Galilean mechanism to biology, exploring the nature of movement and recounting the results of his numerous dissections in his *De motu animalium*. A member of the Florentine Accademia del Cimento, under the patronage of Prince Leopoldo de' Medici, Borelli had worked with Steno on muscles; and he had concluded that animal spirits from the brain caused the fermentation of muscles when they mixed with blood and with lymph.[27] It was a perfectly materialistic way of accounting for voluntary movement. Borelli was a progressive: he is also remembered as the man who took the heart's temperature with a thermometer, showing how erroneous it was to believe that the heart was endowed with an "innate heat," as many still did, including Descartes.

Vitalists took as keen an interest as the mechanists in such functions as voluntary motion and respiration. But their emphasis differed; rather than pursuing the investigation of discrete processes in a machine-like body, they wanted to account for the life of the organism as a whole. They were precisely after the innate forces that mechanists wanted to discard. The most significant exponent among the vitalists was Georg Ernst Stahl, a chemist and physician who eventually served the king of Prussia, Friedrich Wilhelm I, and was also a professor in the Prussian city of Halle. As a chemist, he is remembered for advocating the wrongheaded phlogiston theory, according to which the unknown element in the air that caused combustion—what the Oxford "virtuosi" had never discovered— was an element called phlogiston. He thought it was constitutive of all metals, together with calx, and was released into the air when any given substance was burned. When phlogiston was consumed, he argued, flames died down.

The idea took a while to die under its own weightlessness, as it were. In the second half of the eighteenth century, the chemist and minister Joseph Priestley would hold on to it despite the numerous objections he faced from the so-called "antiphlogistians" and the great chemist Antoine Lavoisier—who would be the first to understand that combustion was instead dependent on the gas oxygen. Priestley even wrote, in 1796, some *Considerations on the Doctrine of Phlogiston and the Decomposition of Water.* By that point he had few allies for his cause. But he was nevertheless an excellent chemist, who tried to accord his deep religiosity with his analysis of matter, offering as a plausible hypothesis the notion that "the Divine Being, when he created matter, only fixed certain centers of various attractions and repulsions."[28] In other words, we were at once fully material and divinely created. It was impossible to think "without an organized body," and our divine origin did not entail that we should have immortal souls. In a sense, life itself was intrinsically divine.

Stahl had believed something very similar. In his view, all matter was infused with a sensitive soul. Without the soul, it was inert. He, too, thought of the body's chemistry in metaphysical terms, expounding his views most famously in his *Theoria medica vera,* published in 1708. As a physician, he had studied fevers and the blood, taking on board Harvey's legacy along with that of Willis and of Malpighi. He had found that, given how potentially lethal fevers could be, they must be due to an overheating or inflammation of the very substance that enabled life. Since the blood was the vital substance of the body, fevers were probably due to blood. While Willis had believed that fermentation or an agitation of the spirits was responsible for the blood's innate heat and for febrile states, Stahl now attributed them to the heat produced by the—mechanical, rather than chemical—friction of fast-moving blood within the body's tissues. But he also agreed with Van Helmont's powerful insight that fevers were attempts by the body to expel the causes of disease. The living body naturally tended to ferment and putrefy, and health was a matter of keeping this tendency in check through what Stahl called "tonic motion," a notion not at all unlike the humoural *krasis*.

Here there were no longer pure or corrupt, balanced or excessive humours: instead, the blood itself was in charge of filtering out toxins and nourishing the organs, and it was this filtering activity that, given the blood's appropriate fluidity, produced humoural secretions. But both Hippocrates and Galen might have recognized the idea as plausible. All these movements of fluids, these shifts in temperature, these secretions and excretions seemed to belong to the humoural body that had been imagined in antiquity. They were innovative, of course, insofar as they were gradually divulging the nature of the nervous system, of sense and motion (as well as reproduction, which was intensely studied during this period). But they were not fundamentally incompatible with the broad, all-embracing humoural model. One can understand how these theoretical shifts were not especially designed to change the way clinicians practiced medical care.

The innovations of the Enlightenment were intertwined with philosophical reflections regarding the nature of life and the relation between body and mind. It was impossible to study the body anew without thinking again about the place of the soul in an increasingly active, decreasingly teleological natural realm. Stahl was an animist, or panpsychist, in the anti-Aristotelian tradition of Girolamo Cardano, Giordano Bruno, and Tommaso Campanella, attached as he was to the idea that spirit, mind, or soul inhered in matter. For Stahl, the Cartesian automaton could not have displayed any "tonic motion," because what ensured the balance in the body's vital force was a kind of soul. Nerves were not hollow, and they did not need to be: the soul was not localized in the brain or in any one place, but existed throughout the body. Without this soul, the body would decompose—it would not have the appropriate functions; nor would it be in the physical state that enabled it to fulfill these functions.

The idea was attractive. But it was nonsense to the Dutch chemist, botanist, and physician Hermann Boerhaave, a professor in Leiden who took from Willis the idea that the brain presided over the body via hollow nerves.[29] Boerhaave was considered the most important physician in Europe in his day. (Samuel Johnson wrote his *Life* for a series of issues of *Gentleman's Magazine* in 1739.) An admirer especially of Hippocrates and Sydenham, Boerhaave was interested primarily in understanding the causes of illness; and like them, he thought it best to investigate the body and the brain, while leaving matters of the soul to religion, not medicine.

7. Nervous Juice

ONE OF BOERHAAVE'S star pupils was the Swiss-born Albrecht von Haller, himself an impressive polymath, eventually a professor of surgery, anatomy, medicine, and botany at the University of Göttingen. Haller was even more of a mechanist than his teacher, whom he is said to have admired as the greatest chemist and doctor since Hippocrates.[30] He researched nerves extensively and found that they were not the cords Descartes had described, and that they were excitable. Nervous states, in other words, were a function of the excitability of the nerves. Muscles, for their part, did not depend on the nervous system for their motive functions and contractibility, but on what he called their intrinsic *irritability*; and it was their intrinsic *sensibility* that made them receptive to external stimuli. No soul at all was required for movement. The living body could be thought of in terms of local, mechanically generated action; and the mind was a function of the nerves.

The notion of irritability was useful to both mechanists and vitalists, and Haller was not the first to conceive of it. A century before, in the mid-1660s, an Englishman and fellow of the Royal Society named Francis Glisson had suggested that the secretion of bile itself was due to the organs' irritation. All organs and all tissue had an autonomous response to events external to them, and it was their irritability, their responsiveness, that enabled the body to live. Glisson was not a full-blown mechanist. But he had given some impetus to the rather anti-humoural, organicist and Van Helmontian idea that organs functioned according to their own autonomous processes and that physiology could be studied only empirically, by a focus on specific parts, rather than by attention to the whole body.

This version of vitalism would be successful for Stahl as well. In the 1730s, it would become popular in the originally mechanistic school of medicine at Montpellier, mainly through the effort of the botanist and nosologist François Boissier de Sauvages. (The term "nosology," from the Greek *nosos*—disease—was used from around that time, to denote the classification of disease.) By mid-century, the doctor Paul-Joseph Barthez had taken up the banner of vitalism, exerting great authority from his position as chair of medicine at Montpellier. Unlike other vitalists, he thought that the "vital principle," thanks to which the body was endowed with motor and sensitive forces, might actually emanate from without, rather than from within the organism—invoking in this way a divine home and origin for an immortal soul. But he also thought that our vital faculties were modifiable in each individual by climate, environment, or lifestyle. The Hippocratic echoes are not fortuitous: Barthez actually venerated Hippocrates as the true father of medicine. Vital forces were the new humours—and inversely, humours, in the minds of those who were loath to think of the body as a machine, were the old vital forces.

The Hippocratic Corpus was rich enough to accommodate these new theories, whether mechanistic or vitalistic, as well as alternatives to them. It befell another star pupil of Boerhaave, the Scotsman Robert Whytt, to suggest one of the most insightful alternatives, positioned somewhere between Stahl the vitalist and Haller the mechanist. In Edinburgh, he too studied the nerves and nervous disease, and he took on board Boerhaave's notion of an "illness of the nerves." All sensation was now due to nervous stimulus. The nerves that caused pleasure could also cause pain, and organs had different degrees of nervous sensibility.

Nerves replaced the old notion of humoural complexion. So one could have a tendency to delicate nerves, for instance; and just as a Hippocratic would have had it, here too, lifestyle could cause disturbances in the organism, which Whytt attributed to what he defined as the "sympathy" between organs and between the brain and each organ. Emotions and mental disturbances had physical manifestations because of this sympathy, including the old phenomena of hypochondria and hysteria—the one suffered by men, the other by women, whose intestine and uterus had a strong sympathetic connection. This was why anxiety, for instance, could cause stomach upset (it triggered the secretion of bile), or why fear and anger caused an accelerated pulse.

Whytt had identified what is still today called the sympathetic nervous system, but he carried quite far—into humoural land—his idea of sympathy, focusing especially on what humouralists had called the hypochondria. Galen might have recognized the physiological map Whytt was drawing, since he himself had discussed sympathy between visceral organs. Moreover, the remedies Whytt recommended for nervous disorders were those that had always been prescribed; their goal was identical to that of humoural prescriptions. They were geared at diminishing excessive nerve sensibility, at relaxing or strengthening particular nerves, and indeed, at rebalancing the whole system. The body was a whole, and Whytt famously disagreed with Haller's much less holistic, extreme mechanistic views over the nature of nerve stimulus. Whytt had tried hard to reconcile this holism with anatomical realities, wondering, as Harvey himself had wondered, where the heart's impetus began. On the basis of the study of cardiac nerves, he surmised that its energy was probably innate, and due to a "sentient principle" rather than to a "mechanical" process.[31] Similarly, although he noticed from experiments on decapitated frogs that reflex actions were connected to the spinal cord, he refused to describe them as entirely automatic. Even blinking and the heartbeat required some sort of overall sentience. Consciousness was not so simple; one could not reduce it to a set of well-oiled connections. There must be such a thing as a vital principle, without which sympathy would not exist.

In a sense, this was true; his "sentient principle" had explanatory power. A major step in the understanding of nerves would be the discovery by another eighteenth-century Italian, the Bolognese Luigi Galvani, of the electricity that ran through them. Galvani found that it was enough to touch the nerves of a dead frog with a metal instrument to trigger muscular contraction. He used an electrostatic machine, of the sort favored by Joseph Priestley, to conduct further experiments, and by the end of the century, in 1791, he published his *Commentary on the Effect of Electricity on Muscular Motion*. His work was important enough that a verb was made out of his name: to galvanize. Stahl's vital force existed, just as did Willis's animal spirits and humoural flows—in the form of an animal electricity that was secreted in the brain, circulated in the nerves, and provoked the irritable muscles to react.

Nerves, at that point, had become as metaphorically laden as humours had been, playing the role that humours had once played in determining health and temperament.[32] In the early 1700s, it was still possible to visualize the organism in hydraulic terms close enough to the humoural model, as Boerhaave had done: he had imagined that particles traveled within fluids through nerves, and that distortions in perception or even melancholy were due to changes in the texture of these fluids and movements of the particles. For Whytt, the nerves themselves had replaced these imagined contents as an explanation for sensation, motion, and perception, because he saw that they were not hollow. With Galvani, nerve function became clearer.

But in either case, the body was invading the soul. The very notion of mental illness was changing just as hypotheses were proliferating regarding the functions of the nervous system and the substances—fluid, solid, or evanescent—of which it was made. Disputes between mechanists and vitalists were particularly charged, since they ultimately concerned the thorny issue of the nature of the soul.

8. The Material Soul

DURING THE 1700S, materialism, along with its counterpart, atheism, had been gaining ground. Anatomy was developing fast. The impressively realistic wax figures collected from the late eighteenth century in the Museo La Specola in Florence, for instance, testify to that: they are three-dimensional *écorchés* that didactically but artistically divulge the body, from bones to skin via organs, veins, nerves, tendons, muscles. Artists and anatomists continued borrowing from each other. The immaterial soul was not swept away by the seeing eye and modeling hand, but it was a fragile presence in a world whose new visibility was celebrated, over and against the world of occult, hidden forces that the vitalists were recycling.

Boerhaave had suggested that the seat of the soul was just where the Scholastics had located common sense in the brain; it was innervated, and not ethereal. Another brilliant pupil of his, the physician and philosopher Julien-Offray de La Mettrie, would go even farther, claiming that matter alone could be studied, and that only physicians were in a position to understand what the soul was. He derided all philosophers, including Descartes, who had offered their merely *a priori* theories on the matter, and he took instead a radically materialist position, in a book entitled *L'Homme-machine (Man a Machine)*, which appeared in 1747. To an extent, it could be seen as a logical continuation of Descartes's own beast-machine thesis—although it was not one that Descartes would have welcomed.

Tab. VII

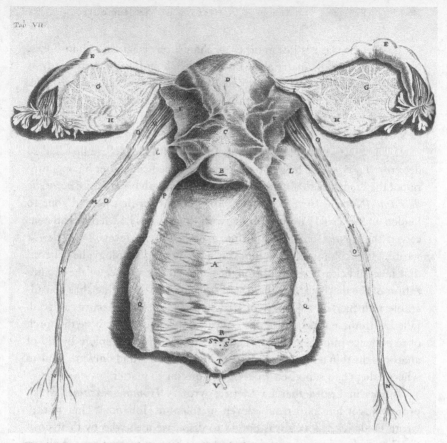

Fig. 26. Regnier de Graaf, *Opera omnia* (2nd ed. Lyon, 1678): *De mulierum organis generationi*, Table 7, p. 122. Regnier de Graaf studied the anatomy and physiology of the male and female human reproductive organs, as well as of the pancreas, and is an early pioneer of the anatomical shift that made possible the passage from old to new humours, from traditional humouralism to modern endocrinology. (Other contemporaries of his who partook in this shift include François de le Boe Sylvius, Regnier's teacher, and Marcello Malpighi.) Regnier still called ovaries "testicles," but he made innovative, precise observations, a sample of which is this impressive engraving representing the female, internal reproductive organs.

La Mettrie was not just another mechanist physician and philosopher. Nor was he the only materialist. But few people in Europe were ready at that point for his radical position, and his opinions got him into trouble. King Frederick II, however, warmly welcomed La Mettrie in Prussia and employed him as his physician. This was just three years before La Mettrie's death in 1751, at age forty-two—apparently from eating an expired pâté. The king wrote his eulogy, recounting how La Mettrie first had to flee from Paris, where he had been practicing medicine, after having infuriated the French clerical establishment in 1745 with his *Histoire naturelle de l'âme (Natural History of the Soul)*. It was then that he had gone to Leiden and received further training with Boerhaave. La Mettrie had conceived the treatise on the soul after a bout of high fever. As Frederick wrote, "For a philosopher an illness is a school of physiology; he believed that he could clearly see that thought is but a consequence of the organization of the machine, and that the disturbance of the springs has considerable influence on that part of us which the metaphysicians call soul. Filled with these ideas during his convalescence, he boldly bore the torch of experience into the night of metaphysics; he tried to explain by aid of anatomy the thin texture of understanding, and he found only mechanism where others had supposed an essence superior to matter."

It was in Leiden that La Mettrie wrote *L'Homme-machine,* which promptly got him into trouble even in tolerant Holland. "This work," wrote Frederick, "which was bound to displease men who by their position are declared enemies of the progress of human reason, roused all the priests of Leiden against its author. Calvinists, Catholics, and Lutherans forgot for the time that consubstantiation, free will, mass for the dead, and the infallibility of the pope divided them: they all united again to persecute a philosopher who had the additional misfortune of being French, at a time when that monarchy was waging a successful war against their High Powers." And so, since "the title of philosopher and the reputation of being unfortunate were enough to procure for La Mettrie a refuge in Prussia with a pension from the king" (as Frederick wrote of himself), La Mettrie "came to Berlin in the month of February in the year 1748," and was promptly welcomed as "a member of the Royal Academy of Science." There he wrote yet another provocative treatise, *L'Homme-plante* (1748). He had no problem with the notion that humans were on a continuum with animals, as well as plants. Reason itself was a part of material nature. It made no sense to define reason as an incorporeal substance.

La Mettrie is not remembered for his medical feats. But his thinking was definitely daring. His starting point was that man was a machine constructed in such a way that it was impossible to have a "clear idea" of it. Therefore, only by "trying to disentangle the soul from the organs of the body, so to speak," could one "reach the highest probability concerning man's own nature, even though one cannot discover with certainty what his nature is." Humouralists had done exactly that—they had looked into the body to understand the soul, at least to a degree. It was true, he wrote, "that melancholy, bile, phlegm, blood etc.—according to the nature, the abundance, and the different combination of these humours—make each man different from another." There was wisdom in the Galenic picture: it simply accorded with the fact that "everything depends on how the machine is running." (La Mettrie, let it be said, was another modern who resorted to the Galenic routine with his patients, bleeding and purging.)

After all, it took but a sudden illness, high fever, stroke, or accident to turn a genius into an idiot or an imbecile into a genius, and for painstakingly acquired knowledge to dissolve into thin air. Soldiers who had lost a leg still felt pain in the absent limb (we call this the phantom-limb syndrome today). "Merely an obstruction in the spleen, in the liver, an impediment in the portal vein" could turn a man of courage into a coward, because "the imagination is obstructed along with the viscera, and this gives rise to all the singular phenomena of hysteria and hypochondria." Our passions were such that "no calm drink or serene abode could soothe or give pleasure to a man devoured with jealousy or hatred." Similarly, it was hard to find rest during an attack of tachycardia (as one calls it today); and one's very will was rendered impotent by opium. Unsatisfied lust could kill the most modest girl; "a hot drink sets into stormy movement the blood which a cold drink would have calmed," and a good meal revived joy "in a sad heart."

Hippocrates, La Mettrie seems to have believed, had clearly been right to think hard about airs, waters, and places. Our physical context mattered, and, "Such is the influence of climate, that a man who goes from one climate to another, feels the change, in spite of himself. He is a walking plant which has transplanted itself; if the climate is not the same, it will surely either degenerate or improve." Quite simply, "the human body is a machine which winds its own springs. It is the living image of perpetual movement." There really was no need to speculate about souls and their seats. We could understand ourselves by looking at the matter we were made of. Guillaume Lamy had written something very similar, and certainly would have approved of La Mettrie's views.

La Mettrie himself was an admirer of Denis Diderot, a Jesuit-educated, anticlerical deist who—unlike La Mettrie—did not abolish the idea of God, but who was one of the greatest freethinkers of the era. His own *Pensées philosophiques* had been burned by the Parliament in 1746, along with La Mettrie's *Histoire naturelle de l'âme;* the members believed that La Mettrie had written both.[33] At once a philosopher, novelist, playwright, and art critic, Diderot certainly was a man who embodied, if anyone ever did, the highest Enlightenment ideals. He was notably the cofounder— with Jean Le Rond d'Alembert—and one of the main coauthors of the monumental *Encyclopédie ou Dictionnaire raisonné des sciences, des arts, et des métiers, par une Société de Gens de lettres.* Against difficult odds and in spite of logistical nightmares, this collective enterprise gave birth to the first modern encyclopedia. It consisted of twenty-eight volumes (seventeen of text, eleven of illustrations), published between 1751 and 1772, and is still an extraordinary source of information.

Diderot was the son of a cutler, and his fascination with artisanry, technology, and physics never ceased. A perusal of the *Encyclopédie*'s vast riches reveals the extent to which matter fascinated its main editor, but also the extent to which new ideas were entangled with their forerunners. The entry on *pierre d'azur* (lapis lazuli), for instance, is categorized not as geology but as *materia medica*—"(mat. med.)"—and describes the stone purely in terms of its purgative virtues, "from above and below," of its use "against melancholy, quartan fever, apoplexy and epilepsy." One reads that copper accounted for the stone's color and for its "corrosive, purgative, and emetic properties"; and that, since there were many better remedies that had exactly the same properties, it had until recently been "used only for the preparation of confectio alkermes" (a cordial based on the alkermes insect).

Entries such as *humeur* and *ténacité des humeurs; evacuant; sens internes; manie; sensibilité, sentiment; passions;* and *pouls* (pulse—an extremely long article) reveal the extent to which novel ideas were embedded within their older, still established forms, and in some ways growing out of them. There was much talk of blood circulation, lymph, nerves, irritability, inflammation, brain fibers, solid or fluid humours, microscopes, mechanism, sensibility. But past authorities were cited. Hellebore appeared as a good cure for mania; there was a list of the alchemists' aqueous solvents; niter and alkali were mentioned as remedies aimed at liquefying the humours; and one read of hot and cold, thick air and phthisis, animal spirits, and yet more purgatives. The attention to traditional knowledge in the *Enyclopédie* was also Diderot's way of covering up under the appearance of impartial erudition the radical program at the heart of his project. But for most people, tradition was still very much present during the self-proclaimed Enlightened days, alongside and in some ways underlying innovation.

9. Modern Humours

THE MERGING of humoural doctrines with modern assumptions about nerves was not limited to the lettered world; it was culturewide. Even in the mid-1800s, one finds for instance a Mrs. Carleton, the author of an *Enquiry into the Nature and Effects of the Nervous Influence* (1845), writing in her *Brief Advice to Travellers in Italy, Addressed to Persons Who Travel for the Purpose of Health, Economy, or Education,* published two years later, that Rome was "the only town [in Italy] perhaps, where *perfect* health may be enjoyed by persons subject to nervous irritation." The air there, she wrote, had

the property of allaying excitation of the nerves, in which it is the direct opposite of Naples. At Naples, the teeth especially are ruined by repeated *fluxions* (as the French call them), and they descend, and seem to walk out of their places—while Rome is the paradise of those subject to tooth-ache from irritability of nerves. Perhaps the tendency of the Roman air to reduce the activity of the nervous and circulating systems predisposes to low fever—but this only occurs when the air is made *too* sedative by its combination with moisture, especially with the heavy evening dews—avoiding this, the dry parts of Rome may be inhabited with impunity even in summer, by those who have *sound* livers; but those who remain should not sleep out of town, the return being considered unsafe. This soft air is so beneficial to some constitutions, that many strangers have settled here for life.

This was nineteenth-century Hippocratism—familiar to us also through the period's fiction, in which characters who fell ill would be subjected to bleedings, offered inhalations, and taken to Switzerland for its better air. One had to avoid bad air, "mal'aria," as Mrs. Carleton was well aware. According to her, "mal'aria" was "a gas that issues from the earth, and that increases in malignancy if it is not continually disturbed and dispersed. . . . The height of a house is a better security than the height of a hill, because an upper story removes us from the *earth,* which is the receptacle of the noxious principle."

This sort of belief takes us straight back to the assumptions that governed anti-plague policies throughout early modernity. Nerves, air, and a vague conception of germs made up the mainstream medical credences. In general, it was still not recommended to take too many baths in the 1700s and 1800s; and when one did have a bath, it was best that the water be lukewarm, rather than hot. Hygiene mattered, but it was a hygiene of restraint, based in part, presumably, on a fear of fluidity, a distrust of humours.[34] No cause had been found that explained why disease was more rife in insalubrious places. Girolamo Fracastoro would not be vindicated until the nineteenth century, with Louis Pasteur, the chemist who understood that microorganisms caused fermentation and infection, and who, in 1885, developed a vaccine against rabies. In the meantime, bleedings and purgings continued.

Although medical care changed during the eighteenth century, along with industrialization, the continuing rise of the middle classes, and the emergence of new ailments, it was not much more comfortable to fall ill then than it had been in the fifteenth century. The omnipresence of ideas that had outlived their theoretical credibility was partly due to the slowness with which new theories were arising to replace them. But the generality of humoural beliefs also explains the ease with which they could feed into an accessible, commonsensical "folk" wisdom, just as "folk" wisdom had contributed to shoring them up. At the higher level, though, humours were no longer a theoretical tool. The body had outgrown them—or so it seemed. Scientific research was developing fast, as modern neuropsychology, medicine, and psychiatry established themselves and became the independent, multifaceted disciplines we know today. The mind was solidly entering the brain, and post-Newtonian physics was sustaining analyses of matter.

But the brain was not just any matter. It was still impossible to wholly apply physics to its study. Mental patients were certainly not mere bundles of nerves, or embodied brains, especially in the eyes of the physicians who tried to study them. They were pathological cases, and they were persons—though what sorts of cases and persons they were was not very clear. The only certainty—and this was perhaps also the problem—was that the old immaterial rational soul now was of no relevance to medicine. Whereas once madness, mania, or extreme melancholy had involved humours that caused bodily disturbances likely to cloud the mind or distort perceptions, but that left intact the immaterial rational soul, now the organs of sense and reason themselves could apparently become ill.

Illness was becoming localized. Psychiatry and medicine were becoming separate fields. Madness had been a pathology of the body; it was now becoming a pathology of the nervous system. It could be a matter of faulty associations, of thought processes gone wrong. John Locke had posited that all our mental processes, including reason, were the outcome of experience, that our knowledge and self-knowledge were the outcome of a dynamic process. In the course of the eighteenth century, so-called sensualist philosophies like that of the Abbé de Condillac were growing out of the Lockean view, which helped understand delirium, say, in terms of the individual's psychological history rather than in terms of the humoural complexion that, earlier, had been read off the individual's medical history. Madness was no longer a matter of heightened passion that needed abating in order for reason to rule again, as humoural models had it. The moral psychology typical of earlier treatises on the passions was insufficient for those whose cognitive faculties had gone awry; but a new form of psychology, at once medical and personalized, could grow on its foundations.

While Willis had inaugurated the practice of studying the brain in order to understand disorders of the mind, or "sensitive soul," one century later a professor of anatomy at Padua, Giovanni Battista Morgagni, was taking the "science of nerves" one step farther. He distinguished between general—that is, humoural—pathologies of the body and localized ones, and actually invented the concept of "organic pathology," the art of analyzing disease by searching for causes in the local disturbances correlated with it. He therefore believed that there were also strict correlations between mental ailments and cerebral disturbances, and he wrote up over 700 such cases. He did confess, in *The Seats and Causes of Diseases Investigated by Anatomy,* first published in Latin in Venice in 1761 (the

English edition appeared in 1769), that he had rarely been able to tell from brain autopsies whether a patient had suffered from mania or melancholy, for instance. But the attempt was made. Neuropsychology was growing into a fully-fledged discipline: once, physicians had taken care of the body and moral philosophers or theologians had taken care of the soul; now, body and soul had met, and required a new sort of physician.

10. Mental Illness

PSYCHIATRY was not quite born as a discipline at this point, however, and neuropsychology was confused. As ever, available treatments were on the whole limited to humoural ones, to the principle of purging the bad in order to favor the good. One can gather this, for instance, from *Select Cases in the Different Species of Insanity,* an account—not a particularly original one—published in 1787 by a "mad-doctor" named William Perfect. Perfect described his cures with great self-confidence, listing the usual bleedings, diets, remedies from *materia medica,* and regimens of mild exercise and exposure to fresh air. These were the cures applied to King George III,[35] whose breakdowns and treatments were widely publicized. Medals were even struck of the lunatic hospitals where he was interned, and of the main doctor in charge of the royal patient, the Reverend Francis Willis (no relation to Thomas).

Willis was a clergyman whose specialty was insanity; he ran one of the many madhouses in the English provinces. His technique was hardly sophisticated, since he considered insanity a form of excitation that had to be contained to be cured, with straitjackets and through psychological submission if need be—apart from the usual purges and bleedings. Other doctors whom he had stopped from approaching the king called him a quack. In fact, the king's recovery in 1789 probably had little to do with Willis, but no one could be sure of that; at any rate it was then that William Pitt, the prime minister, had the doctor's medal struck.

Madness elicited a complex response in the medical and political establishment: benevolence toward those who had lost their minds, medical tyranny over those who should get their minds back, abuse of the abnormal, and fear of the pathologies that could befall us all. But by the late eighteenth century benevolent care was gaining ground in places. Vincenzo Chiarugi was a professor in Florence who held the chairs both of dermatology and of mental pathology (specialization was already on its way). In 1793, he published a treatise on madness, *Della pazzia in genere, e in specie,* and advocated gentleness in treatment rather than the brutality that the clergyman Francis Willis prided himself on. Chiarugi also understood hypochondria in novel terms, wondering whether it really should be considered a form of melancholy. Like madness in general, it was becoming a pathology of the nerves, severed from its historical origins in the actual hypochondria, whose function was no longer that of secreting yellow and black bile. Francesco Sforza would probably not have been treated for a disease of the hypochondria in the eighteenth century.

But nerves served the same metaphorical purposes as humours. Humoural secretions, more biological than mechanical, would perhaps have seemed closer to the mess of life than clean, dry nerves, although as one historian put it, "nerves, like fiddle-strings, needed to be kept in tune. The relaxed nerve, like the slack string, performed flat. By contrast, the overstretched fiber was 'sharp,' scratchy, highly strung. Neurology thus gave a scientific dimension to common talk of being 'up' or 'down,' just as humouralism had made much of being cold, hot, wet, dry, etc."[36] Nerves were even better than humours, in a way: "they were the media of the most delicate sensations," and "not least, they pointed to the brain." Nerves, in other words, were the humours for an age of cerebral supremacy. The rational soul was admirable and, pointedly enough, no longer necessarily immortal. Politics and public language had changed, and metaphysical orthodoxies were less rigid than they had been before.

Things had begun to change earlier, however. In 1729, a governor of Bethlem Hospital, Nicholas Robinson, had published *A New System of the Spleen, Vapours, and Hypochondriack Melancholy, Wherein All the Decays of the Nerves, and Lownesses of the Spirits, Are Mechanically Accounted For.* The book did exactly what the title announced. Nerve fibers and nervous fluid were direct causes of mental illness or "alterations of the mind," whether melancholy or manic, whether the patient was dejected or raging mad. From the late 1600s, a pair of statues by the seventeenth-century Anglo-Dane sculptor Caius Gabriel Cibber had been guarding the entrance to Bethlem: one represented *Melancholy,* the other *Raving Madness,* or mania. Both were figures of insanity in the literal sense, depictions of types unable to live in harmony with the rest of the world, solitary or in search of solitude, and, from the perspective of society, in need of confinement. Insanity was not otherworldly—unless it was connected to possession by the devil, as could still have been the case in Cibber's day—but to be cured of it meant to be brought back into the world.

Insanity always had come in various guises, as the great Burton well knew. But the term also always tended to be applied to those who were unable to act consistently, rationally, and lucidly in the world. Much earlier, in 1592, a lawyer and Master in Chancery, Richard Cosin, had listed the degrees and kinds of insanity that played a part in determining legal responsibility. (He did so in the event that the insanity plea might have been available as defense to William Hacket, one of three men who tried to overthrow the government. But in the end Hacket was hanged.)

These kinds were:

Furor, "an entire and full blindness or darkening of the understanding of the mind, whereby a man knoweth not at all, what he doeth or sayth, and is englished madness or woodnes."

Dementia, "a passion of the minde, bereaving it of the light of the understanding . . . and may be englished distracted or wit, or being beside himselfe."

Insania, "a kind of inconstancie voide in deede of perfite soundnes of minde," who could be "franticke, braine-sicke, cracked-witted, cocke-brained, or hare-brained men, being not altogether unapt for civill societies, or voide of understanding, to perceive what they say or doe, or what is saide unto them: albeit they have many strange conceites, toying fansies, and performe sundry, rash, undiscreete, mad, and foolish parts."

Fatuitas, "the want of wit and understanding, whereith natural fooles are possessed."

Stultitia, "that follie which is seene in such, as albeit they be but simple and grosse witted."

Lethargie, "a notable forgetfulnes of all things almost, that heretofore a man hath knowen, or of their names. . . . This distemperature and weakness commeth by some blowe, sicknes, or age."

Delirium, "that weakenes of conceite and consideration, which we call dotage: when a man, through age or infirmitie, falleth to be a childe againe in discretion: albeit he understand what is said, and can happely speake somwhat pertinently unto sundry matters."[37]

Melancholy was not explicitly included here, but then again, melancholy had a gradation of intensity and was not always a pathology. One can presume that both of Cibber's figures would have contained these types of cognitive disorders, all the bearers of which would have been fit to pass through Bethlem's gates.

During Robinson's reign at the hospital, however, the very notion of pathology was changing, just as were ideas about the best ways to curb it. It was clear to Robinson that "if the Structure or Mechanism of these organs happen to be disorder'd, and the Springs of the Machine out of Tune; no Wonder the Mind perceives the Alteration, and is affected with the Change." Pathology was akin to a fault in the machine, which had to be treated especially insofar as it caused suffering. He thought it inhumane "not to be bold in the Administration of Medicines"—medicines that would be "proportion'd to the Greatness of the Causes," no matter how violent their effects, and that were "capable of making those Alterations in the Fibres of the Brain, necessary to produce a Freedom from those Affections."[38]

Such cures, for better or worse, would soon be forthcoming; but some of them arrived from quarters Robinson might not have expected. At a period when the mind was open to exploration and new experiments were adding mystique to matter, self-proclaimed innovators—sometimes tardily recognized as quacks—were advertising new, revolutionary cures for ailments of all kinds, not only those of the mind.

11. Mesmerism

AS THE EIGHTEENTH CENTURY drew to a close, the new, "magnetic" cures offered by the Austrian physician Franz Anton Mesmer were exciting Vienna's citizens and intriguing, as well as infuriating, the medical world. Mesmer had written his doctorate on medical astrology, and later drew on the notion that our bodies were endowed with an "animal magnetism" that directed an invisible fluid connecting us with planets and with each other. From the mid-1770s, his "séances" were the talk of the town; he became a member of the Bavarian Academy of Sciences and an illustrious man who frequented Mozart. (There are even—satirical—scenes of mesmerism in *Così fan Tutte*.) As the caption of an etching depicting a mesmerist session tells us, his method was capable of curing such ailments as "dropsy, paralysis, gout, scurvy, blindness, accidental deafness." The cure consisted

in the application of a fluid or agent that M. Mesmer directs, at times with one of his fingers, at times with an iron rod that another applies at will, on those who have recourse to him. He also uses a tub, to which are attached ropes that the sick tie around themselves, and iron rods, which they place near the gut of the stomach, the liver, or the spleen, and in general near the part of their bodies that is diseased. The sick, especially women, experience convulsions or crises that bring about their cure. In the antechamber, musicians play tunes likely to make the sick cheerful. Arriving at the home of this famous doctor, one sees a crowd of men and women of every age and state, the cordon bleu, the artisan, the doctor, the surgeon.[39]

Mesmer did not resort to the usual bleedings and purgings. He never opened the body; he left everything to the powerful fluid and its supposed effects on the "very subtle spirit" circulating in our bodies. He did indeed treat the rich as well as the poor with his iron rods and magnetic tub filled with aromatic plants, and he did have excellent results—he was famous for them. There was a catch, though. Since Mesmer's results were unexplained, he had to rely on continued success to maintain credibility—more so than orthodox physicians ever had to. His career in Vienna had come to an end when a wellborn young Viennese pianist he had cured of blindness had subsequently lost her ability to play the piano; this led to her mother's fury, to the girl's relapse into blindness, and to Mesmer's being accused of practicing magic. At this point, he had fled Vienna. In 1778 he arrived in Paris, where his continued self-confidence about a method whose workings seemed designed to mystify was soon alienating as many physicians and academies as it attracted clients, intellectuals, and kings. While Louis XIV was an admirer and supporter, Mesmer was immediately shunned by members of the medical establishment, including the Académie des Sciences. To them, he was a dangerous quack who was not practicing proper medicine.

They were right: Mesmer was not a proper doctor. What he was practicing was, literally, mesmerism. The word today signifies hypnotism. (His name, like Galvani's, was turned into a verb—to mesmerize, or to be mesmerized.) And indeed his initial inspiration for using magnets had come when he once witnessed a priest exorcising a man with the help of a metal crucifix. Mesmer had surmised that the body was connected to the universe by its magnetic fluid, which was channeled by external magnets and metals, much in the way that the sea was affected by the force of gravity—a fashionable concept now that Voltaire had popularized Newton in France. Mesmer even believed that gravity operated within the magnetic fluid, which was also reminiscent of the invisible fluids of the Paracelsians. There was no point in finding out how the mechanism worked, or against what bodily ills exactly—no reason to uncover precise processes and make the invisible visible. Magnetism seemed similar enough to the electricity which Benjamin Franklin had discovered and with which Galvani was experimenting; it was as plausible as phlogiston, and seemed perfectly worthy of Newton's universe, ridden with occult forces as it was.[40] Neither Mesmer's practice nor his *Mémoire sur la découverte du magnétisme animal,* which appeared in 1779 and appealed to these precedents, required any further justification.

Mesmer was tapping into the fascination exerted on the public by these novel, potent, invisible forces, which were useful in the fight against the progress of mechanism. The public in turn bestowed on Mesmer the authority of a physician and the power of a magician who had managed to reconnect the microcosm and the macrocosm. People came in throngs to undergo the tub cure and emerged feeling wonderful. All this provided Mesmer with a tangible fortune, much of it offered at first, in 1781, by the French government, and boosted by a highly visible publicity campaign. Mesmer had charged ten louis per month for the rental of his tubs; but the commercially savvy operation was the creation of the Société de l'Harmonie Universelle by his most zealous follower and propagandizer, Nicolas Bergasse, in 1783. Hundreds of students attended it, and branches sprouted throughout the country. More students and disciples published an extraordinary number of papers, case histories, and pamphlets on mesmerism. Mesmerism offered everything the public needed, especially in a confusing world where the very nature of the air was changing with each experiment: in 1784, phlogiston itself was replaced with Lavoisier's oxygen. Mesmer did not treat wounds, but he did not need to. These were prerevolutionary days: the vogue for the cult of nature and for Rousseau's view of society and knowledge as forces leading us away from nature was at its height, and mesmerism exerted on the body a revolution of its own, reestablishing its order by channeling purely natural forces.

Mesmerism appealed to the founder of homeopathy, Samuel Hahnemann. He was the one who coined the actual term "homeopathy," in opposition to traditional Galenic allopathy. In his work on the matter, the *Organon of the Medical Art*, published in 1810, he wrote that he fully believed in the power on the body of magnetic forces yielded by those "gifted with great nature-power." There was nothing like "the most powerful, good-natured will of a man whose life force is in full bloom" to revive "some persons who remained in apparent death for a long time."[41] Hahnemann accepted the notion of a vital force harbored by the body, which the mesmerist could replace or rejuvenate when someone was ailing. In this he followed Stahl, who was also skeptical about the power of allopathy to cure, and whom Hahnemann even quoted as arguing that "diseases yield to, and are cured by, means that engender a similar suffering *(similia similibus)*."[42]

Magnetism was attractive to those, like Hahnemann, who were in search of "alternative" cures, gentler methods of treatment than those offered by mainstream allopathic practitioners. Mesmer's method seems to have had real effects on the organism; and it certainly did work wonders for psychosomatic illnesses.[43] It is no surprise that it was focused on the region of the old hypochondria, so central to humoural cares and cures. Even though conservatives were right to accuse Mesmer of not toeing the official humoural line, he was actually staying close to its spirit, if not to its letter. Humouralism had made perfect sense of psychosomatic phenomena, after all, and here was a new way of exploiting their power.[44] Ostensibly, Mesmer's sessions worked by creating through elaborate settings—strange sounds, darkness, thick curtains and carpets, flutes, flowing water, mirrors—a "crisis" that dissolved the blockage causing the perceived physical ailment.[45] These organic enactments must have been for the most part psychological events, the outcome of effective hypnosis. But the processes at work were all the more opaque and mysterious.

It was precisely the mystery of these processes that infuriated informed members of medical circles, although established scientists themselves were prone to refer to vital spirits, and the boundary between humoural and alternative cures was fuzzy. Eventually, Mesmer had to leave France as well, after a commission set up to investigate his methods delivered, in 1784, its *Rapport des commissaires chargés par le roy de l'examen du magnétisme animal,* which made it difficult for him or his followers to continue practicing.

The commission was headed by Lavoisier, and one of its five members was Benjamin Franklin.[46] Humouralists had known full well that imagination played an important role in illness and health—they had acknowledged our fundamentally psychosomatic nature—but in the mind of the modernizers who were sitting on the commission, Mesmer was clearly abusing imagination. If he was using hypnosis, then he was influencing the psyche while pretending to cure the body. The members of the commission believed that "animal magnetism could well exist without being useful, but it cannot be useful if it does not exist." In other words, it was the supposed nonexistence of the phenomenon that they thought problematic.[47]

The charge was loud: Mesmer left his many pupils and disciples behind in 1785, wandered around Europe, took refuge in England from the French Revolution, and spent the rest of his life in Germany.

12. The Birth of Psychiatry

IN THE HANDS of the young Freud, the technique that Mesmer had used so profitably would eventually provide insights into the makeup of the unconscious mind. But it did not further neuropsychological research in any major way. A number of contemporaries of Mesmer, however, had been making headway in their investigations of the brain, and the late eighteenth and nineteenth centuries are populated with figures who were contributing to unraveling the structure of the central nervous system, progressively establishing the foundations of our modern neurosciences. A history of the humours, of course, is not a history of the modern neurosciences, and would have to stop at its opening. But notions that arise out of the humoural mind-set have perdured, testifying to the extent to which early modern culture is profoundly inextricable from our modernity.

The questions central to neuroscientific pursuits had arisen out of the work on nerves that people like Haller, following the example of Willis and Glisson, had been pursuing in the eighteenth century. But it was in the application of these findings to psychology that humours could still be found lurking. Melancholy, notably, survived as a full-fledged, often mysterious syndrome, even though it had outgrown the atrabilious physiology of which the concept was initially a product and had become a multifaceted pathology.

Modern psychiatry was borne on the back of concepts inaugurated by vitalists and mechanists, and it accompanied the changes that the care of the mentally ill was undergoing in the late 1700s. Many people still consider Philippe Pinel a founding father of psychiatry. He was indeed a modernizer who, in the midst of the French Revolution, in 1793, had decided to liberate from their customary restraints the patients at the insane asylum he headed at Bicêtre. He had realized that coercive treatments were not helpful, and he looked for other methods.

But for a while at least, he also believed that the hypochondria played a central role in madness. He, too, had once been intrigued by mesmerism, at a time when he still spoke a humoural language. In 1786, in an article entitled "Observation on a Nervous Melancholy Degenerated into Mania," he had recounted the case of a dear friend of his who had died a few years before, whom he described as having a "bilious, sanguine temperament." The reality of magnetism had been impossible to prove, just as the report drawn up by the commission entrusted with the investigation of Mesmer had concluded. But Pinel agreed that illnesses had their origin in the viscera or hypochondria—the "epigastric and hypogastric areas," densely innervated and therefore likely to carry the "subtle matter" supposedly responsible for magnetism.[48] Mania was related to melancholy and hypochondria, and its seat, he thought, was in the gut, as it had been for the humouralists.

Later, Pinel moved away from mesmerism and magnetism. He became well known for his work on patho-anatomy. In 1795 he was named head of La Salpêtrière, the hospital in Paris which had been created by Louis XIV in 1656 to house women deemed indigent, sick, poor, or mad, and in which all possible cases of mental and psychological distress could be found when Pinel arrived. Pinel began to engage in philosophical reflection on the passions, in line with the classical literature on the topic from Cicero and Galen on, and to think about madness in moral rather than purely physiological terms. He developed the notion of "moral insanity," and his *Traité médico-philosophique sur l'aliénation mentale ou la manie*, published in 1800, testified to his dedication to understanding each mental patient as a case *sui generis*, whom the doctor could attend to properly only by being a good Hippocratic. Doctors should now register the patient's symptoms, and form a diagnosis and prognosis on the basis of careful observation.

Pinel had also been the translator into French of the work of William Cullen, one of the famous exponents, with Whytt, of the medical school of Edinburgh, noteworthy in the second half of the 1700s. Cullen had been the first to classify diseases according to their symptoms and presumed causes. Diseases were a matter of "neuralpathology" and could be febrile, nervous, wasting, or local; these categories in turn had their own subcategories. Cullen supposed he could recognize what family a disease belonged to by inferring its nature from its symptoms. He believed that sensory disturbances and disturbances of appetite and peripheral motion should go under the heading of local diseases. The disciplines of neurophysiology and neuropathology did not yet exist, so this hypothesis could not be explored in a significant way; and although neuroanatomy was developing, it was still vague.

Cullen is still remembered as the man who first coined the term "neuroses," close in meaning to Thomas Willis's notion of "diseases of the brain and nervous stock" and to the "nervous disorders" of Cullen's compatriot Whytt. It was also akin to what Sydenham had understood as "hysteric disorders," the female malady that was the counterpart of male hypochondria.[49] Yet under the heading of neuroses Cullen included not only hysteria, predictably enough, but also a whole range of illnesses one would not today associate with neuroses, from apoplexy to dyspepsia, from tetanus and epilepsy to diabetes.

In *First Lines in the Practice of Physic,* published in 1784, Cullen explained that neuroses were "all those preternatural affections of sense and motion . . . which do not depend upon a topical affection of the organs, but upon a more general affection of the nervous system." He thought, indeed, that "almost the whole of the diseases of the human body might be called *nervous,*" not merely in the "loose" sense of "hysteric and hypochondriac disorders, which are themselves hardly to be defined with sufficient precision," but in the broader sense given initially by Willis and pursued by Whytt.

Delirium, for Cullen, was a "false judgment that produced "disproportionate emotions," and insanity was a kind of delirium, an "affection of the mind" that, given the connection between mind and body, also must depend on the "state of our corporeal part." It was still conceptually connected to the old melancholy, then—and treated much as it had been for centuries, with cold baths, purges, and hellebore. Old categories were being recycled by now. Jean-Étienne-Dominique Esquirol, a pupil of Pinel and a cofounder with him of the psychiatric clinic, described love-melancholy, for instance, as a manic state in which "amorous ideas" had overwhelmed and altered reason. But this was at a time when hysteria and madness were no longer so easily definable and when their nosology expanded to include, for example, cases where madness was defined as a complete but not permanent loss of self-conscious reason, and cases of mania without delirium.[50]

It was evidently rather difficult to identify what part of the brain was involved in the higher intellectual functions that seemed to shut down in states of madness or to be denatured in cases of mania. But according to Cullen, one could at least observe "that the different state of the motion of the blood in the vessels of the brain has some share in affecting the operations of the intellect." And so the brains of those who had recovered from insanity probably were free of any "organic lesions." It seemed, in the end, that "the state of intellectual functions" and the capacity to form accurate representations or ideas of the world depended on the state of what Cullen termed "Nervous Power, or, as we suppose, of a subtle very moveable fluid, included or inherent, in a manner we do not clearly understand."[51]

The "subtile, moveable" fluids had always existed, of course—as humours, which in a literal sense are just fluids. Their precise nature, as we have gathered by now, had not been known in previous centuries; their qualities had been summed up in Aristotelian terms—according to their basic temperature, acidity, quantity, and humidity—as long as Aristotelianism had prevailed. Cullen, who characterized melancholy as "partial insanity without indigestion" and mania as "universal insanity," nevertheless did not have a clear idea of what the qualities and constituents of the newly conceived "nervous power" might be.

In a sense it might as well have been called spirit, *pneuma*—or magnetic fluid. Robert Burton had written that spirit was "a most subtle vapour, which is expressed from the blood, and the instrument of the soul, to perform all his actions; a common tie or medium between the body and the soul, as some will have it; or as Paracelsus, a fourth soul of itself."[52] There was in fact a straight line connecting the humours that still prevailed in Burton to the iatrochemistry that purportedly rejected them, to the holism of vitalist creeds and to *libertin* materialism. It was a line that drew together all those who had proclaimed the embodiment of soul, or mind, often against the grain and, at least in the seventeenth and eighteenth centuries, at the risk of their freedom and even their lives.

But what was the soul, or mind? Were these terms equivalent? Could the study of matter yield better knowledge of either of them? Was that a question one could even ask? Were the new neurologists contributing to answering this question? Could they answer Pinel's moral questions? No one really knew. The dissatisfaction and fears that the rapidly evolving sciences provoked in the latter eighteenth century and throughout the nineteenth century found expression in Romanticism and in the search for a new sort of soul and spiritualized nature. This search begat movements like vitalism and mesmerism, but also highly individual, idiosyncratic programs or ideologies, and even early science fiction like Mary Shelley's *Frankenstein*. Extreme mechanism built factories, but it did not soothe fragile souls or repair broken bodies.[53]

While the sciences were increasingly specialized and effective, magical beliefs failed to disappear. Sexuality—especially female sexuality and the associated specters of new sorts of "uterine fury" and hysteria—became associated with excessive fluidity and the loss of reason, which, whether it was indeed embodied or not, was still considered the mark of our divine nature. Natural theology was back, especially in England, and the argument from design—the notion that nature's perfection could only be the product of a divine "intelligence"—had powerful advocates.

Answers were being forged within the sciences, however, and some of these answers, after mesmerism, became widely popular. One attempt at understanding the nature of the soul or mind was to divide it into functions and to search for their localization in the brain, as Willis and Morgagni had done, rather than to assume that the soul was immanent throughout the body, in the terms which Stahl, for instance, had chosen. Descartes had not entirely succeeded in burying the ventricular theory of the Scholastics, according to which memory, imagination, and common sense had each been housed in their own, respective brain ventricle. Ventricles themselves, granted, had not housed these functions since Vesalius. But the functions themselves still existed. Ventricles were out, but cortical structures, those that Willis had begun to study, were now in.

13. Brain Localization

THE DRIVE to localize specific mental functions in specific parts of the brain was now increasing tremendously. The Austrian physician and neuroanatomist Franz Joseph Gall is perhaps most famous for having developed the ill-fated theory known as phrenology, according to which each mental function was precisely localized in such a way that one could tell character, temperament, and moral qualities from the contours of the skull. It was, in a way, a visible, solid equivalent of humoural determinism—and of the physiognomics favored in the sixteenth century by, for instance, Giacomo Della Porta, who had thought that facial features were excellent indicators of character and had written a well-known treatise on the matter.

Gall divided the skull into twenty-seven parts. There were areas for improbably specific functions such as the sense of satire, kindness and benevolence, affection, guile, arrogance, circumspection, the organ of religion, the memory of words and the desire to study, the sense of places, and so on. Gall's colleague and follower from 1800 was Johan Spurzheim, with whom he successfully toured Europe, lecturing and demonstrating. They settled in Paris in 1807, where Gall opened a medical practice; and they were the coauthors of the first two volumes of an eventually hefty six-volume *Anatomie et physiologie du système nerveux en général, et du cerveau en particulier,* which came out in 1810. But the collaboration came to an end soon after; Spurzheim left Paris for England in 1812.

It was in fact Spurzheim, not Gall, who later adopted the term "phrenology" (science of mind, in Greek). Gall remains associated with it, but he disliked the term, favoring instead "organology" to emphasize that his starting point was the brain's physiology, not *a priori* assumptions about what sort of relation the brain bore to the mind. He did believe, however, that an understanding of cerebral physiology would lead to that of brain function and of visible anatomy. As he wrote, "The functions or faculties . . . are in direct proportion to the development of the organs appropriate to them." Furthermore, "the brain is composed of as many specific systems as distinct functions it performs," and there were "anatomical facts by which we have established that the nerves arise in various places and from various masses of grey substance, and that the various specific systems of the brain are brought into being in the plurality of the fascicles,

Fig. 27. Johann Kaspar Lavater, *The four temperaments*, from *Physiognomische Fragmente zur Beförderung der Menschenkenntnis und Menschenliebe* (Leipzig and Winterthur, 1778), vol. 4, p. 352. Lavater is remembered principally for his enormously popular work on physiognomics. It is based in part on the seventeenth-century precedent of della Porta, but the system he developed is parallel to the cranioscopy favored by phrenologists.

layers, and convolutions."⁵⁴ This was the basis for his belief in "organology." If Spurzheim's phrenology was what became best known and popular, that is most notably because it was simplistic enough to be understood. Anyone could feel the bumps of a friend's skull and, phrenological map in hand, list qualities and faults. Some major practitioners, like George Combe, even combined it with mesmerism, calling the result "mesmeric phrenology."⁵⁵ In fact, phrenology also became infamous, for its adherents tended to believe that the neurologically determined traits that surfaced on the skull were inheritable—a thought that would lead to the

Fig. 28. A phrenology figure (possibly ca. 1820s).

fallacious and tendentious reductionism of eugenics.

Humours had been too general for their inheritability to pose a problem of this sort. But the geographical determinism inaugurated with the Hippocratic *Airs, Waters, Places* could seem somewhat biased (especially the notion, present in Hippocratic discourses and in contemporary Greek literature on the Persian wars, that Asians were "softer" than Europeans). The very idea of a complexion could be interpreted in an overly deterministic vein, and the current meaning of the term—the color of skin—bears witness to the dangers inherent, at least potentially, in the belief that organic form is the direct, clear translation of soul, character, type, or quality.

Ideology was not the only impetus in the search for localization, however. The notion that cerebral areas must be divided in some way according to their respective function had taken off by the nineteenth century, but not without opposition, especially from Pierre Flourens. He was a physiologist who, in 1822, gave a lecture to the Académie des Sciences in which he took Gall to task, arguing that brain functions were not localized in particular places but distributed throughout the cerebral structure—just as they had been associated with the brain's four ventricles in scholastic psychology. In 1825, Jean-Jacques Bouillaud, a French physician who was a founder of the Société Phrénologique instituted after Gall's death, had contradicted Flourens's position in favor of that of Gall, reporting to the Académie Royale de Médecine his empirical finding that patients who had lost their capacity to speak often presented a lesion in the brain's frontal lobe. This was strong evidence for a correlation between the cerebrum and at least one of the mental functions listed by Gall.[56]

But localization had a bad name at that point because of its association with phrenology, which Flourens, in particular, had damned, and Bouillaud's lecture met with opposition from the anti-locationists. He had nevertheless established over 100 correlations between frontal lobe lesions and speech failure, either through memory loss or through motor difficulties impeding word pronunciation. The evidence was too powerful to ignore. By 1860 Bouillaud was an established figure who contributed to keeping the localization debates alive.

Notions that are still present in our assumptions about brain function were being mooted during those decades, and fervently discussed. The map of the brain was coming into focus—literally bit by bit. It was increasingly possible to speak of the mind as brain, and to offer scientific evidence for this—a far cry from Descartes's disembodied soul, or so it seemed. Locationism grew stronger in spite of continued attacks, and despite the relegation of Gall to the dustbin of quackery. In 1861, it was the turn of Paul Broca, secretary of the recently founded Société d'Anthropologie, to enter the debate on the side of the locationists. He presented evidence from patients with what he called "aphemia"— renamed "aphasia" in 1864—which showed that the faculty of speech was located within the brain's convolutions in the left frontal lobe, near the primary motor cortex, and was not visible from the cranium surface, as phrenologists had believed. In 1870, the Germans Gustav Fritsch and Eduard Hitzig demonstrated in a paper, *On the Electrical Excitability of the Cerebrum,* that speech was primarily a motor function rather than an intellectual function.

The debates went on. Eventually, the cerebral area that Broca had identified in his aphasic patients was named Broca's area. In 1874 Carl Wernicke identified that an area in the brain's left temporal lobe, near the primary auditory cortex, was associated with language comprehension: it was named Wernicke's area. It was now clear that either of these areas was implicated in strokes leading, broadly, to receptive or to expressive aphasia. The areas still bear the names of these scientists, whose insights would mark a turning point in the understanding of severe neurological malfunction: both Broca and Wernicke had provided crucial clues to the study of the normal cerebral correlates of language—the highest cognitive function, always considered to be what differentiated humans from "brutes."

But motor and cognitive functions now seemed to be perilously close to each other. The capacity to communicate, to take part in society and lead a normal life had been shown to be, at least in part, a clear-cut function of an area in the brain. It was as if the pulley effect that Descartes had imagined operating between, say, feet and brain, really did exist, but included the faculty of speech and reason as well. La Mettrie had been right to describe our soul in all its fragile materiality. In spite of the continued influence of Haller's notion of irritability and of vitalist traditions, the mind and brain were becoming one; the old rational soul was no longer a central concern, and the definition of life itself was changing. At the same time the brain and body seemed increasingly split apart.

Given the complexity of the issues at stake in these studies, many scientists at this early point in the history of the modern neurosciences became involved in examining the assumptions underlying their work. Questions regarding the scope, the status, and even the history of the neurosciences increasingly emerged, in parallel to the mainstream scientific research that was developing so rapidly and impressively. This was the case with the fields of biology and natural history. The microscopic world was expanding, yielding more and more of its secrets. But its mysteries, too, were multiplying.

Practitioners of the experimental method in modern medicine were also growing self-conscious, especially after the appearance of the lucid *Introduction à l'étude de la médecine expérimentale* by the French physician Claude Bernard in 1865. He argued there that facts about anatomy did not necessarily give a clear picture of physiology, and he warned against the tendency to explain one by means of the other. Anatomy could not possibly yield a complete picture of the "phenomena of life." Ever since the "great Haller"—as Bernard called him—had developed his physiology on the reductive basis of irritable and sensitive fibers, "the humoural or physico-chemical part of physiology, which one cannot dissect and which constitutes what one calls our internal environment, has been neglected and thrown into the shadows."[57]

Claude Bernard thought it might therefore be useful to take seriously an idea proposed by the distinguished French zoologist and comparative anatomist Henri de Blainville, according to whom the brain should be understood not so much as "the organ of thought" but rather as "its *substratum*." One could then, thought Bernard, consider all organs to be the substratum of their functions. So for instance, "although one may understand how a secreted liquid flows through a gland's ducts, one cannot have any clue about the essence of the secreting phenomena, and may just as well say that the gland is the substratum of the secretion."[58]

This was a conceptually ingenious way of accounting for the relationship between a structure and a function. It is significant that such philosophically rich reflections in support of a functionalist thesis constituted an introduction to experimental medicine, by a brilliant doctor who had conducted important research on a range of topics pertaining to physiology—from hepatic and gastric functions to pathology, the nervous system, and toxicology. The more one examined the world, Bernard seemed to be saying, the more acute became the need for ascertaining analytically what constituted evidence. Seeing was never enough: one had to gauge first what it was that one saw.

Humoural medicine had grown on a similar ground of opacity. At first, to be sure, its basic tenets had developed out of empirical evidence—it had been supported by the use of techniques and remedies that, on occasion, had turned out to be empirically appropriate. But medical care had developed for the most part in spite of the theoretical confusion bequeathed by humouralism. And yet, it was precisely the awareness of the fundamental gap between theory and practice that had allowed someone like Claude Bernard to recognize the complexity of the relationship between the functioning of our organism and known, potentially or actually visible anatomical forms.

14. Hypnosis, Hysterics, Neurosis

THE GAP between theory and practice was equivalent to the gap between mind and brain, or between physiological humoural events and mental psychological events. One could venture to say that humours had accounted for mania and melancholy similarly to the way in which the substratum of Blainville had accounted for the brain's relation to thought. For the humouralists, the psyche (in the word's old sense of soul) had certainly not been a set of functions localized in brain areas studied by neuroscientists. Rather, it was spread throughout the body, as the body's substratum. (Galen himself, after all, had written a treatise claiming *That the Powers of the Soul Follow the Body's Mixtures*.) Theories of brain localization on their own were not yet contributing much to the effort to understand the psyche (in the word's modern sense of mental structure), temperament, or moral imagination; and so there was still a need for the holistic physiology that humouralism had accommodated, and that people like Stahl, Whytt, and even Haller had defended.

The milieu of neuroscientists who were investigating the anatomical-physiological basis of cognitive functions was actually open to the sort of work that enabled the psyche itself to be taken inside out, so to speak. Mania and melancholy continued to be studied. Hypnosis was taken up quite seriously by James Braid, a surgeon in Manchester who was fascinated with the mesmerism sessions he attended there; he was the one who had decided to name the technique hypnotism, or hypnosis. He was certain that it was naturally possible for the nervous system to be put into a hypnotized state, and that there was nothing quackish about the physician's capacity to induce this "nervous sleep," or "lucid sleep" as the mesmerists had referred to it. He himself had achieved it in patients, without any recourse to magnetic fluid.

In 1843 Braid published his *Neurypnology; Or, the Rationale of Nervous Sleep, Considered in Relation with Animal Magnetism,* in which he explained "that the phenomena of Mesmerism were to be accounted for on the principle of a derangement of the state of the cerebro-spinal centers, and of the circulatory, respiratory, and muscular systems" that were induced "by a fixed stare, absolute repose of body, fixed attention, and suppressed respiration." The state "depended on the physical and psychical condition of the patient," rather than "on the volition, or passes of the operator, throwing out a magnetic fluid, or exciting into activity some mystical universal fluid or medium."[59] Here was an excellent cure—but its effectiveness had nothing to do with magnetism. It had to do, rather, with the nervous system. Braid had found a way of accessing the psyche via the "derangement" of the body's central systems, which Mesmer had induced: once the patient was hypnotized, the doctor could effectively induce mental states by pure suggestion.

Hypnosis was soon to grab the attention of Jean-Martin Charcot, an illustrious physician at La Salpêtrière who eventually was its first chair of neurology. He upheld the localization theory and studied aphasia, epilepsy, and, with the help of photography, hysteria. He thought at first that he had found a new disease, which he called hystero-epilepsy and which manifested itself as a rather spectacular convulsive fit: the patient lost sensory awareness and sometimes fainted. Soon, though, it emerged that something quite different was happening—autosuggestion. Charcot had been able to cure the symptoms with magnets and electricity, but his colleagues convinced him that the cure had nothing to do with the implements, and everything to do with the patients' expectation. In the end, hystero-epilepsy, for all its dramatic effects, had been created by the doctor. The effects were not symptoms of real pathologies. (Moreover, Charcot's infamous photographs of his female patients in the grip of such fits tended to be theatrical depictions of staged or brutally, voluntarily induced "hysteria.")[60] Hysteria could be reproduced by means of hypnosis. This meant that hysterical symptoms were manifestations of some sort of trauma.

In the winter of 1885–1886, a twenty-six-year-old neurologist, Sigmund Freud, was working in Paris with Charcot on hysteria. Like many before him, he was intrigued by the phenomenon and by the hypnotism that Charcot excelled in. When he returned to Vienna, Freud began to treat hysterics and neurasthenics—patients who had suffered what one still calls a "nervous breakdown." In 1888 he wrote two articles for a medical dictionary—"Aphasie," on aphasia; and "Gehirn," on the brain. Aphasia, as Freud rightly acknowledged, was usually the consequence of a cerebral trauma, and could affect either motor (usually expressive) functions or sensory (usually receptive) ones. But he added that it was not "always the consequence of a material brain process; rather, neuroses like hysteria and neurasthenia may also produce aphasic disturbances." In the first case, the patient was literally speechless, "mute." In the second case, the patient forgot and confused words. In both cases, though, the prognosis was relatively optimistic: the patient could improve.[61]

Freud was about to turn to psychology, and from there, to the foundation of what he thought of as the new science of psychoanalysis. His article on the brain helps one understand the rationale for his switch away from the study of the brain itself. In the section on its physiology, which followed a precisely documented section on its anatomy, he described the brain as "that organ which converts centripetal excitations, supplied by the sensory pathways of the spinal cord and through the gateway of the higher senses, into purposive and coordinated centrifugal movement impulses." Moreover, he wrote, "there exists the fact, inaccessible through mechanical understanding, that simultaneously to the mechanically definable excited state of specific brain elements, specific states of consciousness, only accessible through introspection, may occur. The actual fact of the connection of changes in the material state of the brain with changes in the state of consciousness, even though [this fact is] mechanically incomprehensible, makes the brain the organ of mental activity."[62]

Psychic life was at once dynamic and localized. It was material, but not fully explicable in mechanical terms. Instead, introspection—our self-consciousness—might lead us to reexperience the brain activity involved in such emotional crises as hysteria and neurasthenia. Freud began to study the traumas whose memories seemed to be released when the nervous system was in its "hypnoid state," as he called it. Hysteria, he ventured, was the somatic manifestation of trauma, a case of the mind entering the body that Freud called conversion. Hysteria occurred when repressed anguish surfaced like a message in a bottle, or indeed like humours within a hydraulic organism. Freud posited the existence of an unconscious driven by the libidinal humour, the locus of repressed conflicts between the parts of a new tripartite soul—the ego, the id, and the superego, vaguely corresponding to the old appetitive, sensitive, and rational souls. And he abandoned the use of hypnosis, realizing that he could achieve the same results without it. Psychoanalysis was born, and with it the venture to cure melancholy and mania—two facets, perhaps, of the same "moral insanity" Pinel had described, sometimes amenable to physiological treatment but also reversible through introspection.

Some people have argued that the era of psychoanalysis is over, its authority partly dispelled by a mistrust of the methods and assumptions on which it was born, its practice partly outshone by physiological cures—cheaper, quicker, and seemingly more radical than talking cures. But eras are never tidy. And beyond Freud's notion of an unconscious as a dark place too unpleasant to spend time in without the constraint of a couch; beyond his attention to the worlds that inhabit us beneath our immediately graspable drives, needs, and appetites; beyond his emphasis on the value of introspection for an ethical life—beyond all that, the notion of an unconscious that we somatized in various, sometimes extreme ways was a return to humoural form. It was an acknowledgment that our organisms were traversed by stuff invisible to the conscious eye, that our innards were as present to our sleeping, dreaming, neurotic, or maddened selves as our skin was smooth. And it was a recognition that, however much we shoved the terror at our bloody origins beneath the garb of our conscious, thinking minds, the mystery of our primal, humoural embodiment returned, violently or surreptitiously. We have been born into this particular notion of an unconscious for over 100 years now. But it is possible that its specifically Freudian form is no longer necessary for it to signify a powerful reality that humoural medicine had already recognized.

The world has changed since the early days of brain science. Our self-understanding has been changed by the rapid growth of psychoanalysis, psychiatry, neuropsychiatry, neuropsychology, neurophysiology, neuropathology, neuropharmacology, molecular biology, genetics, biochemistry, cognitive psychology, evolutionary biology, developmental psychology, and so on, as well as by novel kinds of mechanism and vitalism, like behaviorism and artificial life. The world beneath our skin has changed. We see dimensions that were probably unimaginable even to the microscope-wielders of the seventeenth century. We understand how our bodies work in a way that could not have been possible before the discovery of genes. Imaging techniques show us our own living brain, within our skull, without any bone grinding or spilling of blood, without the need for technological metaphor that until recently fueled the comparison of our minds to computers. We are able to register the movements of what once were "nervous fluids," "vital spirits," and "animal spirits" coursing through veins, arteries, nerves, and neurons, jumping across our synapses and making us what we are. The dispute between "locationists" and "holists" has shifted, and although modularity is the new locationism, the most convincing account of the conscious brain that is emerging matches the sort of picture that humouralism had given of the body. As the neuroscientist Gerald Edelman has put it, "the long-standing argument between locationists and holists dissolves if one considers how the functionally segregated regions of the brain are connected as a complex system in an intricate but integral fashion."[63]

But few neuroscientists heed the entreaty not to fall for the temptation to localize our functions in order to explain them. Doctors also have trouble accommodating their knowledge of individual organs with a sense of how interconnected our functions are. Medicine chops us into bits. It has become so specialized that Hippocratic doctors might not recognize today a profession whose goal was the care of illness through the understanding of the whole patient. Localized medicine can work wonders, of course, thanks to the findings of microbiology, imaging machines, diagnostic techniques, and advanced technologies. But ever since we realized that our hearts were pumps and not the seats of vital souls, our medicine has become increasingly mechanistic, focused on soulless pulleys, easily forgetful of our complex humours.

And yet, an entire industry sustains the Hippocratic conceits of diet, exercise, and air for the pursuit of health. Diets have replaced all-round regimens. Medicines that help us combat melancholy and curb mania have replaced some of the old *materia medica*. The culture of self-help has replaced guides to passions—still necessary, for the introspection advocated by the ancients and by their inheritors is as hard to maintain as it ever was. We do have much more control over our bodies and environments than we used to. But the search for humoural balance continues.

VII

Science: Contemporary Humours

(Twentieth and Twenty-first Centuries)

—*Doctor, my brain hurts!*
—*We'll have to have it out.*

MONTY PYTHON[1]

*Any conception of the world must include some
acknowledgment of its own incompleteness: at a minimum
it will admit the existence of things or events
we don't know about now.*

THOMAS NAGEL[2]

1. The Neurological Self

SOLDIERS WHO, on the many battlefields of the eighteenth and nineteenth centuries, survived shots to the head and then lived with brain damage, instead of dying, became sources for brain research. The idea central to neuropsychology—that to understand mental function, one must study the brain and nervous system—emerged from the observation of patients who had lost a part of their brain function. Odd subjects became objects of scientific scrutiny, as if the scientists needed to notice what could go wrong in order to identify what it took to function normally, or in order to define, even, what it is to be normal. Today, brain-damaged patients can still help neuroscientists reconstruct the genesis of our functions: the capacity to speak, count, write, and read; to make sense of sounds, sights, and words; the capacity to remember places and names or recognize faces; the emotional response to situations; vision and the perception of depth and color; the sense of self through time; or the sense of body image and proprioception.

Given how reliant we are on the brain's ordinary functioning, cases of brain dysfunction can be spectacular for those who witness them, as was for instance that of the man who failed to recognize faces and even "mistook his wife for a hat," as Oliver Sacks recounted in a book that bears that title. Major damage of this sort differs widely from the ordinary dyscrasia that melancholics suffered in the days of Timothy Bright, say. But at that point melancholy also included the more radical disturbances such as those described by Sacks, like that of a man afraid to walk, lest he break the leg he thought was made of glass; another man thought the clouds would fall onto his head, and another believed he was made of butter and so was unwilling to sit next to a fire, for fear of melting. Today these sorts of delusions are treated as neurological cases. Our perception of what constitutes a precise physical illness changed as discoveries about the body's systems accumulated.

It was relatively simple in Broca's day to correlate visible damage (usually from an accident or a stroke) with a broad but still unread area of the brain, and with a dramatically visible disability. We have a clearer sense of the brain's complexities today, so it is no longer simple to determine the causes and mechanisms of a neurological illness with the exactitude commensurate to the task. Neurological dysfunctions are confusing—for those who live with them, of course, but also for those who witness them— because they concern the deepest layers of subjectivity itself. The brain can be understood only insofar as it is the mind, and vice versa.

But the mind is larger than the brain. The loss of one function can be compensated for with a hypertrophy of another function, for instance. A person who becomes blind might develop a capacity to hear sounds no seeing person will pay attention to, and a manic crisis can result in the heightened lucidity and artistic inspiration that are said to precede an epileptic fit (hence the Hippocratic association of manic creativity and epilepsy). Our subjectivity and consciousness, in other words, are more than an assemblage of functions, and more than the sum of identifiable parts. Yet, as Oliver Sacks put it in *The Man Who Mistook His Wife for a Hat*, "Traditional neurology, by its mechanicalness, its emphasis on deficits, conceals from us the actual life which is instinct in all cerebral functions— at least higher functions such as those of imagination, memory and perception. It conceals from us the very life of the mind."

Melancholy was a broad concept that revealed rather than concealed the life of the mind. Timothy Bright had described it as a case of vapors "annoying" the heart and "passing up to the brain," where it "counterfeits terrible objects to the fantasy, and polluting both the substance, and spirits of the brain, causes it without external occasion, to forge monstrous fictions." Since the heart had "no judgment of discretion in itself," it gave "credit to the mistaken report of the brain" and broke "into that inordinate passion, against reason." This was a damnable state in those who relinquished self-control.

Mania and hysteria still exist as names for conditions—though their old versions tend to resemble more common nervous breakdowns and endogenous depressions, broadly put. The contemporary equivalents of melancholy, from its natural to its adust forms, range from those we are usually able (at least partially) to control and steer, such as ordinary blues, unhappy romantic passion, passing depression, or the mild existential pain that suffuses our everyday lives, to the pathologized conditions that might have been seen in Bedlam. Today, these extreme conditions, from obsessive compulsion to schizophrenia, have their own aetiology.

All these identifiable states of mind can be correlated with substances that play the explanatory roles once given to humours: neurotransmitters, enzymes, and hormones, identified, named, and understood with increasing precision. But our mental life has also been pathologized to a higher degree than was possible when the only available treatments were those of the regimen and *materia medica,* and when our fits, foibles, fantasies, and fears were perceived to be the outcome of personality types, on a continuum with milder, ordinary mood swings and humourally determined states. In order to balance humours that have run amok, we know that it is possible to resort to a new, powerful pharmacopoeia and thereby relinquish autonomous control over them, in order to regain balance when reason alone is powerless to do so. Melancholy once named a condition that, at best, one could console; today, that condition tends to be experienced as an illness or negative state that has to be cured, neutralized, eradicated.

With our new humours, syndromes that were once either broadly conceived as possession by spirits, demons, and devils or explained, humourally and more prosaically, as disturbances in the courses of spirits in the three seats of the Scholastics' soul (liver, heart, brain) are much more specific and diversified. But whatever their aetiology, they still point—as they did in the old system—to the fragility of our embodied selves. They testify to the dependence of our very subjectivity and sense of self on mere concoctions. The modern humoural self can still be transformed from within. Our innate temperaments can still turn on us, while reason, the rational mind whose highest functions remain as mysterious as the rational soul once was, observes the chaos.

This is the case, for instance, with Tourette's syndrome, named after Georges Gilles de la Tourette, the pupil of Charcot who first described it. During a Tourette crisis, cognitive functions accelerate massively, leading to tics, sometimes insults, compulsions, repetitive behavior, and so on; but ordinary consciousness is not wiped out. The full person remains, unable to control what is happening, anxious, ashamed, confused. All of us can feel possessed by the very processes at work in our minds, thanks to which we are emoting, cognizing, functioning creatures in the first place. Emotional distress can color everyday perceptions. The passions, as Galen, Aristotle, and Plato knew, can take over the will or reason, and remind us that we are not disembodied, unified, singular souls. On an ordinary level, women know well, for instance, how "chemical" the blues that precedes menstruation can feel, how it is at once intense, suffusing one's whole worldview, and inconsequential once one learns how to recognize its brevity and very particular, clear-cut causality (it is due for the most part to a sharp decline in progesterone and estrogen). But it does take a certain detachment, the recognition of *what,* in chemical terms, that powerful feeling arises out of—black bile, hormonal change, whatever—in order for the rational mind not to follow the siren song of the blues. "That is what it means to be gloomy," as J. M. Coetzee eloquently puts it in his novel *Slow Man:* "at a level far below the play and flicker of the intellect (*Why not this? Why not that?*) he, *he,* the *he* he calls sometimes *you,* sometimes *I,* is all too ready to embrace darkness, stillness, extinction. *He*: not the one whose mind used to dart this way and that but the one who aches all night."[3]

2. The Pharmaceutical Self

NEWSPAPERS CERTAINLY do not neglect the humours. A well advertised scientific publication or a concentration of research on an attractive theme by a few teams of scientists will emerge on the front page—for instance, that falling in love is "all in the brain." The neurotransmitters dopamine and norepinephrine can indeed help us account chemically for the love-melancholy Ferrand was such an expert on: addiction is characterized by an increase in dopamine levels in the brain. One widely reported set of experiments with functional MRIs (which measure levels of oxygen in the blood flowing in the brain), performed on subjects who claimed to be in love, showed that parts of the brain that were particularly activated overlapped with some that are also activated in the experience of euphoria that accompanies the use of drugs—especially cocaine—or gambling. Parts of the brain responsible for the production of the hormones vasopressin and oxytocin, which, as the anthropologist Helen Fisher reported, are involved in the formation of strong bonds between individuals (in prairie voles, actually, but in humans, too, perhaps by way of olfactory signals), were also activated. In the first of their two articles reporting their experiments on the love-struck subjects (focused on romantic love—the second set of experiments, conducted a few years later, compared romantic with maternal love), Andreas Bartels and Semir Zeki exclaimed their fascination "that the face that launched a thousand ships should have done so through such a limited expanse of cortex."[4]

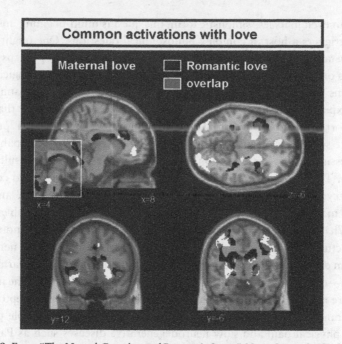

Fig. 29. From "The Neural Correlates of Romantic Love," *NeuroImage* 21(3), 2004, Figure 3: Common activations, by scientists Andreas Bartels and Semir Zeki. This image shows the areas in the brain activated when the subject viewed his/her beloved ("romantic love"), and when the subject viewed his/her child ("maternal love"), and especially the high degree of overlap between the two.

The experiences we consider most fundamental to human life, and to our sense of its value, are all being scrutinized in similar ways—not only language and love but also religious belief and the perception of music and of the visual arts. In 2003 the *New York Times* ran an article by Sandra Blakeslee, "Humanity? Maybe It's in the Wiring," which began with the claim that "neuroscientists have given up looking for the seat of the soul, but they are still seeking what may be special about human brains, what it is that provides the basis for a level of self-awareness and complex emotions unlike those of other animals."[5] The quest for the correlates of emotions and the sense of free will, of what makes us what we are, involves the study of the "circuitry" of the brain, as the article pointed out, rather than of the "specific locations" of functions that neuroscientists were looking for up until the 1950s.

But even in its less simplistic form, this quest is mind-boggling, insofar as it triggers a host of complex philosophical questions. One might ask, for instance, whether these correlations are not redundant: perhaps they tell us only what we already know, confirming our phenomenological intuitions and, in the end, saying little about the contents and meaning of our experiences. One might also ask what it is in the brain exactly that we are correlating with the mind—as the philosopher Jerry Fodor put it, "what sort of brain mechanisms might implement a mind" capable of processing sensory information in the way it does, with the results we know. These sorts of questions, posed by many philosophers and cognitive psychologists, are important to ask as discoveries emerge, and as brain maps become more precise; but they are not easily answerable.[6]

What is clear so far, though, is that advances in the understanding of the actual ailments, disorders, diseases, and syndromes of the nervous system can result in advances in our understanding of how the pharmacopoeia—old and new, natural and synthetic—may actually affect the mind. We are able to produce drugs that work on highly specified receptors in the brain, which regulate neurotransmitters and not only are able to reduce physical pain and slow neurodegenerative diseases such as Parkinson's, but also can change our minds, alleviate our anxieties, or reduce our fears. Common pain relievers like aspirin, ibuprofen, paracetamol, naproxen, nimesulide and other nonsteroidal anti-inflammatory drugs (NSAIDs) aside, medications that work on serotonin and norepinephrine receptors are perhaps the most familiar, given how frequently they are prescribed in the western world, especially in the United States.

Pharmaceutical companies are thus a far cry from the apothecaries of old; but the historical mistrust between apothecaries and the medical corps has not entirely disappeared. Like apothecaries, pharmaceuticals have their own concerns and concoct medications not necessarily in line with the primary concerns of doctors; and like apothecaries, they also seek medical approval, while responding to market forces. There is much investment today in research on pills for anxiety, eating disorders, hysteria, obsessive-compulsive disorders, addiction, phobias, and depression, and in the production of millions of monoamine-oxidase inhibitors (MAOIs) and selective serotonin reuptake inhibitors (SSRIs). Mood—and personality, as Peter Kramer notably described it in 1993 in his *Listening to Prozac*—is humoural enough that it can take but a pill to reroute it, to open or close the neuronal pathways that allow it to take over, replete with its animal spirits. Appropriate doses and drug combinations are not easy for doctors to gauge, but people otherwise oblivious of scientific jargon have become familiar with the initials that apply to their prescribed drugs, designed to boost spirits or calm nerves, to increase energy or decrease appetite. Our desires can be regulated, our will steered—to some extent. From Valium to Prozac, from Zanax to Viagra and beyond, each decade since the 1950s has seemed to have its most discussed designer pill.

Nonmedical psychotropic drugs have been in use for millennia, for religious or recreational purposes. But our neuropharmacopoeia was unimaginable before neurophysiology was born—before the brain became the mind, before pharmaceuticals could target particular receptors in the brain, and also before psychiatry had become a discipline with its own catalog, the *Diagnostic and Statistical Manual of Mental Disorders (DSM)*, regularly updated and now in its fourth edition. In its earlier editions, the *DSM* followed the categories devised by the nineteenth-century pioneers of neuropathology, dividing mental ailments into "organic" ones, which left cerebral lesions found by these early researchers at autopsy; and "functional" ones, which left no cerebral lesions and were merely visible in behavior. The new edition differs insofar as "the basis of the new intellectual framework for psychiatry is that all mental processes are biological, and therefore any alteration in those processes is necessarily organic," in the words of the neurobiologist and Nobel laureate Eric Kandel.[7]

But the organic basis is not always clear, and symptoms are not always easy to differentiate. The current *DSM* orders the "disorders" that have become recognized under headings—anxiety disorders, childhood disorders, eating disorders, mood disorders, personality disorders, psychotic disorders, and substance-related disorders, among others (a category called "other disorders" includes Tourette's syndrome). It gives careful diagnostic criteria and differential diagnoses for post-traumatic stress disorder, separation anxiety, hyperactivity disorder, cyclothymic and dysthymic disorders, antisocial personality disorder, anorexia nervosa, autistic disorder, or schizophrenia.

To label a state is indeed to put it in legible order, to interpret it according to visible symptoms—and often, naming a condition is a way of establishing a diagnosis. This is the case with schizophrenia, for instance, which was coined as a diagnostic category in the early twentieth century by the Swiss psychiatrist Eugen Bleuler.[8] The more syndromes we recognize, the more targeted the possible treatments or palliative care, and so the more effective one may become at dispelling the often acute pain that can characterize serious conditions such as schizophrenia, extreme anxiety, or manic depression. But with this attention to syndromes comes a tendency to indulge in also treating as "disorders" those states whose pathological status is sometimes unclear, such as hyperactivity in children or post-traumatic stress syndrome in adults. We tend to pay medically frantic attention to milder states, or mild versions of hyperactivity or stress—and are quick to turn the possibly normal into the necessarily pathological. Certainly, it is a great relief to be able to put a name to a set of psychological (or physical, for that matter) problems or pains. But much effort would be needed to understand our sufferings in new terms, and thus to enrich with new concepts the nosology that originated in the eighteenth century with Boissier de Sauvages's *Nosologie méthodique.*[9] Insofar as nosology is the classification of diseases according to recognizable and recognized symptoms, it is a list of the state of knowledge at any point in time; and that list is never final.

The criteria for identifying deep depression have always existed—it is the old adust melancholy, which does have the characteristics of a constitutional illness. In an article in the *New York Times,* Peter Kramer took up the humoural history of depression, referring to the Hippocratic definition of melancholy as an illness "that caused epileptic seizures when it affected the body and caused dejection when it affected the mind," as he put it; and also referring to the view expressed in the Aristotelian *Problemata,* XXX that depression was an excess of black bile, otherwise present in moderate amounts in all creative people. Kramer believes that it was the generality of the symptoms ascribed to melancholy by the Greeks that eventually gave it, especially from Ficino on, an ambiguously noble status. As he wrote, "Depression is common and spans the life cycle. When you add in (as the Greeks did) mania, schizophrenia and epilepsy, not to mention hemorrhoids, you encompass a good deal of what humankind suffers altogether. Such an impasse calls for the elaboration of myth. Over time, 'melancholy' became a universal metaphor, standing in for sin and innocent suffering, self-indulgence and sacrifice, inferiority and perspicacity."[10]

For Kramer, this has led us to underplay the sheer negativity of deep depression and the extent to which it is an actual illness, one that not only pertains to mind and brain but also is "linked with harm to the heart, to endocrine glands, to bones. Depressives die young—not only of suicide, but also of heart attacks and strokes. Depression is a multisystem disease, one we would consider dangerous to health even if we lacked the concept 'mental illness.' " Yet, as Kramer also acknowledged years before, in *Listening to Prozac,* the humoural tradition recognized a sliding scale of melancholy states, from a moderate nonpathological condition to extreme illness. In other words, the line is not always clear between illness amenable to psychiatric care and the "neurotic" mind-set that Freud, in particular, defined in the form still familiar to us. The difference between deep, dangerous depression and ordinary blues is real; but that does not make the blues, and our awareness of it, any easier to define.

Nor is neurotic anxiety—the tendency to fret over small, unimportant things, to worry about what people think, to feel anxious without any identifiable reason—a clear-cut condition. But it is probably the closest modern incarnation of the natural sort of melancholy described by Burton and his predecessors: common enough, potentially creative, and possibly more intense in those more likely to create. Unlike fully fledged depression, it has never been an illness. Inadequate self-knowledge was always an unfortunate state, but it was curable and reparable with the right sort of moral advice, of the sort one finds in treatises on the passions and in the moral psychology of the classical tradition. Today, self-help books offering the way to self-knowledge are the offspring of the old treatises on the passions, but they are not the only ones.

Freud had brought in the "talking cure," and despite cries that its days are over, there is a vast panoply of such cures, from counseling to traditional psychoanalysis and various forms of psychotherapy, all aimed at naming and placing emotions in order to control them better—just as Galen and the moral psychologists advised. It takes the cognitive, not the pharmaceutical route, relying on the capacity for reason to access emotion through the exercise of speech and silence. It helps us shake hands, so to speak, with the hidden humours that may govern our reactions, but also with our nonmaterial, non-humoural thoughts—all products of the brain, although hardly experienced as such by our conscious, embodied selves. Because our thought processes are so complex, because our self-consciousness is so ineffable and multiple, and because reason and emotion are so entwined, the definition of what constitutes the "normal," "ordered" brain—rather than the "disordered" one—tends to be rather elusive.

3. Brain Images and Body Image

WILLIS AND WREN, however, would have welcomed neuroimaging techniques. Together with the study of the microscopic world that constitutes us—the genetic and molecular substrate of life—they have been yielding some remarkable insights. Willis might even say that they are making our corporeal soul visible. Undergoing an fMRI exam can be a strange experience. It can trigger the sense that one's rational soul will be visible too, that silent thoughts and fantasies are somehow being "read"—including the very thought that one's thoughts are being read. It can generate a heightened awareness of the many interconnected cognitive and emotive operations, sensations and responses that are concentrated within each second of one's waking life. Of course, what the machine can in fact "see" is limited, by definition. Although it certainly reveals hidden depths, it cannot tell us what our fears, hopes, fantasies, desires, aversions, or abstract thoughts are. It can only register activity; and it requires subtle, careful interpretation to signify anything.[11]

But this and other imaging techniques have in fact been helping us revise Cartesian preconceptions about how we think and feel, showing, notably, the extent to which our emotions are conjoined with our thoughts. There have been many significant discoveries regarding the nature of thought and emotion in the past years. For example, John Allman, at the California Institute of Technology, found "spindle neurons" located in the anterior cingulate cortex of the human brain: these are activated whenever we experience a strong emotion or try to resolve a complex task that requires elaborate decision making. Significantly, these spindle neurons develop from infancy on—and we seem to share them only with great apes, but even then just to a limited extent.[12] They could be seen as one of the structures Willis was looking for when he resolved on the intercostal nerve as that which distinguished humans from animals—not entirely unlike the move made by Descartes when he designated the pineal gland as the seat of the soul.

At any rate, this sort of finding delivers a picture of the brain that corresponds to something we might already know about ourselves: our decision making is not a matter of logical calculation or of disembodied abstract relations. Our sense of self is also embodied, dependent on the "body image" that the brain dynamically creates. One writer on consciousness, Israel Rosenfield, argued that without memory relating to one's embodied self—where memory is identity within a spatiotemporal continuity—there would be no sense of self. As he put it, "the body image becomes conscious by reference to itself."[13] There are cases of brain damage that result in disturbances in the sense of self, stories (some from the early days of neuroscientific research but recast by the likes of Rosenfield and Sacks) in which patients lost the sense that their hand was theirs, or that their house was their own. Ever since Paul Schilder's *The Image and Appearance of the Human Body*, first published in 1950, a heir to the late nineteenth-, early twentieth-century contributions of Henry Head, work continues on the foundations of "body image" and on what Wernicke himself had called the "somatopsyche."[14] More recently, small regions in the brain called the insulae, one in each hemisphere, have also been located, by Arthur Craig at the Barrow Neurological Institute, as probably significant centers for the transformation of sensory information into feelings and conscious sense experience. Blakeslee reported him describing the insulae as "a system that represents the material me."

Perhaps this is true. But such claims can be confusing because one cannot really "locate" in one place the embodied self, and because the "mind," although fully dependent on the brain, cannot easily be conceived as identical with it: the "material me" may seem little more than an abstract idea when it emerges out of a theory about the functions of microscopic neurons, based on laboratory experiments and computer images.

Still, studies of the neurophysiology of emotion, in progress since the 1970s and particularly intense since the 1990s, do help us recapture what had been lost in orthodox interpretations of Cartesian dualism. These studies have been replacing the disembodied, immaterial "me" with the humoural "material me." Now we are the ceaselessly processing, perpetually emoting, neuron-firing, chemically altering, electrically charged material mass of the brain and nerves, the innervated body in constant dialogue with a brain ceaselessly involved in "feedback" with the body, the neuronal body image without which the brain would not be "mine."

As early as 1890, the American psychologist William James, in his monumental *Principles of Psychology*, described emotions as primarily embodied, in terms not unlike those used to describe humoural flows: "a purely disembodied human emotion is a nonentity," he wrote.[15] In his view, the physical manifestation of the emotion, such as crying, even preceded the emotion—here, sadness—rather than following it. We were somatic beings, through and through. Humouralists always knew that; but our humours are even more spectacular, more multiple than what the Hippocratics and especially the Galenists had imagined.

It took a while for neuroscientists and then for cognitive scientists to pay attention to the emotions, to take advantage of the transformation—partly due to James—of moral psychology into investigative psychology, of an ethical tradition into a scientific one. But the notion of an "embodied self," in which emotions are constitutive of the whole "rational" being and necessary to its very existence, clearly became established toward the 1990s. Antonio Damasio opened his first book, *Descartes's Error*, with the famous neuropathological case of Phineas Gage, a railway construction foreman in Vermont. In September 1848, an iron rod pierced Gage's cranium after he accidentally provoked an explosion on site. Gage recovered, after successful treatment, but his personality changed radically: he became irascible, impatient, aggressive, antisocial, and incapable of making decisions. Damasio found that damage to the ventromedial prefrontal cortex, suffered by Gage as well as by one of his own patients, would account for these strange incapacities in an otherwise perfectly intelligent person: it is an area involved in processing emotions and fixing what he called "somatic markers," the body's communications to the prefrontal cortex that produce the "gut" feelings informing decisions we take when we make choices. Perhaps this is what accounts for the concupiscible and irascible appetites of the Scholastics—the tendency to run away from pain and toward pleasure. Without the capacity to make these elementary distinctions, we would also be unable to make wise, "rational" choices.

4. The Emotional Self

WE ARE HARDLY, if at all, aware of those gut-level emotive responses without which we seem unable to function, and which characterize the normal nervous system—or, as one may prefer to call it, the humoural constitution of an organism endowed with a tripartite soul. Our basic emotions—the old primary passions—are functions of our evolved physiology: for the most part, they could be called evolved responses to environmental impacts and threats. But in order to study them, we first have to resort to looking outside our own minds. Ever since

Charles Darwin published *On the Origin of Species* in 1859, and ever since he argued, especially in the *Descent of Man,* which appeared in 1871, that our basic emotions were not distinct from those of animals, psychology has become as comparative as anatomy was in Renaissance Padua.

Using rats, the neuroscientist Joseph LeDoux has carried out extensive research on fear, one of the basic emotions. Others might be anger or rage, maternal attachment, and sexual attraction. (The psychologist Paul Ekman, for his part, famously listed the "basic emotions" as anger, disgust, fear, happiness, and sadness.)[16] Le Doux's experiments led him to understand the centrality of the brain's almond-shaped amygdala, located in the medial temporal lobe, for the elaboration of emotions. As he explained, the emotional stimulus (say, perceiving a snake) is first processed in the sensory thalamus; the percept then takes one or both of two routes: straight to the amygdala, and to the amygdala via the sensory cortex, thanks to which we can cognize what is going on, but always slightly after we begin to react.[17] Quick, rather unpoetic reactions crucial to immediate survival—like fear—are those that have bypassed the cortex. When the Emperor Hadrian gouged out his servant's eye with a stylus, he was, one might say, acting on the amygdala's cue, overtaken with choleric passion. Only afterward, once the gesture had been initiated, and perhaps already accomplished, could he realize what it was he had done. Likewise, the sophisticated emotions we feel and express emerge out of the responses triggered when the cortical pathway is involved, and so when we are able to recognize what is going on.

It is notoriously difficult to reconstruct the work of the nonconscious processes without which there would be no consciousness. It is indeed the job of cognitive science to identify, in LeDoux's words, how "the mental representation of the apple that you consciously perceive is created by the unconscious turnings of mental gears."[18] Emotions are generated and processed at the nonconscious level, within complex neural systems, just as are cognitive operations and perceptions. But when we perceive an object, the operations going on in the brain are quite different from those that happen when we are evaluating this object emotively. The brain even seems able to "know that something is good or bad before it knows exactly what it is."[19] Emotional evaluation precedes full cognition—at least, in situations that would have required, evolutionarily speaking, the capacity for such knowledge. In this sense, emotions in their raw state are cognitive tools. Without them we would be like Phineas Gage: we would not understand the place of the self within time and space, or within the realm of intentions, causes, and effects.

These cognitive emotions are not quite the same as the crude passions and "agitations of the spirits" that people like Thomas Wright worried might distract fine young men from the path of virtue. Once we respond emotionally to a fearful, disgusting, or desirable sight, or to a troubling piece of news, or to a beautiful piece of music, other emotions follow, less immediate and less reactive. It is through them that we consciously know the world. A reaction of anger at a politician's lie, for instance, can be strong but mediated by preexisting analysis, beliefs, judgments, hopes, and fears; and in turn, it will mediate one's judgment of that politician's credentials, perhaps affecting how one later chooses to vote, leading to a shift in one's political principles, and so on.

Emotions are not fixed responses of course, and they can turn sour, too—suffused, one might say, with yellow, blackened, or blackening bile. As LeDoux wrote, virtually echoing the creed of moderation advanced by Aristotle and then taken up within the early modern literature of moral psychology, "When fear becomes anxiety, desire gives way to greed, or annoyance turns to anger, anger to hatred, friendship to envy, love to obsession, or pleasure to addiction, our emotions start working against us." And—unknowingly perhaps—echoing the Hippocratics, he continued: "Mental health is maintained by emotional hygiene, and mental problems, to a large extent, reflect a breakdown of emotional order. Emotions can have both useful and pathological consequences."[20]

Emotional hygiene would require the capacity—also emotionally mediated—not to take the step from appropriate emotional response to inappropriate passion. It would require us to give up our addictions—not to start smoking again, not to think about an impossible infatuation, not to buy yet another pair of shoes. We are not only passionate, appetitive creatures whose responses to hormonal, humoural fluxes and outbreaks are immutable: if we call ourselves rational, it is because of the second-order awareness we have of the emotions one might fear the fMRI can expose. This is one of the ways in which human passions, ranging as they do from sensations to emotions to feelings, differ from animal passions. Once we are aware of an emotion, its content changes. The appetitive soul is amenable to the influence of the rational soul. We are capable of experiencing general states of happiness or anxiety, but also of knowing that we are experiencing these states. It has also been shown that, by visualizing one's heartbeat on a computer screen, one can inflect it, regulate its pace through the conscious control of breathing (just as practitioners of meditation do). The technique has been offered recently to stressed executives and school pupils.[21] Meanwhile, philosophers continue to study the relationship between passion and awareness, emotion and reason, reaction and action.[22]

We may be rationally aware of the extent to which we are made of the fluids that are secreted throughout the body, running through us, from the enteric, or automatic nervous system in the old hypochondria, erstwhile seat of the "appetitive soul," to the central nervous system, seat of the old "rational soul." Yet we are built in such a way that we cannot wholly escape them, and are profoundly conditioned by them and by the "brain's humours." This expression was actually favored by a neuroendocrinologist, Jean-Didier Vincent, who also called the brain an endocrine gland. The brain is protected by the brain-blood barrier. It is penetrable only by the sense-data transmitted by the nervous system and by the humours— the word now used, in fact, to denote "the substances secreted by the cells and by the fluids that transport them."[23] Our hormones, enzymes, and neurotransmitters are always flowing, just as they flow in other animals. Controlled by the small but crucial hypothalamus at the base of the brain, by the pituitary gland, and by the adrenal gland, they determine our sexual characteristics, our appetites, and much else. They trigger each other, prompting and inhibiting secretions, ensuring survival within the external environment and the stability of the "inner environment" (it was Claude Bernard who coined the expression *milieu intérieur*) against its imbalances and aggressors. They work to conserve the organism's homeostasis through complex chemical operations that involve the viscera, those "guts" that humouralists associated with the hypochondria.

Without these humours, one would not live, eat, drink, or reproduce. Their impact overrides both emotion and will, determining the very conditions that enable us to have passions in the first place. Simplifying the enormously complicated, intensely humoural endocrine and neuroendocrine world depicted by Vincent, a man might be in love but impotent perhaps because of an excess of prolactin, a hormone secreted in the pituitary gland; and without luliberin, a neurotransmitter connecting the hypothalamus to the mesencephalus, one might not even fall in love at all. Nor, for that matter, would one ever experience love-melancholy. Perhaps there would be no tome by Burton, and perhaps one would not even know what it means to long. These, surely, are puzzling thoughts.

Yet they are puzzling also because it is easy to forget ourselves within compelling scientific accounts, regardless of whether or not they are truly descriptive of our embodied, physical experience: explanations all too easily replace the phenomena they are explaining. We tend to live through our conceptualizing minds. As Antonio and Hanna Damasio put it: "We spend a good part of our lives attending to the sights and sounds of the world outside us, oblivious of the fact that we (mentally speaking) exist in our bodies, and that our bodies exist in our minds."[24] When we read about what is going on in the brain when we are afraid, or falling in love, there is a fair chance that we will gain some important insights; but it is also probable that these intellectual insights will not change anything about our actual experience of fear or of falling in love. A map of the brain, or indeed of the humours, can only identify features of the landscape; the experience, meaning, and value remain for the emotional traveler to discover. Scientific explanations can help us identify mental pathologies, like acute forms of depression, and develop tools, such as antidepressants, with which to treat resulting ailments. But the definition of what constitutes a mental ailment amenable to physical treatment is not always clear, and although chemical treatment can succeed where psychotherapy fails, the reverse is also true. Chemistry cannot tell us all there is to know about what goes on in a depressive individual, or in any mind; nor can it account for the sense of self. The "explanatory gap" between a scientific theory and actual experience remains identical through time, whether the scientific theory is based on humours or on hormones.

What doctors or scientists tell us about our makeup, internal structures, and mechanisms does not necessarily correspond to the way we represent ourselves to ourselves. For instance, one might argue that most people in western or westernized societies believe that colds are caused by external microbes taking advantage of a body whose resistance has been weakened through the action of internal or external factors. Although microbiologists might be the first to admit how little is really understood about the world of cells, genes, and molecules that they investigate in such detail, their explanations for these events concern activity at the microscopic level. Yet the experience of the cold has nothing microscopic about it: a sneeze is always a sneeze; and the belief that it is due to a "bug," however true, is identical in structure to the belief that it is caused by certain movements of the animal spirits. This might seem a trivial observation; but it is not as self-explanatory as it may seem when what is being observed is neither a sneeze nor a sniff nor a cough but the makeup of human physiology and psychology.

Because of this constancy in the structure of intuitive explanation, time travel is relatively easy for those who seek to understand the history of ideas about human nature. Of course, it would be rash to believe that a contemporary patient's meeting with Galen would be free of misunderstandings. The two would probably agree that Hadrian should and could have controlled his anger. The Roman doctor, however, would know much more about Aristotle than the average GP, or indeed than today's average patient; but he certainly would not understand what the time traveler meant by "virus" or by the notion that the symptoms of a cold were caused by a substance external to the human body. And, crucially, he would be baffled if he were asked to take the patient's blood pressure.

5. New Temperaments

A ROMAN TIME TRAVELER, however, might be surprised to find out that the notion of temperaments is still alive in a form he would recognize. The psychologist Jerome Kagan of Harvard has worked on temperaments for years, concentrating especially on our reactivity before the external world, on our propensity to fear it, on our evolved capacity for "fight or flight," and on the emotions that, in evolutionary terms, have ensured our survival.

These "gut" reactions depend to a large extent on the amygdala, which are particularly involved in inhibition and arousal, our main drives, or, to use scholastic terminology again, concupiscible and irascible appetites. In some ways they are the new hypochondria; Kagan would probably think so. In a book aptly titled *Galen's Prophecy,* he argued—in line with views originating in the 1960s—that Freud, in turning everyone into an equally psychologically burdened, potentially phobic individual, had effectively put an end to the notion that some people are temperamentally more prone to phobia than others. Kagan traced the revival of the idea of temperament to the psychologist Hans Eysenck, who was certainly responsible for pursuing the notion that individuals vary in emotional reactivity and that these variations exist at birth, within the endocrine and nervous systems. It is true enough that all children are not equally fearful, or playful—or indeed choleric, phlegmatic, melancholic, or sanguine.

Kagan tried to understand the nature of these differences, in an effort to find out whether or not types of behavior were physiologically ingrained, as the Hippocratics had thought. He studied the levels and courses of enzymes, hormones, and neurotransmitters, observing, for instance, that infants with a high level of the enzyme dopamine-beta-hydroxylase (DBH), which is essential for the transformation of dopamine into norepinephrine, are "unusually sensitive to lights, sounds and new foods at five months of age; at one year they have an intense dislike of unfamiliar foods."[25] According to Kagan, hormones help account for the difference between easy and fussy young eaters. And the humours go on determining the dominant inner season beyond this stage, of course: among children aged four "who display high degrees of motor activity and frequent crying to visual, auditory, and olfactory stimulation, about one-half are highly fearful when they are one and two years old," perhaps because norepinephrine "increases the excitability of the amygdala and its projections to the corpus striatum, cortex, and sympathetic nervous system." Inversely, children who exhibited an unusually low level of DBH had "severe conduct disorder," in part because "children with very low norepinephrine levels might experience minimal fear over violating parental and community standards, in part because of lower sympathetic reactivity."[26] Our nervous systems are not all equally "irritable," to borrow Stahl's word. It takes a particularly high reactivity that is later fostered in a violent environment to become an antisocial ruffian.

In his belief—shared by the physiologist Ivan Pavlov, a founder of behavioral therapy—that even dogs could be characterized as sanguine, choleric, phlegmatic, or melancholic, Kagan has been building on Eysenck's notion that individuals could be divided into types, especially introverts and extroverts, according to the degree of excitability of the nervous system. For Eysenck, there were states of inhibition: the choleric type was not inhibited in the least; and the phlegmatic was not particularly excitable but inhibited, unlike the melancholic, who was at once underexcited and overinhibited.

This is a tradition in which it is possible to hold the view, explicitly analogous with humoural theory, that the biological makeup of the organism is strongly indicative of an individual's psychology. If, as Damasio

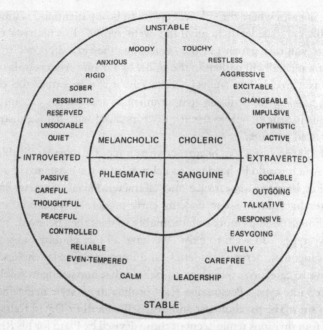

Fig. 30. Diagram of personality types from Hans Eysenck, *Personality and Individual Differences* (New York: Plenum, 1985).

and others like Edelman have established, the body is represented in the brain and particularly in the somatosensory cortex, then it is quite plausible to read bodily types in mental terms—and vice versa—as the humouralists did. Furthermore, the typological labeling Kagan advances may serve to normalize, so to speak, individuals who have been tagged with a psychiatric disorder.

But the problem with typological, as opposed to psychiatric labeling is that it can also be used by political authorities to identify psychological or psychopathological tendencies on the basis of physique and body build, as was once done with phrenology and physiognomy.[27] Categories in general are helpful for those who seek further psychological understanding, but they can also be misused and become dangerously reductive. The discipline of psychiatry, for its part, is based on the capacity to categorize states into identifiable pathologies, and debates about its practices still rage. Medical ethics, sociological studies, and even political decisions engage with these issues. It is difficult to define what is a mental illness, especially in cases where the subject refuses to be a patient and is unwilling or unable to consider that anything went wrong. Psychiatrists cannot deny free will to a patient; at the same time, they are supposed to judge whether a mentally ill patient in the midst of a manic crisis is or is not behaving as a free-willing subject. Obversely, it can be rather too easy to create a nosology of ordinary temperament to account for extraordinary cases: this is certainly of no use for doctors faced with manic, distressed, or depressed patients.

Still, this nosology can be attractive as a register of generalizable variations. And it would be hard not to acknowledge the existence of a correspondence between appearance and character. Facial characteristics do translate character in some way, as early modern physiognomists well knew, though perhaps more unfathomably so than they believed; and personality "types" do seem to exist. We have all met chubby phlegmatics, poised sanguines, nervous cholerics, and introverted melancholics. Most of us love to categorize ourselves, to be told that our strangest foibles can be tucked into types. Personality tests proliferate in style magazines and on Web sites. The psychologist and best-selling author David Keirsey has created four divisions reminiscent of those devised by Plato for his *Republic* —guardians, artisans, rationalists, and idealists. Within these, Keirsey fits vocations or professions, focusing on activity rather than on character to devise his personality types. This system derives in part from the Jungian Myers-Biggs Type Indicator, widely used since its creation in the 1940s. The somatotypes devised by the psychologist William Sheldon in the same period—ectomorphic, endomorphic, and mesomorphic—are inventive variations on the genre, derived from the classifications of the German psychiatrist Ernst Kretschmer. At a "folk" level, astrological signs are another sort of nosology, serving in our idle and not so idle hours as shortcuts to our most broadly defined "personality traits" and "tendencies." We easily recognize the categories in tests and charts that classify intelligence, character, qualities as a lover, self-confidence, anxieties, cleaning habits, favorite music, colors, and whatnot.[28]

But here, humours disappear behind culture. It is rather at the microscopic level that scientific research seems to be corroborating the humoural intuition, which always made sense as a general framework. Yet is it right to say that, from our sophisticated vantage point, we are showing, 2,500 years later, that Hippocrates had rather good insights, even though we are the better scientists? Only to a point. We do know much more than the Hippocratics did; but the remarkable echoes between two so very different approaches to the study of the body, so far apart historically, indicate that there are truths about the body that perdure beyond the theories available to describe it, and also that the efficiency of the maps we use to understand ourselves is not necessarily determined by their age.

6. Mind Over Matter

OUR VARIOUS HUMOURS are keys to the map of our psyche. This map indicates how we imagine our inner world, how we conceive of the relation between our rational, conscious thoughts and the brain that produces them. Our imagination is powerful. And this is why its power over our humours has always been acknowledged—why gazing at representations of illness could be thought to induce illness, and why music could help melancholics. Yet, while our scientific theories allow us to step outside ourselves in order to look back in, our nonconscious minds are active chemists that can affect our bodies from within.

We have barely begun to comprehend the processes underlying psychosomatic phenomena. But there does exist a promising new field, psychoneuroimmunology, which involves studying these mechanisms to help us understand how our states of mind affect our immune system and therefore our capacity to resist illness. Doctors, however, are rarely equipped with the biopharmaceutical knowledge they would need to predict how a treatment will affect the individual patient. Those who also work as research scientists are often sorely aware of the gap between their two practices. Patients rely on their doctors' knowing more than the patient knows, on their having the medical expertise that will justify their being trusted. Without a minimal trust, no effective medical transaction would happen. But in fact doctors are the first to admit that they must often work in a half-light. They must prescribe medications on the basis of faith in the underlying pharmacological research—a modern version of the confidence earlier physicians had to have in the sometimes shady apothecary.

Most of us allow ourselves in any case to swallow milligram after milligram of substances whose exact actions can seem mysterious even to expert clinicians. We generally accept a treatment, whether chemical or herbal, either because we empirically "know" it works, or because an expert or magazine article has advised its use. One may have heard that echinacea reinforces the immune system, believe this piece of information, and act on it by taking extracts of it at the onset of a cold, but without knowing exactly how the substance functions. The *materia medica* is a repository of this sort of empirical knowledge, a catalogue of remedies that, in their crude or distilled form, can be powerfully medicinal. The humoural rationale underlying them no longer applies, and botanical chemistry provides us with much more satisfactory accounts for the workings, say, of witch hazel as an astringent. But of course not all ancient treatments make empirical or even humoural sense: those whose pedigree leads back to magical thinking must have owed whatever potency they had to the phenomenon of placebo.

There is a real difference in effects between a powerful medication and an incantation, or between an analgesic and a banal mint.[29] But there is also a difference between a substance that has a placebo effect and one that does not.[30] Various theories have been advanced to explain the placebo effect. We tend to refer to it pejoratively, to describe medications that have been shown not to have any demonstrable clinical value. But the pejorative connotation is not always called for: we also indulge in the placebo effect and take it for granted in our everyday lives, much more than we are aware of. We all resort to "mind over matter" thinking when we are ill, and are told that "negative" thoughts will only make things worse. In Gaetano Donizetti's 1832 opera *L'elisir d'amore,* a mountebank sells a potion to the lovestruck Nemorino who, after drinking it, does find his powers of attraction radically augmented. The proximate cause of the potion's effectiveness is in fact external circumstance, but also Nemorino's firm belief in its virtues.

Mountebanks are rarely seen today, but quackish claims to cures of everything remain easy to sell. There are stories of people afflicted with severe back pain who claim to get significant relief from magnets; and even more people claim to be helped by homeopathy. These were always "alternative" cures: once they were anti-humoural; today they are still uncorroborated by scientific evidence that they actually work in any specific way. That is not to say that their effects are not real, or even that they might not be found one day to exert identifiable actions on the body. But the only available way so far to account for the staying power of techniques of this sort is the placebo effect—mysterious enough in itself. There is also testimony about cures of deadly cancers through a tenacious belief in the possibility of survival. Some people believe in the efficacy of collective prayer. In the 1920s–1950s, Max Gerson, reacting to the limited efficiency of allopathic and invasive treatments of cancer, and despite the opposition of most of the medical establishment, promoted alternative cures, now offered by the Gerson Institute, consisting of old-fashioned coffee enemas and a fruit and vegetable diet.[31]

There is no evidence that the huge doses of vitamin C recommended by the chemist Linus Pauling to ward off colds and even to reduce cancer (Gerson also recommended its use) actually work in either case, but we certainly make them work when we have colds. Vitamin C is an essential nutrient, but Pauling's own laboratory found that "high doses of the vitamin could reduce the average number of symptomatic days per cold from 7.8 to 7.1"[32] That is an embarrassing figure. But no one seems to mind; Pauling himself did not. Similarly, some parents swear to the effectiveness of homeopathy on their unknowing children—the children's unknowingness is proof to them that the treatment is not placebo. But again, there has been no evidence so far to back up their claim, or at least, no successful randomized control trial (RCT), the official experimental protocol without which no medication can be licensed.

The placebo effect has been shown to take place in some precise cases. For instance, studies of the brains of patients with Parkinson's disease—characterized by an insufficient production of dopamine by the basal ganglia—showed that the basal ganglia responded when they were injected with a saline solution. The response was not as high as when they were injected with actual medication, but it was significantly higher than in patients who were not given anything.[33] The body responds to conscious cues in indirect, often unpredictable ways. As humouralists well knew, happiness does seem to induce health, insofar as it promotes a better immune system, while the obverse, as Kramer observed, is also true—it is possible for depression to cause serious immunodeficiency, if not death. Psychosomatic illnesses seem to be the flipside of positive placebo effects, and they are no less real for their symptoms' being triggered or favored by mental states. The messages that are chemically transmitted between blood cells and neurons are complex enough that our thoughts—whatever a thought *is*—can cause and be affected by changes in our physical or emotional state. Some cancer patients undergoing chemotherapy, which frequently induces nausea and vomiting, have been known to feel nauseated and vomit as they approach the hospital.

We usually experience the placebo effect during what is called the acute phase response to the initial outset of a disease or in the immediate aftermath of an injury. This is the equivalent of the inflammation typical of humoural diagnoses, marked by the "swelling, redness, heat and pain"—*tumor, rubor, calor,* and *dolor*—that humoural doctors knew as signs of physical or mental stress. Classical symptoms include nausea, headaches, fatigue and fever, digestive trouble and appetite loss, swellings and aches.[34] (The mechanisms involved in the acute phase response are also involved in depression and anxiety, whose symptoms are aetiologically akin to inflammations, and not unlike infections).[35] Placebos have been found capable of relieving these symptoms to a degree, by the release of opiates into the organism. Our immune system reacts in complex ways to external inputs, and Pavlovian conditioned reflexes operate at a nonconscious level; they are as potent as they are little understood. There have been a number of experiments testing our responses to such factors as brands of medicine, colors of pills, status of the medical authority prescribing the drug, and so on. In all cases, beliefs of patients have been seen to play a role in shifting their perception of pain.[36] Companies of all kinds make use of this tendency: a brand can seem to affect the taste of a drink as easily as it may justify the price of a pair of shoes.

Of course symptoms do not tell the whole story of illness, malaise, depression, or injury. Doctors cannot rely on the beliefs of patients to be of help, though they sometimes must take these beliefs into account, and probably do to an extent, not always consciously. A positive cure is not the same as one that is based on belief in a shaman or priest, or in a treatment that is a placebo. But the distinction between positive cure and belief is not always clear. The *materia medica* can be derided today as listing mere placebos, and bloodletting can be scoffed at as bogus.[37] But of the thousands of remedies gathered over millennia, some were truly effective, not just placebos. Bloodletting itself worked against high blood pressure, or as an anti-inflammatory.

Whether the issue is placebo or "alternative" therapies—either homeopathic or based on ancient herbal allopathy—fights can be vicious between those who believe that something not proved to work actually does work and those who believe that anything that has not been proved to work simply cannot work. Such fights take place at the highest levels of medical practice, between advocates of "alternative" therapies and practitioners who think only "conventional" therapies make sense. But the boundary between conventional and alternative has always been fuzzy; and ironically, both sides implicitly acknowledge the power of belief.

7. The Regimen Returns

HEALTH IS A state in which the body is a source of pleasure rather than pain. Medical students must know the healthy organism before studying illness, which is a dysfunction, a dyscrasia, an imbalance, a humoural excess or lack—a blip in the overall, subtle *krasis*, in the combination of the body's components. When we are ill or stressed, whether in mind or body or both, we have the immediate sense that something in us has been thrown off-balance. Like sick animals, we might instinctively try to find cures with which to regain our equilibrium. Life is optimally lived when all the body's operations run smoothly, but there are probably as many possible dysfunctions as there are operations.

Even in health, our "internal environment" is never in a constant, static state. Our states of balance are as transitory as the body's inner seasons, punctuated by crises that can occur in their midst—our very own floods, storms, and fires. We preserve our equilibrium partly by fighting against such assaults: from within, with inflammations and fevers; and from without, through conscious decisions about food, sleep, exercise, and medication. This is why the *Regimen* was so popular for so long— and why it remains popular in its modern versions. Whether beset by illness or not, we are always searching for "balance," or prodded to seek it by a whole industry that reminds consumers how much they should long for it. Food, spas, cosmetics, healing techniques and mineral waters are sold on that principle, and year after year diets are generated that are supposed to balance weight, water retention, metabolism—and life.

Indeed, belief does not only influence the way we use medicines; it is a force potent enough to operate blindly in the market for diets, too. A few decades ago, a pineapple-only diet was in vogue that promised graduated, effective, permanent slimming. Its adherents did lose weight as promised, not unsurprisingly, although many ended up severely compromising their health. Not all diets are quite so ill-conceived or dangerous, but to become a fad, a diet must persuade its audience that it will work even if nothing else has worked until then. Its authority must be unquestionable, its guidelines as clear and assured as the list of ingredients and cooking directions in a recipe.

The need for change, for a palliative remedy against possible future ills, can turn food groups into medications, into potential panaceas or possible poisons. Dietary "news" often gets front-page attention. It may focus on the dangers of butter, red meat, sugar, hydrogenated fats, and french fries; or it may extol the virtues of red wine, dark chocolate, whole grains, grape seed oil, virgin olive oil, mustard greens, papaya, or seafood. The earnest prescriptions and proscriptions shift like musical chairs, with each change of tune about benefits and side effects. Broccoli fights against cancer. Vitamin E is a cause of cancer. Pomegranate juice has "antioxidant" properties, and is advertised on the New York subway as the fluid in a hospital drip, more glamorous as a grim lifesaver than as the nectar squeezed out of that wondrous fruit (perhaps the one forbidden in Eden). Pizza eaters have fewer cancers than others. The list could go on. Misused statistics transform facts about foods into certainties about health.

That is because of an increasing demand for facts. The accuracy of media-bred information tends to be far less constant than are potent our taste buds' inclinations, but in an effort to pursue health and longevity, more and more people in the developed world spend increasing amounts of time and money evaluating the pros and cons of foods—and of so-called food supplements, too. Markets and bodies of legislation have emerged out of this need to evaluate what we eat and drink in terms of health outcome. The *Regimen* has become big business. The chemical components of food and drink appear on all U.S. packaging, even of mineral water. And the connection between our body's chemistry and the chemistry of food is used by marketers intent on selling healthfulness, as well as by advertisers keen on stressing the sexiness of dark chocolate or, indeed, the life-prolonging qualities of pomegranate juice.

But food is also a crucial and legitimate concern in a world increasingly focused on health issues, just as it must be true that each meal is an occasion to respond to our body's needs. The more information we ingest, so to speak, along with the food we eat, the more we are aware, even subconsciously, of what *becomes* of the food after it has been chewed, of its effects on the body beyond the taste buds: we are evaluating it in light of the consequences of ingestion on digestion, not of the sensation of taste. Regardless of the countless and ever renewed rumors, right or wrong, about foods' vices and virtues, our physical lives, from circadian rhythms to appetite, are made of the constant inner processes studied by scientists and imagined by humouralists. Of course, even as we produce concoctions at the stove, or imagine dishes by contemplating them in cookbooks, we are—thankfully—unaware of most of these inner processes. The culinary imagination is not only a matter of thinking how bad for those with water retention or hypertension would be that extra bit of salt in the soup, or of how wonderfully healthy that steamed fish would be for everyone. But it is also a sense that a dish is a set of chemical metamorphoses, all of which will ultimately have an effect, however hard to gauge, on our physical and mental state.

That "we are what we eat" seems true enough. And this is of course a central tenet of the theory of humours, where it was clear to all that bodily temperature and humidity could be adjusted by the appropriate foods, herbs, and plants, and where diet was one of the three modes of medical intervention—along with pharmacology and surgery. Now that our humoural history has reached the contemporary world, it is worth returning to those earlier humoural times—specifically to the intensely aromatic and overwhelming cooking one might have experienced then. For a long time, cookbooks relied on the *materia medica* and were partly medical. They served as explicit guides for the pursuit of the *Regimen*, offering recipes for a humourally balanced life in the household, and ultimately for an ethical, morally balanced life in society. Food was as profound as that. The emphasis on its moral dimension was still prominent in nineteenth-century cookbooks such as that of Pellegrino Artusi, *La scienza in cucina e l'arte di mangiar bene,* which established the "bourgeois" cuisine of early-twentieth-century Italy and revolutionized the heavy, spicy cooking of earlier times.[38]

The first printed medical cookbook in which we can witness this cooking was by a rather shady, politically active fifteenth-century Italian humanist called Bartolomeo Sacchi and known as Platina, who eventually became the librarian to Pope Sixtus IV. Entitled *De honesta voluptate et valetudine—On Right Pleasure and Good Health*[39]—and first published around 1474, his book reads like a guide to general and culinary pharmaceuticals. It was written not for medical specialists, but for well-to-do eaters and the politically prominent men of Platina's day, and it appeared in numerous editions, as well as translations from the Latin into Italian, German, and French. It combined standard advice from the *Regimen* about exercise, diet, and so on, with the botanical and pharmaceutical descriptions found in such treatises as the *Circa instans*. One learned here, for instance, that savory "moves the urine and, when drunk in wine, arouses the lethargic who are sleeping as if dead." Surely this is not the sort of information we expect to find in our cookbooks today.

But it was in the book's second half, much of it drawn from *De arte coquinaria* by the influential Maestro Martino, chef to Cardinal Trevisan, that recipes in the form of directions for the use of ingredients, many still recognizable in today's Italy, were bundled with medical and moral advice on how to balance the humours and ensure good digestion. Such advice was necessary, given how heavy the food could be. Here one could read, for instance, that beef was generally "of a cool and dry nature, being very hard both to cook and to digest. It offers gross, disturbed, and melancholic nourishment. It drives a person toward quartan fever, eczema, and scaly skin disease, but veal is eaten more safely because it is almost of medium nourishment, and so the tables of the nobility seek it frequently, with no harm." Hot and moist foods were fine, especially for those in good health; others had to counterbalance lacks and excesses as much as they could. But it was important for food to taste good, and Martino, Platina, and their colleagues would have disapproved thoroughly of the protein powders sold in health shops to those of us eager to boost or control our diet.

Galen himself had written a number of treatises on nutrition—one on the powers of foods, another on the humours derived from foods, and yet another on dieting, taken up by the medieval compilers and translated into Arabic and then Latin.[40] It might well have come in handy: ancient Roman cuisine, most famously epitomized in *De re coquinaria* by the writer known as Marcus Gavius Apicius, could be very rich. Roman cooking was also spicy: pepper, ginger, cumin, poppy seeds, herbs, and the smelly asafetida were commonly used. Fish sauce, honey, and *agresto* or verjuice (the wine made from sour grapes) were essential ingredients in a Roman kitchen. Offal was popular, and ostrich was a luxury item.

What is remarkable in Galen's short manual for dieters is the extent to which some of his advice and analysis of ingredients remains relevant.

True, he did believe that apples, pears, and cucumbers produce a "phleg-matic humour" especially when eaten raw; that lamb should be avoided because of its excessive humidity; and that only "men of an asinine na-ture" should eat horses, asses, and deer, just as only men of a wild nature should eat bears and lions. But many people might well recognize Galen's point that whoever wants to follow a diet should "not even touch" flour, though the dieter could safely take a modicum of semolina, cooked in water or in honeyed or otherwise sweetened wine. Milk whey was thin-ning, although milk itself was fattening, as were soft-boiled eggs. Semo-lina puddings with dill, leeks, and some liquefying substances like marjoram, pennyroyal, savory, pepper, and thyme, on the other hand, were only moderately fattening.[41]

There was always more to food than its nutritional value. Quality mattered. Much Roman cooking would probably seem rather good to us (there are a number of contemporary cookbooks based on Apicius). The Indian Ocean trade had ensured since antiquity the presence in Europe of some eastern spices. But the Romans, Galen included, did not yet have ac-cess to the vast quantities of spices that later Europeans brought from the East, and then from the New World in the West.

Spices made Venice a rich city during the Middle Ages. They would lead Venetians and other Europeans to circumnavigate the world, to es-tablish an international trade and precious transportation routes, and to go to war. The common belief that spices were necessary to cover up the taste of putrid meat and fish is not quite accurate. It is true that spices considered hot and dry were thought to counteract the excessive humidity that seemed to account for putrefaction, and therefore could be used without fear that the rot might kill. It is also true that flesh gone bad could be quite well covered up by pungent spices. But people did not actually expect to eat rotten meat or smelly fish, and in any case spices were costly enough that one would have thought twice before wasting them in such a way. In the absence of refrigeration, salt was used as a preservative to en-sure the provision of some animal protein during the winter, and in these cases spices would have been used to counteract the saltiness. Otherwise, spices simply enlivened dull dishes, and boosted their medicinal powers.[42] Spices were expensive because they were both luxurious ornaments that brought pleasure, and necessary ingredients for the pursuit of health.

Of course, all substances were potentially medicinal or even druglike, in that they had humoural effects. Shakespeare's Falstaff, in *Henry V*, knew this well enough when he sang the praises of sherry:

> A good sherris sack hath a two-fold operation in it. It ascends me into the brain; dries me there all the foolish and dull and curdy vapours which environ it; makes it apprehensive, quick, forgetive, full of nimble fiery and delectable shapes, which, delivered o'er to the voice, the tongue, which is the birth, becomes excellent wit. The second property of your excellent sherris is, the warming of the blood; which, before cold and settled, left the liver white and pale, which is the badge of pusillanimity and cowardice; but the sherris warms it and makes it course from the inwards to the parts extreme: it illumineth the face, which as a beacon gives warning to all the rest of this little kingdom, man, to arm; and then the vital commoners and inland petty spirits muster me all to their captain, the heart, who, great and puffed up with this retinue, doth any deed of courage; and this valour comes of sherris.[43]

But the "valor" of sherry was short-lived. Drunkenness is soon slept off. Spices, on the other hand, like herbs, were carefully prescribed for their long-term properties, listed in the *materia medica*. Ginger, for instance, was always used as a digestive, though that did not make cloves, also endowed with digestive properties, any less precious when they arrived on the market. The history of spices is in part the prehistory of the pharmaceutical industry. To use spices today is, to some extent, to engage in humoural medicine, to tap into the stuff of which empirical "folk" knowledge was made. It also reveals the extent to which the spice trail allowed cuisines to merge, borrow, and develop. The spices most prized in Renaissance Europe had made their way there from the Middle East, via the Arab conquests, and remain widely used in Moroccan cooking, for instance. But they originally stemmed from further east, and they are still staples in Indian cuisine: curcuma, cardamom, fennel seeds, fenugreek seeds and leaves, cumin, coriander seeds and leaves, mustard seeds, poppy seeds, cinnamon bark, chili peppers (only after the Portuguese brought them to Cochin from South America), and of course pepper, cloves, nutmeg, and ginger.

8. Full Circle

THERE IS LITTLE better even today than an infusion of fresh ginger to soothe a bout of nausea. In humoural terms, ginger is "warming." We would say, in more precise modern terms, that it has potent anti-emetic properties. It also has an antispasmodic effect. In many parts of the world, it is used against coughs, and generally is a good anti-inflammatory. Eucalyptus inhalations can unclog a blocked nose, cutting through phlegm, both literally and humourally speaking. A bath of lavender and orange flower essences can help promote a good night's sleep: both plants are re-laxants. In India, turmeric, or curcuma (related to ginger) is often used on wounds and as an anti-inflammatory to treat joint pain or a sore throat, and to aid digestion.

Ethnobotany is a developing field today. Mainstream research into the chemistry of herbs and plants confirms some ancient intuitions, and shows some others to be myths. At any rate, the botanical world has been ex-ploited for as long as biotechnology has existed, yielding some well-known pharmaceuticals. (Many new ones are emerging from the Amazon forest, for instance.) The most familiar is aspirin, or acetylsalicylic acid. It was synthesized in the late nineteenth century on the basis of salicin, the active substance in the bark and leaves of willow trees that had been in use since Hippocratic days as a true pain reliever. More recently, an ingredient in the turmeric root has been found—curcumin—that seems to stop the spread of breast cancer into the lungs and the development of colorectal cancer.[44] Until recently digitalis was used in medications as a heart stimu-lant; it is found in foxglove, a plant that was prescribed by humouralists for dropsy (water retention), which can be a symptom of impending heart failure.

But while herbal remedies have become a veritable industry, and mil-lennial knowledge of plant lore is considered enviable, fewer pharmaceu-ticals are derived from the botanical world than was the case when aspirin was synthesized or when digitalis was processed.[45] Research on curcumin has become mainstream, though it is not yet used as an official medication against cancer—it takes time for a "folk" remedy to be transformed into an authorized drug. But not many of the old "simples"—pharmacologically active substances in their raw, natural state—are deemed worthy of the in-vestment necessary for this transformation. Meanwhile, despite their popularity, practitioners of herbalism are still considered "alternative" doctors, and, just like apothecaries in the past, they have their own, independent certificates. The same applies to practitioners of aromatherapy, homeopathy, and acupuncture. All such "holistic" practices are really alternatives to mechanism in particular; but their status is in fact quite unclear.

Indeed, some of these traditions stem from western empirical traditions, such as alchemy, that were never particularly "mainstream" in any case. When we use disinfectants and antibiotics to kill the germs that cause an illness, we are using today's equivalents of the humoural pharmacopoeia and are in line with humoural and contemporary mainstream medicine. When we use homeopathy, or take vitamins and supplements to promote our defenses against an illness rather than medications to attack its cause, we are, in a way, following the rules of alchemy. But an "alternative" practice such as the ancient art of aromatherapy would have been more acceptable to Hippocratics and official Galenists, who used distilled plant essences, than it is to the medical establishment today. What is considered established shifts in light of medical discoveries and beliefs.

It is significant culturally, medically, and historically that the most popular of the "alternative" medicines today are those from China and India, each based on humouralism, just as fully as is the Hippocratic-Galenic tradition. They have taken root in the West. So have the now increasingly "mainstream" physical practices such as yoga and tai chi based on the integration of mind and body, meant to render us more attentive to our embodiment and rebalance ourselves, as Hippocratics would have wished us to do. And the Galenic tradition still informs much of the language used to discuss clinical conditions in Iran, where, of course, it arrived millennia ago.

The history of the Indian and the Chinese medical cultures, in particular, and of their mutual influences, would require another book. But a history of humours in the West cannot end without opening the door at least to India, with whose traditions it is most entwined. There, a medical system developed called ayurveda, translated as "knowledge of life" or "science of longevity." It is based on passages within the Rig Veda (the oldest of the collections of hymns, or Vedas, from which Hinduism developed), but it was written over centuries and well into the fifth century AD.[46] Since it is based on an ancient cosmology, the correspondence between microcosm and macrocosm is as essential here as it has been in Europe, although there are five elements rather than four (space is the fifth), and there are three temperaments, called *doshas*.

These temperaments are *vata,* corresponding to air and ether (space); *pitta,* corresponding to fire; and *kapha,* corresponding to earth and water. All of us are a combination of the three, but to varying degrees. Someone who has excessive *vata* will have different dietary needs from someone who is high in *kapha.* Here too, health is a matter of balance between the three, pursued through diet. Ayurvedic cooking is based on the counterpoint between the elements of taste, the *rasa,* which can be sweet, sour, salty, pungent, bitter, or astringent; the foods' heating or cooling properties, called the *virya;* and their effects on the body in the aftermath of digestion, called the *vipak,* which can be sweet, sour, or pungent. There is a sophisticated art to using the Indian spices that were so dear to the early modern European world. Cuisine and medicine are as allied in ayurveda as they were in Platina.

The similarities between the Indian and the Greek medical traditions are striking, and there are various historical theories to explain them. Reciprocal borrowings must have occurred at various stages. Indians and Greeks must have met in the pre-Socratic Greece of the late sixth century and the fifth century BC. The emphasis in the ayurveda and in the Hippocratics alike on the centrality to the organism and to the genesis of illness of *pneuma,* wind, air, or breath might attest to that (there was a Hippocratic treatise called *On the Winds).*[47] It has long been believed that Plato's *Timaeus,* so influential in western culture, may have translated, in part, some Indian concepts. (No one has demonstrated this incontrovertibly, however.)[48] Certainly, herbs and spices always traveled across the Indian Ocean during antiquity, and Roman cuisine benefited from these eastern imports—especially pepper, which Hippocratic Greeks already knew of and used for pessaries.[49] In 326 BC, Greeks arrived in northern India with Alexander the Great. And, most recently, during the Mogul period, Arabic and Persian works entered the country, especially those by and on Avicenna's *Canon.* Greek medicine merged somewhat with local tradition, and was named Unani, Arabic for "Ionian"—that is, Greek. Mahatma Gandhi inaugurated the first Unani academy in 1921—it has since become a large company, Hamdard, and spawned a university in Delhi.[50] The ayurvedic tradition itself is alive and increasingly popular in the West, in the form of personality tests, spas, cookbooks, and schools.

It is possible that, regardless of their origins and names, these traditions are attractive precisely because they are "alternatives," slightly mysterious in the eyes of westerners. What those who turn to such practices perhaps fail to realize is that mystery has never entirely disappeared from western concepts of the body. Our sciences pinpoint microscopy, but there are still too many parts to discover for the macroscopic picture to be fully understood. We are more than the sum of our parts. We are fascinated with accounts of ourselves which translate into ordinary language the invisible layers we are made of, and which give us simple rules to live a better life; and this fascination leads to a return to our historical roots.

In our need to understand what we are made of, we resort to structures and traditions that seem to match our intuitions. Doctors know how wide remains the gap between the seen and the unseen; no amount of computerized, hands-off imaging of our innards will replace the presence of the attentive Hippocratic physician. We do not really know what it would mean to bridge, at last, practice and theory, scientific knowledge and "folk" belief. It has never been done. We have complex insides and potent emotions, passions, and tempers, too humoural, too fluid to be easily fixed by a set of answers, a picture on the screen, a pharmaceutical substance.

The old humours are gone, but they still serve as useful, suggestive, and malleable images. Other theories have replaced the humoural system, but its explanatory structure has remained. However much we may now know about the body, the brain, and the mind, the gap between what is known and what is not known must, in the end, remain identical at any given time. That is because theories are always at a distance from the reality they describe, and from the processes—embodied and physiological, abstract and mental—that produce them. Whenever we try to seize ourselves, we get locked into a hall of mirrors. A gap appears between the thought or theory and the reality—ourselves—we are struggling to describe, a gap that is no smaller today than it was in ancient Greece, Renaissance Rome, or nineteenth-century Paris. Humours, and their contemporary equivalents, act as a bridge between the theories we devise and the mental functions that enable us to devise these theories—but without filling in the gap. The historical retelling of explanations people adopted in other times and other places simply reconstructs other, older halls of mirrors.

And so, as a product of the effort to understand ourselves, the humoural system is a reflection of our own self-consciousness. It reminds us that we must content ourselves with approximative explanations, and that the best doctors and the best scientists are those who, like Socrates, know how little they know. The history of humoural certainties is really the underside of our present perplexities.

Endnotes

NOTES TO CHAPTER I

1. Galen, *On the Humours,* in Mark Grant, *Galen on Food and Diet* (London and New York: Routledge, 2000), p. 14.
2. Elisabeth Young-Bruehl, "Psychoanalysis and Characterology," in *Where Do We Fall When We Fall in Love?* (New York: Other Press, 2003), p. 306.
3. On the parallels between Greek and Indian traditions in this regard, see Jean Filliozat, *The Classical Doctrine of Indian Medicine* (Delhi: Munshi Ram Manohar Lal, 1964), especially pp. 196–257.
4. Aristotle, *Metaphysics,* trans. W. D. Ross (London, c. 1908), I.5.
5. The Hippocratic Oath reads as follows:
   "I swear by Apollo the Physician and Asclepius and Hygeia and Panaceia and all the gods and goddesses, making them my witnesses, that I will fulfill according to my ability and judgment this oath and this covenant:
   —to hold him who has taught me this art as equal to my parent and to live my life in partnership with him, and if he is in need of money to give him a share of mine, and to regard his offspring as equal to my brothers in male lineage and to teach them this art—if they desire to learn it—without fee and covenant;
   —to give share of precepts and oral instruction and all other learning to my sons and to the sons of him who has instructed me and to pupils who have signed the covenant and have taken an oath according to the medical law, but to no one else;
   —I will apply dietetic measure for the benefit of the sick according to my ability and judgment;
   —I will keep them from harm and injustice.
   —I will neither give a deadly drug to anybody if asked for it, nor will I make a suggestion to this effect.
   —Similarly I will not give a woman an abortive remedy.
   —In purity and in holiness I will guard my life and my art.

—I will not use the knife, not even on sufferers from stone, but will withdraw in favor of such men as are engaged in this work.

—Whatever houses I may visit, I will come for the benefit of the sick, remaining free of all intentional injustice, of all mischief and in particular of sexual relations with both female and male persons, be they free or slaves.

—What I may see or hear in the course of the treatment or even outside of the treatment in regard to the life of men, which on no account one must spread abroad, I will keep to myself holding such things shameful to be spoken about.

—If I fulfill this oath and do not violate it, may it be granted to me to enjoy life and art, being honored with fame among all men for all time to come; if I transgress it and swear falsely, may the opposite be my lot." (trans. Ludwig Edelstein.)

The Declaration of Geneva (signed in Sydney in 1968) reads thus:

"At the time of being admitted as a member of the medical profession:

—I solemnly pledge myself to consecrate my life to the service of humanity;

—I will give to my teachers the respect and gratitude which is their due;

—I will practice my profession with conscience and dignity; the health of my patient will be my first consideration;

—I will maintain by all the means in my power, the honor and the noble traditions of the medical profession; my colleagues will be my brothers;

—I will not permit considerations of religion, nationality, race, party politics, or social standing to intervene between my duty and my patient;

—I will maintain the utmost respect for human life from the time of conception; even under threat, I will not use my medical knowledge contrary to the laws of humanity;

—I make these promises solemnly, freely, and upon my honor."

6. On Hippocrates, see Jacques Jouanna's authoritative *Hippocrates*, trans. M. B. DeBeroise (Baltimore and London: Johns Hopkins University Press, 1999).

7. Guido Majno, *The Healing Hand: Man and Wound in the Ancient World* (Cambridge, Mass., and London: Harvard University Press, 1975), pp. 116–140.

8. *Places in Man*, in *Hippocrates* VIII, trans. Paul Potter (Loeb, Cambridge, Mass., and London: Harvard University Press, 1995), Loeb: vol. V, pp. 19–21.

9. *The Sacred Disease*, in *Hippocratic Writings*, ed. G. E. R. Lloyd (London: Penguin, 1983), p. 238. Note that for G. E. R. Lloyd, the rationalist Hippocratic author was also guilty of "bluffing" in his assurance that mental illness was due to excessive phlegm or bile. Although, as he wrote, the author's "ruling out of references to the divine or demonic, is a release from one mystification," the new theory was "achieved at a cost of the substitution of another of a different kind, at least when the theorists' own proposed explanations were quite unsubstantiated and imaginary." See G. E. R. Lloyd, *The Revolutions of Wisdom: Studies in the Claims and Practice of Ancient Greek Science* (Berkeley, Los Angeles, London: University of California Press, 1987), p. 28.

10. There is a passage on blood vessels in Aristotle's *History of Animals,* which Aristotle himself attributes to Polybus and which turns out to be identical to one in *Nature of Man.*

11. See Ruth Padel's powerful *In and Out of the Mind: Greek Images of the Tragic Self* (Princeton, N.J., and Chichester: Princeton University Press, 1992). Here, pp. 12–18.

12. Ibid., pp. 68–69.

13. As Padel writes, "What is in us is obscure, like earth," ibid., p. 71.

14. *Nature of Man,* VII, in *Hippocrates* IV, trans. W. H. S. Jones (Loeb, Cambridge, Mass., and London: Harvard University Press, 1931: 1998), p. 23.

15. *Diseases,* I, 2, in *Hippocrates* V, trans. Paul Potter (Loeb, Cambridge, Mass., and London: Harvard University Press, 1988); Vol. V, pp. 102–103.

16. *Regimen* III, 81, in *Hippocrates* IV, p. 411.

17. *Regimen in Acute Diseases,* 66, in *Hippocratic Writings,* p. 204.

18. *A Regimen for Health,* 4, in *Hippocratic Writings,* p. 273.

19. *Prognosis,* 14, in *Hippocratic Writings,* p. 177.

20. *Regimen in Acute Diseases,* 53, 54, 56, in *Hippocratic Writings,* p. 200.

21. *Epidemics,* III: vii, in *Hippocratic Writings,* p. 118.

22. On Asklepios, see James E. Bailey, "Asklepios: Ancient Hero of Medical Caring," *Annals of Internal Medicine,* 124:2 (1996), pp. 257–263.

23. Majno, *Healing Hand,* pp. 201–205.

24. Pliny, *Natural History,* trans. W. H. S. Jones (Cambridge, Mass.: Loeb, 1956), Book 29, 1–2. See Jouanna, *Hippocrates,* pp. 18–19.

25. Pliny, *Natural History,* Book 30, xi.

26. Celsus, *De medicina* (London: William Heinemann, 1971), I, 15. On Celsus and the Roman encyclopedists, see Majno, *Healing Hand,* pp. 353–381; Majno quotes this passage from Celsus, p. 354.

27. Galen, *On the Doctrines of Hippocrates and Plato (De placitis Hippocratis et Platonis),* ed. Phillip de Lacy, III, 5 (Berlin, 1978).

28. Ibid., I, 9.

29. See Padel, *In and Out of the Mind,* p. 58.

30. Galen, *De placitis,* I, 6.

31. Ibid., VI, 6, 2–24.

32. Ibid., VII, 3.

33. Ibid., II, 3.

34. Galen, *On the Natural Faculties,* trans. Arthur John Brock, II, 9 (Cambridge, Mass. and London: Loeb, 1916, 2000).

35. Galen, *On black bile (Peri melainē cholē): De la bile noire,* trans. and ed. Vincent Barras, Terpsichore Birchler and Anne-France Morand (Paris: Gallimard, 1998), p. 95.

36. See Maria Michela Sassi, *The Science of Man in Ancient Greece* (Chicago and London: University of Chicago Press, 2001), pp. 157–158.

37. The exact title is *That the Soul's Faculties Follow the Body's Temperaments*—in Latin, *Quod animi mores corporis temperamenta sequantur.*

38. From such works as the *Parts of Animals* and the *History of Animals.*

39. Galen, *On the Natural Faculties,* II, 9.

40. The destruction was probably a result of a fire Julius Caesar set to spare his fleet and get himself out of a dangerous impasse during the Roman civil wars. Cae-

sar's account of the battle against Achillas—who was "trying to occupy Alexandria"—mentions only that "he [Caesar] burnt all those ships and the rest that were in the docks": see Julius Caesar, *The Civil Wars*, trans. A. G. Peskett, III, 111 (Cambridge, Mass.: Loeb, 1914). But Plutarch, who knew Alexandria well, tells us that the fire "spread from the dockyards and destroyed the great library": see Plutarch, *Lives*, trans. Bernadette Perrin, "Caesar," XIX (Cambridge, Mass.: Loeb, 1919).

41. Plutarch, *Lives*, "Demetrius."

NOTES TO CHAPTER TWO

1. George Eliot, *Middlemarch* (Oxford and New York: Oxford University Press, 1986), p. 118.
2. Tertullian, De ieiunio I.1: cited by Peter Brown, *The Body and Society: Men, Women and Sexual Renunciation in Early Christianity* (New York: Columbia University Press, 1988), p. 77.
3. These included *Epidemics, Nature of the Child, Prognostics, Fractures, Aphorisms* (a collection), and a collection of obscure sayings entitled *On Humours*.
4. Agnellus of Ravenna, *Lecures on Galen's De sectis*, Seminar Classics 609 (Buffalo, N.Y., Seminar Classics 609, 1981), pp. 39, 67. (Latin text and translation.)
5. John of Alexandria, *Commentary on Hippocrates' Epidemics VI—Fragments; Commentary of an Anonymous Author on Hippocrates' Epidemics VI—Fragments,* ed., trans., notes John M. Duffy (Berlin, Akademie Verlag, 1997), pp. 82–85.
6. Paulus of Aegina, *Treatise of Medicine* (3 vols.), trans. and comm. Francis Adams (London: Sydenham Society, 1894), I, 95.
7. Ibid., III, 4.
8. The author of this translation is Istafan ibn-Basil.
9. Abu Zakariyya' Yuhanna ibn Masawayh, *Le livre des axiomes médicaux (Aphorismi),* ed. of Arabic and Latin texts, trans. into French by Danielle Jacquart and Gérard Troupeau (Geneva: Droz, and Paris: Champion, 1980), n. 55, pp. 158–159.
10. Ibid., n. 41, pp. 146–147.
11. Ibid., n. 62, pp. 166–167.
12. Ibid., n. 56, pp. 160–161.
13. Ibid., n. 21, pp. 126–127.
14. This relies on Roy Porter, *The Greatest Benefit to Mankind: A Medical History of Humanity* (New York: Norton, 1999), pp. 83–84, 93–95. On Leviticus, see Mary Douglas, *Leviticus as Literature* (Oxford and New York: Oxford University Press, 1999).
15. The physician was Ya'qub al-Kaskari. See Peter E. Pormann, "Theory and Practice in the Early Hospitals in Baghdad—al Kaskari on Rabies and Melancholy," *Zeitschrift für Geschichte der Arabisch-Islamischen Wissenchaften,* 15 (2002–2003), p. 206.
16. Galen, *On the Use of Breathing,* III, 9.
17. Galen, *On the Natural Faculties,* II, 3.
18. See Vivian Nutton, "God, Galen, and the Depaganization of Ancient Medi-

cine" (The 1999 York Quodlibet Lecture), in Peter Biller and Joseph Ziegler (eds.), *Religion and Medicine in the Middle Ages* (Woodbridge, Suffolk; Rochester, NY: York Medieval Press, 2001).

19. See Peter Brown, *The Body and Society,* esp. pp. 18–19, 77–79.

20. Isaac Israeli, *The Book of Definitions,* in *A Neoplatonic Philosopher of the Early Tenth Century: His Works Translated with Comments and an Outline of His Philosophy,* ed. and trans. A. Altmann and S. M. Stern (Oxford: Oxford University Press, 1958), p. 51.

21. Israeli, "The Book on Spirit and Soul," ibid., p. 109.

22. Constantinus would also translate a work by a disciple of Isaac Judaeus, Abu Ja'far Ahmad ibn Abi Khalid ibn al-Jazzar—who was born in al-Qayrawan, would join a Sufi cell, examine patients for free, and write his own treatise on fevers as well as one on sexual diseases. It was his influential book on pharmacology—the *Treatise on Simple Drugs*—that would be translated by Constantinus as the *Liber de gradibus.*

23. The phrase is in *Epidemics* VI. See Richard Klibansky, Erwin Panofsky, and Fritz Saxl's classic study on the history of melancholy, *Saturn and Melancholy: Studies in the History of Natural Philosophy, Religion, and Art* (London: Thomas Nelson, 1964); the edition used here is the Italian one (recently reissued), *Saturno e la melancolia: Studi su storia della filosofia naturale, medicina, religione e arte* (Turin: Einaudi, 1983, 2002), pp. 46, 79–80.

24. Pliny, *Natural History,* trans. W. H. S. Jones (Loeb, Cambridge, Mass., and London: Harvard University Press, 1956), Book 25, xxii.

25. Al-Kaskari took this from Galen's *De locis affectis.*

26. Paulus of Aegina, III, 14, trans. Francis Adams, *The Seven Books of Paulus Aegineta* (London, 1844), I, p. 383; cited in Pormann, "Theory and Practice," pp. 213–214.

27. Pormann, "Theory and Practice," pp. 214–215. The origin and use of the terms *birsam* and *sirsam* are confusing: *sam* means inflammation in Persian, *bar* is breast, and *sar* is head; but ar-Razi, for one, thought they denoted the same condition. See Manfred Ullmann, *Islamic Medicine* (Edinburgh: Edinburgh University Press, 1978), p. 29.

28. Chaucer mentions ar-Razi in his list of doctors in the *Doctor of Physick:* "Well know he the old Esculapius, / And Dioscorides, and eke Rufus; / Old Hippocras, Hali, and Gallien; / Serapion, Rasis, and Avicen; / Averroism Damascene, and Constantin; / Bernard, and Gatisden, and Gilbertin."

29. Its Latin title was the *Continens;* the translator was Faraj ibn Salim, whose name was Latinized as Farragut.

30. See Charles Burnett, *Arabic Medicine in the Mediterranean,* at http://www.muslimheritage.com.

31. Ibid.

32. Avicenna, *The General Principles of Avicenna's Canon of Medicine,* ed. and trans. Mazhar H. Shah (Karachi: Naveed Clinic, 1966), Part I, Vol. 1, Ch. 3: section II, p. 30. The translation of O. Cameron Gruner, in *A Treatise on the Canon of Medicine of Avicenna, Incorporating a Translation of the First Book* (London, 1930), reads: "Allah most Beneficent has furnished every animal and each of its members with a temperament which is entirely the most appropriate and best adapted for the performance of its functions and passive states.—The

proof of this belongs to philosophy and not to medicine. In the case of man, He has bestowed upon him the most befitting temperament possible of all in this world, as well as faculties corresponding to all the active and passive states of man. Each organ and member has also received the proper temperament requisite for its function. Some he has made hotter, others colder, others drier, and others moister."

33. Avicenna, *General Principles*, Part I, Vol. 1, Ch. 1: Section I, p. 17.
34. Moses Maimonides, *On Asthma*, ed., trans. and annotated Gerrit Bos (Provo, Utah: Brigham Young University Press, 2002), p. 6. (A parallel Arabic-English text of Vol. 1 of Maimonides's complete medical works.)

NOTES TO CHAPTER THREE

1. Chaucer, *Canterbury Tales:* Prologue, ed. Scott German. Online at http://www. canterburytales.org/canterbury_tales.html
2. Montaigne, *Essais* (1580), "Of the Resemblance of Children to their Fathers."
3. See Michel Mollat, *Les pauvres du Moyen Age* (Paris: Hachette, 1999).
4. See David Freedberg, *The Power of Images: Studies in the History and Theory of Response* (Chicago, Ill., and London: University of Chicago Press, 1989). A notable relic was the so-called "true cross" which the emperor Theodosius had given to Saint Ambrose in the eleventh century, and supposed fragments of which could be found in countless medieval churches.
5. The translation had been executed by a theologian who had also translated a few of Galen's treatises, William of Moerbeke.
6. See http://www.piar.hu/councils/ecum12.htm, entry 22.
7. See Nemesius, *On the Nature of Man*. It was translated by the physician Alfanus in the mid-eleventh century, before he became bishop of Salerno.
8. Mondino made use of Burgundio da Pisa's Latin translation of the Arabic, shortened version of Galen's *De usu partium*, which he had commented on.
9. In the sixteenth century, *De spermate* was still printed along with authentic works by Galen. The sole vernacular translation from the Latin was into Middle English, in the fifteenth century.
10. Galen, *De locis affectis*, V, 8. Quoted in Manfred Ullmann, *Islamic Medicine* (Edinburgh University Press, 1978), p. 35. The description was picked up by Rufus of Ephesus and thence copied by Aetius of Amida.
11. A large number of documents about Alderotti—author of one of the earliest commentaries on Avicenna's *Canon*—have survived. These documents were extensively studied by Nancy Siraisi, and presented and discussed in her *Taddeo Alderotti and His Pupils: Two Generations of Italian Learning* (Princeton, N.J.: Princeton University Press, 1981).
12. Siraisi, *Taddeo Alderotti*, p. 130.
13. Ibid., p. 282.
14. Geoffrey Chaucer, *The Nun's Priest's Tale of the Cock and Hen, Chantecleer and Pertelote*, ed. Sinan Kökbugur (Librarius, http://www.librarius.com/cantales.htm), v. 102–149.
15. See Guido Majno, *The Healing Hand: Man and Wound in the Ancient World*

(Cambridge, MA, and London: Harvard University Press, 1975), p. 188: "Imagine cutting the veins of a patient who has already half bled to death through his wound, plus the combined effects of enemas, purges, and vomiting, plus the side effects of poisoning with hellebore, and all of this on a starvation diet."

16. Some scholars have thought that Trotula is the name of a female Salernitan physician from the eleventh or twelfth century, but no such Trotula has been found.

17. "On the Regimen of Pregnant Women," in *The Trotula*, ed. and trans. Monica H. Green (Philadelphia: University of Pennsylvania Press, 2002), p. 77.

18. Cited in Katharine Park, "Medicine and Society in Medieval Europe, 500–1500," in Andrew Wear, *Medicine in Society: Historical Essays* (Cambridge and New York: Cambridge University Press, 2002), pp. 59–90.

19. See Katharine Park, *Doctors and Medicine in Early Renaissance Florence* (Princeton, N.J.: Princeton University Press, 1985).

20. The discovery of the vacuum in the seventeenth century is due to Evangelista Torricelli and Robert Boyle's experiments with the air pump at Oxford during the 1650s and 1660s, over and against the Aristotelian belief that "nature abhors a vacuum."

21. See http://www.fda.gov/bbs/topics/answers/2004/ANS01294.html. See also John Colapinto, "Bloodsuckers," *New Yorker* (July 25, 2005), pp. 72–81.

22. See his *Libellus de conservatione sanitatis* (published in 1475). Cited in Carlo Cipolla, *Miasmi e umori* (Bologna: Il Mulino, 1989), p. 86n, from Gonario Deffenu, *Benedetto Riguardati* (Milan, 1955).

23. See Nancy Siraisi, *Medieval and Renaissance Medicine: An Introduction to Knowledge and Practice* (Chicago, Ill., and London: University of Chicago Press, 1990), pp. 115–118.

24. Ibid., p. 116.

25. See Majno, *Healing Hand*, p. 414. Majno tells us that the compound could still be found listed in the German and French pharmacopoeia in the late nineteenth century. In his *A Traveler in Italy* (Methuen, NY: Dodd, Mead and Co., 1964; Da Capo Press, 2002), H. V. Morton reported having found a Venice pharmacy, *Testa d'Oro*, which still sold theriaca. (Thanks to Susan Gainer for the reference.)

26. Galen, *De simplicium medicamentorum temperamentis ac facultatibus*; Pliny the Elder, *Historia naturalis*; Theophrastus, *Historia plantarum*.

27. Another translation from an Arabic text (by the twelfth-century Serapion the Younger) was the *Herbolario volgare*, widespread in the Renaissance and useful as a short, schematically illustrated guide to the nature and medical properties of the hundreds of plants that had been catalogued up until then.

28. The author of the *Circa instans* was a famous Salernitan physician called Matthaeus Platearius. The first French edition was printed in Besançon in 1488.

29. John Gerard: 1597; Thomas Johnson: 1633. The Flemish herbal (in Latin) was by Rembert Dodoens (1583); the German one was by Jacob Dietrich von Bergzabern.

30. Matthaeus Platearius, *Le livre des simples médecines*, from MS. 12322, Bibliothèque Nationale de Paris, adapted by Ghislaine Malandin (Paris: Éditions Ozalid, 1986).

31. In works such as *De simplicium medicamentorum temperamentis, De locis affectis*, and *De compositione medicamentorum secundum locum*.

32. Shakespeare, *Romeo and Juliet,* Act II, Scene 3.

33. See Jean Starobinski, "Le passé de la passion: Textes médicaux et commentaires," *Nouvelle Revue de Psychanalyse,* 31 (1980), pp. 51–76.

34. *A Regimen for Health, 5,* in *Hippocratic Writings* ed. by E. R. Lloyd (London: Penguin, 1983), pp. 273–274.

35. See the English text of the *Regimen sanitatis Salernitatum: A Salernitan Regimen of Health* at http://www.godecookery.com/regimen/regimen.htm.

36. *Regimen,* in *Hippocrates* IV, trans. W. H. S. Jones (Loeb, Cambridge, Mass., and London: Harvard University Press, 1988), II, LVI.

37. Ibid., II, LXII.

38. Ibid., III, LXII.

39. See Galen, *De differentiis febrium.* Al-Majusi and his translator Constantinus would refer to it.

40. Girolamo Fracastoro, *De sympathia et antipathia rerum, liber unus: De contagione et contagiosis morbis et eorum curatione, libri III* (Venice, 1546).

41. Cipolla, *Miasmi e umori,* pp. 60, 86, n. 30.

42. This account largely relies on Robert S. Gottfried, *The Black Death: Natural and Human Disaster in Medieval Europe* (New York: Free Press, 1983).

43. Ullmann, *Islamic Medicine,* p. 93.

44. Marchione di Coppo Stefani, *Cronaca fiorentina,* Rerum Italicarum Scriptores, vol. 30, ed. Niccolo Rodolico (Città di Castello, 1903–13), Rubric 643, at http://www.fordham.edu/halsall/med/marchione.html.

45. Movable type was a new technique in the West, although it had already been in use in China for perhaps as long as a millennium.

NOTES TO CHAPTER FOUR

1. Thomas Walkington, *The Optick Glasse of Humours* (1607; 1631), p. 136.

2. François Rabelais, *Gargantua and his son Pantagruel (Le Tiers-Livre,* 1546), trans. Sir Thomas Urquhart of Cromarty (1693: 1894), 3, XXXI: "How the physician Rondibilis counselleth Panurge."

3. Giorgio Cosmacini, "La malattia del duca Francesco," in *Carteggio degli oratori mantovani alla corte sforzesca (1450–1500),* ed. Franca Leverotti, Vol. 3: 1461 (Rome: Ministero per i beni e le attività culturali. Ufficio centrale per i beni archivistici, 2000), pp. 23–26.

4. This is Cosmacini's account, ibid.

5. Ibid.

6. Marsilio Ficino, *Liber de vita: Three Books on Life,* trans. and ed. Carol V. Kaske and John R. Clark (Binghamton, N.Y.: Medieval & Renaissance Texts & Studies, in conjunction with the Renaissance Society of America, 1989), I, 10.

7. Dante, *Divine Comedy,* Canto VII: "Fitto nel limo dico: 'Tristi fummo / ne l'aere dolce che dal sol s'allegra, / portando dentro accidioso fumo: // or ci attristiam ne la belleta negra' // Così girammo de la lorda pozza / grand'arco, tra la ripa secca e'l mézzo, / con li occhi vòlti a chi del fango ingozza."

8. See Jean Starobinski, "Tradition de la mélancolie" (doctoral thesis, 1960), in Yves Hersant, *Mélancolies: De l'antiquité au XXème siècle* (Paris: Robert Laffont, 2005), pp. 550–559.

9. *Hippocratic Writings*, ed. G. E. R. Lloyd (London: Penguin, 1983), *Aphorisms*, XXIII, LXVI.

10. See Raymond Klibansky, Erwin Panofsky, and Fritz Saxl, *Saturn and Melancholy* (London: Nelson, 1964); the edition used here is the Italian one, *Saturno e la melancolia: Studi in storia della Filosofia naturale, medicina, religione e arte* (Turin: Einaudi, 1983, 2002). Recent anthologies on melancholy include: Jennifer Radden, ed., *The Nature of Melancholy: From Aristotle to Kristeva* (Oxford: Oxford University Press, 2000); Patrick Dandrey, *Anthologie de l'humeur noire: Écrits sur la mélancolie d'Hippocrate à l'Encyclopédie* (Paris: Gallimard, Le Promeneur, 2005); Hersant, *Mélancolies: De l'Antiquité au XXème siècle*. See also the catalogue, edited by Jean Clair, of the exhibition *Mélancolie: Génie et folie en Occident* (Paris: Gallimard, 2005), which took place in 2005–2006 at the Grand Palais in Paris and at the Nationalgallerie in Berlin in 2006.

11. For an epochal, critical discussion of this view, see Klibansky, Panofsky, Saxl, *Saturno e la melanconia: Studi in storia della Filosofia naturale, medicina, religione e arte* (Turin: Einaudi, 1983, 2002), pp. 302–321, 328–342.

12. The association of air and melancholy has a rich classical history; see Ruth Padel, *In and Out of the Mind* (Princeton, N.J. and Chichester; Princeton University Press, 1992), pp. 88–98.

13. Galen, *On the Usefulness of Parts*, I, 233. Galen also wrote a treatise on black bile: see Galien, *De la bile noire*, ed. and trans. Vincent Barras, Terpsichore Birchler, and Anne-France Morand (Paris: Gallimard, 1998).

14. Alcabitius drew on Ptolemy and other Arabic sources, such as Abu Masar and al-Kindi. See *Al-Qabisi (Alcabitius): The Introduction to Astrology—Editions of the Arabic and Latin texts and an English translation*, eds. Charles Burnett, Keiji Yamamoto, Michio Yano (London and Turin: The Warburg Institute, Nino Aragno Editore, 2004). See also Klibansky, Panofsky, Saxl, *Saturno e la melancolia*, pp. 119–125.

15. See William R. Newman, *Promethean Ambitions: Alchemy and the Quest to Perfect Nature* (Chicago, Ill.: University of Chicago Press, 2004), pp. 37–41.

16. Ficino, *Three Books on Life*, II, 11.

17. Ibid., III, 6.

18. Ibid., I, 26.

19. Ibid., III, 22.

20. Ibid., I, 12.

21. Ibid., I, 10.

22. Plato, *Republic*, 617b–c.

23. Plato, *Timaeus*, 14/47d: "So much of music as is adapted to the sound of the voice and to the sense of hearing is granted to us for the sake of harmony; and harmony, which has motions akin to the revolutions of our souls, is not regarded by the intelligent votary of the Muses as given by them with a view to irrational pleasure, which is deemed to be the purpose of it in our day, but as meant to correct any discord which may have arisen in the courses of the soul, and to be our ally in bringing her into harmony and agreement with herself; and rhythm too was given by them for the same reason, on account of the irregular and graceless ways which prevail among mankind generally, and to help us against them." (Translation Benjamin Jowett)

24. Aristotle, *Metaphysics*, 985b–986a.

25. Ficino, *Three Books on Life*, III, 22.

26. Ficino, *Sopra lo amore o ver'Convito di Platone: Comento di Marsilio Ficini Fiorentino sopra il Convito di Platone*, ed. G. Ottaviano (Milan: Celuc, 1973), VI, 6. (The text was first composed in Latin in 1469; the Italian translation was executed by Ficino himself and published posthumously in Florence in 1544).

27. But see Klibansky, Panofsky, Saxl, pp. 264–266.

28. See Angela Voss, "Marsilio Ficino, the Second Orpheus," in *Music as Medicine: The History of Musical Therapy since Antiquity*, ed. Peregrine Horden (Aldershot: Ashgate, 2000), pp. 154–172.

29. Ibid., p. 161.

30. William Shakespeare, *Twelfth Night; or, What You Will*, Act II Scene IV.

31. Samuel I: 16, 23.

32. Ficino, *Three Books on Life*, I, 10.

33. Martianus Capella, *The Marriage of Philology and Mercury: Le nozze di Filologia e Mercurio*, ed. and trans. Ilaria Ramelli (Milan: Bompiani, 2001), IX, 926.

34. Brandolini wrote the piece for Giovanni de' Medici, who became Pope Leo X at the time when he presented it to him, in 1513.

35. Raffaele Brandolini, *On Music and Poetry (De musica et poetica)*, trans., intro., and notes by Ann Moyer and Marc Laureys (Tempe, Ariz.: Medieval and Renaissance Texts and Studies, Arizona Center for Medieval and Renaissance Studies, 2001), Vol. 232, 32v–33r.

36. A Spanish treatise, *Musica practica*, by Bartolomeo Ramos de Pareja, appeared in 1482; it described in nearly scientific detail these neat, all-encompassing correspondences.

37. See Kristin Lippincott, "Two Astrological Ceilings Reconsidered: The Sala di Galatea in the Villa Farnesina and the Sala del Mappamondo at the Caprarola," *Journal of the Warburg and Courtauld Institutes*, 53 (1990), p. 185. In 1934, Fritz Saxl, Warburg's heir at the helm of what by then had become the Warburg Institute, calculated (with the help of another scholar, Arthur Beer) Chigi's birth date on the basis of the ceiling. The result was either November 30 or December 11, 1466. Further calculations then narrowed down the range of possible dates to November 29 or 30 of that year. Chigi's baptismal record was published in 1984 by the historian Ingrid Rowland, revealing that Agostino was "baptized on the thirtieth day of November 1466 and was born on the twenty-ninth day of the said month at the hour 21½."

38. Ibid.

39. Paracelsus, *Selected Writings*, ed. Jolande Jacobi, trans. Norbert Guterman (Princeton and Chichester: Princeton University Press, 1951/1988), p. 21: from *Das Buch Paragranum, letzte Bearbeitung in vier Büchern* (1530).

40. Ibid.

41. Ibid., pp. 19–20. Galen had denied that categories of taste, as opposed to qualities, were of any medical use.

42. The first English translation would appear a century later, in 1658.

43. Giambattista della Porta, *Magiae naturalis libri XX* (Naples, 1584); translated into English as *Natural Magick* (London, 1658; facsimile reprint New York: Basic Books, 1957), pp. x, 15.

44. Ibid., VIII, 3.

45. Vivian Nutton, "Introduction" in Vesalius, *De humani corporis fabrica*, ed. Daniel Garrison and Malcolm Hast, at http://vesalius.northwestern.edu/flash.html.

46. The reasons for Vesalius's pilgrimage are not clear. The explanation most frequently given is that it was a penance imposed by the emperor to satisfy a call by the Spanish Inquisition for Vesalius's execution on trumped-up but at that time typical charges of Protestant heresy.

47. The College of Physicians still exists under that name; it describes itself (on its Web site) as a "charity that aims to ensure high quality care for patients by promoting the highest standards of medical practice." See http.//www.replondon.ac.uk/college/.

48. It was printed by Marco Fabio Calvo, a friend of the painter Raphael.

49. In 1528, there appeared in Paris the 1322 Latin translation by Niccolò da Reggio of Galen's *De usu partium*. The next year, the first reliable translation into Latin of Galen's important *Anatomical procedures (De Anatomicis administrationibus)* came out. It was begun by a Greek translator, scholar and friend of Thomas Linacre's, Demetrius Chalcondylas, and then corrected by Jacopo Berengario da Carpi. It was then revised again by Vesalius's teacher Guinther, in view of the printing of the complete Galen in Latin.

50. Benvenuto Cellini, *Autobiography*, trans. John Addington Symonds (New York: Collier Press, Kessinger Publishing, 2005), ch. 28, pp. 53–54.

51. Nutton, "Introduction," *De humani corporis fabrica*, at http://vesalius.northwestern.edu).

52. Ibid.

53. On fugitive sheets, see Andrea Carlino, *Paper Bodies: A Catalogue of Anatomical Fugitive Sheets, 1538–1687,* trans. Noga Arikha (London, Wellcome Institute for The History of Medicine: *Medical History*, Supp. n.1g.,1999).

54. *Gli amori degli dei* is the work of two Mannerist painters: Perino del Vaga (a pupil of Raphael and of Ghirlandaio) and the eccentric, great visionary Rosso Fiorentino. On its connection with Estienne, see Carlino, *Fugitive Sheets*, p. 26.

55. The *Fabrica* was printed in Basel by Johannes Oporinus, who had until recently been a professor of Greek and Latin there.

56. *Fabrica*, 2r–2v, at http://vesalius.northwestern.edu.

57. Fuchs's *New Herbal* was illustrated by three artists: Albrecht Meyer, Heinrich Füllmaurer, and Veyt Rudolff Speckle, and it was also printed in Basel by Johannes Oporinus.

58. *De humani corporis fabrica ex Galeni & Andreae Vesalii libris concinnatae epitome,* 1551.

59. *Fabrica,* 2v.

60. Ibid., 3r.

61. See the point made here by the editors of the online edition of the *Fabrica*.

62. Valverde published the stunning *Anatomia del corpo humano* in Rome in 1560. Eustachi observed and illustrated in color organs such as the uterus, the inner ear—hence the Eustachian tube—the adrenal glands, and the kidneys, composing a *Tabulae anatomicae* that was not printed until 1783.

63. *De subtilitate*, published in 1550; and *De rerum varietate*, published in 1557.

64. Anthony Grafton, "Introduction" to the classic English translation by Jean Stoner of *The Book of my Life: De vita propria liber*, 1575 (New York: New York Review Books, 2002), p. xii.

65. See Nancy Siraisi's book on Cardano's medical career, *The Clock and the Mir-*

ror: *Girolamo Cardano and Renaissance Medicine* (Princeton, N.J.: Princeton University Press, 1997).

66. Anthony Grafton showed this in his book on Cardano's astrological career, *Cardano's Cosmos: The Worlds and Works of a Renaissance Astrologer* (Cambridge, Mass., and London: Harvard University Press, 1999, new edition 2001).

67. Shakespeare, *King Lear*, Act I, Scene 2.

68. The historian Elisabetta Mori was the first to study these letters. I owe the information about Isabella to her research, on which I gratefully rely for this account. See her "L'onore perduto del duca di Bracciano: dalla corrispondenza di Paolo Giordano Orsini e Isabella de' Medici," *Dimensioni e problemi della ricerca storica*, No. 2 (2004), pp. 135–174.

69. Eucharius Rösslin's *Rosegarten*—its full title was *Der Swangern Frawen und hebammen roßgarten*—was published in 1515, the first treatise of its kind. It would remain in print for 200 years.

70. Galen, *On the Natural Faculties*, II, 9.

71. Bedlam was first founded as the Order of Saint Mary of Bethlehem in the mid-thirteenth century, and was a hospice for the insane from 1377.

72. Robert Burton, *The Anatomy of Melancholy* (New York: New York Review Books, 2001), First partition, section 3, number 2, subsection IV.

73. This comes from the French physician Jean Fernel (1497–1558), who in 1548 published *On the Hidden Causes*, a commentary on Galen's *De morbus causis*.

74. Burton, partition I, section 2, member 3, subsection V.

75. Ibid., partition III, section 2, member 3 (no subsection).

76. Jason Pratensis, *De cerebri morbis* (Basel, 1549).

77. Burton, *Anatomy of Melancholy*, Partition III, section 2, member 3.

78. Ibid., partition I, section 3, member 1, subsection II

79. Ibid., partition I, section 1, member 2, 4

80. Thomas Walkington, *Optick Glasse of Humours* (London, 1607, 1631), pp. 131–132.

81. The first, rather different edition of Ferrand's work had appeared earlier, in 1610. The second edition was translated into English by Edmund Chilmead and appeared in Oxford in 1640.

82. Galen, *De placitis Hippocratis et Platonis*, ed. Kühn, vol. V, p. 429; reference provided in Jacques Ferrand, *La mélancolie érotique* (Paris, 1610, 1623): *A Treatise on Lovesickness*, trans., ed., and critical intro Donald A. Beecher and Massimo Ciavolella (Syracuse, N.Y.: Syracuse University Press, 1990).

83. Ferrand, ch. 7.

84. Ibid., ch. 8.

85. In their extensively annotated edition of Ferrand's text, Beecher and Ciavolella show that Ferrand seems to have poached this description from another, earlier work by Jean Liébault (c. 1535–1596), *Trois livres appartenans aux infirmitez et maladies des femmes* (1598), the translation into French of an Italian work by another physician, Giovanni Marinelli. Liébault was a physician who had been born in Dijon; as it happens, he was married to the daughter of Charles Estienne.

86. Ferrand, *Mélancolie*, ch. 33.

87. Ibid., ch. 31.
88. Pharmaceuticals are outlined in ch. 32.
89. Ferrand, *Mélancolie*, ch. 35.
90. Ibid., ch. 30.
91. Burton, *Anatomy*, partition III, section 2, member 5, subsection II.
92. Shakespeare, *A Midsummer Night's Dream*, Act V, Scene 1.
93. Burton, *Anatomy*, partition I, section I, member 1, subsection V.

NOTES TO CHAPTER FIVE

1. Molière, *Le malade imaginaire*, Act III, Scene 5 (1673).
2. Thomas Browne, *Religio medici* (London, 1643), part I, section 43. Quoted from London, 1835 edition, by Guido Giglioni, *Immaginazione e malattia: Saggio su Jan Baptiste van Helmont* (Milan: FrancoAngeli, 2000), p. 9, n.1.
3. Giacomo della Porta wrote an influential treatise on physiognomy. The Bolognese natural philosopher Ulisse Aldrovandi described new species and monsters. So did Ambroise Paré in his *On Monsters and Marvels*. The multivolume *Histoires prodigieuses* was published over a couple of decades in the second half of the sixteenth century.
4. Michel de Montaigne, *Essais:* "Of the Resemblance of Children to their Fathers."
5. Campanella's text was first published in Italian in 1602 as *La città del sole: Dialogo poetico;* the Latin edition appeared in 1613–1614.
6. Tommaso Campanella, *La Città del Sole: Dialogo Poetico—The City of the Sun: A Poetical Dialogue,* trans. and ed. Daniel J. Donno (Berkeley, Los Angeles, and London: University of California Press, 1981), p. 57. (Bilingual edition.)
7. See Germana Ernst, "The Sky in a Room: Campanella's *Apologeticus* in defense of the pamphlet *De siderali fato vitando,*" trans. Noga Arikha, *Culture and Cosmos,* 6:1 (Spring–Summer 2002), pp. 45–54.
8. In 1620, Maffeo Barberini had even written a poem in Galileo's honor, *Adulatio perniciosa.*
9. Burton, *The Anatomy of Melancholy* (New York: New York Review Books, 2001), partition II, section 2, member 6, subsection III.
10. Ibid., partition II, section 2, member III.
11. Ibid., partition I, section 3, member 1, subsection III.
12. Aristotle outlined these dissections in *On the Motion of Animals* especially.
13. Aristotle, *On the Generation of Animals,* III, X, 760b.
14. Galen, *On the Usefulness of Parts,* I, 303.
15. Ibid., I, 32–33.
16. Ibid., I, 324.
17. Michael Servetus, *Christianismi Restitutio,* in Charles Donald O'Malley, *A Translation of His Geographical, Medical, and Astrological Writings with Introductions and Notes* (Philadelphia, Pa.: American Philosophical Society; and London: Lloyd-Luke Medical, 1953), pp. 201–208.
18. Manfred Ullmann, *Islamic Medicine* (Edinburgh: Edinburgh University Press, 1978), p. 69.
19. William Harvey, "Movement of the Heart and Blood in Animals: An Anatomi-

cal Essay," in *The Circulation of the Blood and Other Writings,* trans. Kenneth J. Franklin (London: Dent, 1963), p. 63.

20. Ibid., p. 174.

21. Ibid., p. 175.

22. Francis Bacon, *Novum organum,* published in 1620 as the second part of *The Great Instauration.*

23. The French followed suit in 1666, when the minister Jean-Baptiste Colbert gathered the natural philosophers who were to make up the Académie Royale des Sciences—although Louis XIV did not give it a formal charter until 1699.

24. Harvey, *The Circulation of the Blood,* p. 150.

25. Harvey, *Movement,* p. 92.

26. Harvey, *Circulation,* p. 165.

27. This was the view of the seventeenth-century lawyer Gérauld de Cordemoy.

28. Another set of objections was from Gassendi, to whom Mersenne, a frien of both him and Descartes, had sent the manuscript of the *Meditations.* Gassendi wondered why Descartes had bothered with the evil genius and the complicated demonstrations that followed, rather than just settling with the skeptical idea—continuously developed from its classical sources over at least a century, and notably by Montaigne in the 1500s—that it was our nature to have limited knowledge and that we had to make do with the probable, at least for the sake of pragmatism. (This objection seemed so beside the point to Descartes that he chose to leave it out of the first edition.)

29. The long article on animal minds is under the entry "Rorarius," for Girolamo Rorario, a sixteenth-century papal legate who in 1544 had written a tract *That Brute Animals Use Reason Better than Men,* arguing exactly that, just as Montaigne had done. It had been published, in Paris, only 100 years later, in 1648.

30. On the Mary Toft case, see Dennis Todd, *Imagining Monsters: Miscreations of the Self in Eighteenth-Century England* (Chicago, Ill., and London: University of Chicago Press, 1995).

31. See Walter Pagel's seminal book, *Joan Baptista Van Helmont: Reformer of Science and Medicine* (Cambridge: Cambridge University Press, 1982). My section on Van Helmont is largely based on Pagel's account.

32. Tommaso Campanella, *De sensu rerum et magia.* Campanella developed his theory of fevers in another, later treatise, *Medicinalia,* published in 1635.

33. Daniel Sennert, *The weapon-salves maladie; or, a declaration of its insufficience to performe what is attributed to it,* trans. out of his 5th booke, pt. 4, chap. 10 *Practicae Medicinae* (London, 1637). See Tonnio Griffero, "Immagini contagiose: Malattia e cure magnetiche nella *philosophia per ignem* di Johann Baptist van Helmont," *Rivista di Estetica* (2000), 15, XL, 19–45.

34. Van Helmont, *De magnetica vulnerum curatione* (Paris, 1621).

35. Digby was a Royalist who negotiated with Cromwell on behalf of Catholics, and a Catholic who turned Anglican for a few years, but he remained chancellor to Queen Henrietta Maria after the Restoration. His wife, Venetia, was painted by Van Dyck on her deathbed. There was a rumor, never substantiated, that Digby had poisoned her.

36. Thomas Sydenham, "Theologia Rationalis," in *Thomas Sydenham (1624–*

1689): His Life and Original Writings, ed. Kenneth Dewhurst (Berkeley, Los Angeles, and London: University of California Press, 1966).

37. Galileo wrote about the reversed telescope to his young patron of the *Accademia dei Lincei,* Federico Cesi, encouraging him and his friends to begin investigating the world of fleas and plants. David Freedberg has explored this in his study of the "Lynxes," or *Lincei,* in *The Eye of the Lynx: Galileo, His Friends, and the Beginnings of Modern Natural History* (Chicago, Ill., and London: University of Chicago Press, 2002).

38. John Locke, *An Essay Concerning Human Understanding* (originally published in London, 1690), ed. Peter H. Nidditch (Oxford: Clarendon, 1972), II, xxii, 12.

39. This is reported by Robert Burton, *The Anatomy of Melancholy,* partition I, section 1, member 1, subsection 2.

40. One of its exponents was Franz de le Boë, or Franciscus Sylvius, a French Harvcian based in Leiden, who also continued research on the role of acid in digestion.

41. See Benjamin Woolley, *Heal Thyself: Nicholas Culpeper and the Seventeenth-Century Struggle to Bring Medicine to the People* (London and New York: HarperCollins, 2004).

42. Cited in KTL Fu, "The Healing Hand in Literature: Shakespeare and Surgery," *Hong Kong Medical Journal,* 4, (1998), pp. 77–88: p. 78.

43. See Roy Porter, *Quacks: Fakers and Charlatans in English Medicine* (Stroud, Gloucestershire: Tempus, 2000), pp. 52–53.

44. Porter, ibid., refers to a few lines from this passage—ch. 2 of Defoe's *Journal of the Plague Year* (London, 1722).

45. Richard Lower, "Tractatus de corde (1669)," trans. Kenneth J. Franklin, in *Early Science in Oxford,* ed. R. T. Gunther (Oxford, For the Subscribers, 1932), Vol. 9, pp. 171, 188–192. Also in Richard Hunter and Ida Macalpine, *Three Hundred Years of Psychiatry, 1535–1860* (Hartsdale, N.Y.: Carlisle, 1982), pp. 185–186.

46. Samuel Pepys, *Diary,* Vol. VIII, *1667,* ed. Robert Latham and William Matthews (London: Bell and Hyman, 1974), pp. 543, 554.

47. Their efforts were first recounted in detail in a now classic book by Robert G. Frank, Jr., *Harvey and the Oxford Physiologists: A Study of Scientific Ideas* (Berkeley, Los Angeles, and London: University of California Press, 1980). They have been much studied since then.

48. Boyle published the account of his famous experiments with the air pump in 1662, in *The Spring and Weight of the Air.* For the creation of his air pump, he took the cue from the Italian chemist Evangelista Torricelli's discovery of air pressure, following on the original attempt by a German natural philosopher and ideator of the barometer, Otto von Guericke. Another collaborator was the multitalented Christopher Wren, astronomer, mathematician, and eventually one of England's greatest architects, commissioned by Charles II to build or rebuild at least fifty churches in London, including Saint Paul's Cathedral, after the Great Fire of 1666.

49. Richard Lower, *Diatribae Thomae Willisii de febribus vindicatio* (Amsterdam,

1666). *Tractatus de corde* (London, 1669), trans. Kenneth J. Franklin in R. T. Gunther, *Early Science in Oxford*, Vol. 9 (Oxford, for the subscribers 1932).

50. See *The Correspondence of Marcello Malpighi*, Vol. V, 1693–1694, ed. Howard B. Adelman (Ithaca, N.Y., and London: Cornell University Press, 1975), pp. 1958–1960.

NOTES TO CHAPTER SIX

1. Laurence Sterne, *Tristram Shandy*, I, 1 (1760).
2. William James, *The Principles of Psychology* (Cambridge, MA, and London: Harvard University Press, 1981), Chapter 9: "The Stream of Thought," p. 239.
3. Descartes's *Of Man* accompanied the anti-Aristotelian *The World*, his treatise of physics, whose imminent publication in 1633 he had stopped at the last minute when he heard of Galileo's condemnation—he did not want to upset the Church, or indeed, to get himself into trouble.
4. Nicolas Malebranche, *De la recherche de la vérité: Où l'on traitte de la nature de l'esprit de l'homme, et de l'usage qu'il en doit faire pour eviter l'erreur dans les sciences* (Paris, 1674–1675), II, I, v.
5. Juan Luis Vives, *The Passions of the Soul: The Third Book of De Anima et Vita*, ed. Carlos G. Norena (Lewiston, NY: E. Mellen Press, Studies in Renaissance Literature, 1990, v. 4, p. 2). Vives, a friend of Erasmus, was born in Valencia but died in Bruges.
6. The episode has never been forgotten. It was recycled in films and novels, and tourists today can take a tour of the places in the town where the main events happened. Michel de Certeau analyzed the episode in *The Possession at Loudun* (Chicago, Ill., and London: University of Chicago Press, 2000), originally *La possession de Loudun* (Paris: Gallimard, 1970).
7. Thomas Wright, *The Passions of the Mind in Generall* (London, 1601), II, 2.
8. Ibid., I, 2.
9. Ibid., II, 2.
10. Ibid., II, 3.
11. See Sheila Barker, "Poussin, Plague, and Early Modern Medicine," *Art Bulletin* (December 2004), pp. 659–689.
12. Matteo Naldi, *Regole per la cura del contagio* (Rome, 1656), p. 9. Cited in Barker, "Poussin," p. 661.
13. On the medical, intellectual, and philosophical context of the study of the brain by Willis and others, see Robert Martensen's insightful *The Brain Takes Shape: An Early History* (Oxford: Oxford University Press, 2004). Willis's career as a forerunner of neurology has been ably traced by Carl Zimmer in *Soul Made Flesh: The Discovery of the Brain—and How It Changed the World* (New York: Free Press and London: William Heinemann, 2004).
14. For a full account of this, see Zimmer, *Soul Made Flesh*.
15. Thomas Willis, *De anima brutorum* or *The Soul of Brutes, Which is that of the Vital and Sensitive of Man* (London, 1672 Latin, 1683 English); reproduced in Richard Hunter and Ida Macalpine, *Three Hundred Years of Psychiatry, 1535–1860* (Hartsdale, NY: Carlisle Publishing, Inc., 1982), p. 190.

16. Martensen, *The Brain Takes Shape*, p. 165.

17. Willis, *The Soul of Brutes*, p. 1.

18. Martensen, *The Brain Takes Shape*, p. 138. And see Rina Knoeff, "The Reins of the Soul: The Centrality of the Intercostal Nerves to the Neurology of Thomas Willis and to Samuel Parker's Theology," *Journal of the History of Medicine and Allied Sciences* (2004), 59:3, pp. 413–440.

19. Lucretius, *De natura rerum*, IV, 822–823.

20. Guillaume Lamy, *Explication mechanique et physique des fonctions de l'âme sensitive* (Paris, 1677). (The title means "Mechanical and Physical Explanation of the Functions of the Sensitive Soul.") Published with *Discours anatomiques* in edition by Anna Minerbi Belgrado (Paris: Universitas; Oxford: Voltaire Foundation, 1996).

21. Jonathan Swift, "A Discourse Concerning the Mechanical Operation of the Spirit. In a Letter to a Friend. A Fragment" (London, 1704), in *A Tale of a Tub and Other Works* (Oxford: World Classics, Oxford University Press, 1986) p. 132.

22. Ibid., p. 140.

23. Ibid., p. 141.

24. Thomas Tryon, *A Treatise of Dreams & Visions . . . To Which Is added, a Discourse of the Causes, Natures, and Cure of the Phrensie, Madness, or Distraction. By Philotheos Physiologus* (London, 1689). Reproduced in Hunter and Macalpine, *Three Hundred Years of Psychiatry*, pp. 233–235.

25. See Henry Harris, *The Birth of the Cell* (New Haven and London: Yale University Press, 1999).

26. Molière, *Le médecin volant*, Scene IV "Hippocrate dit, et Galien par vives raisons persuade qu'une personne ne se porte pas bien quand elle est malade. Vous avez raison de mettre votre espérance en moi; car je suis le plus grand, le plus habile, le plus docte médecin qui soit dans la faculté végétable, sensitive et minérale." (Trans. author.)

27. Lymph had been first identified in the mid-sixteenth century by Gabriele Fallopia, a physician from Modena who had seen in the vivisected bodies of condemned criminals a "yellowish substance" in intestinal veins connected to liver and lungs. The lymph would be analyzed in depth a century later, especially by the Dane Thomas Bartholin. Fallopia, incidentally, was the physician who described accurately the so-called Fallopian tubes.

28. Joseph Priestley, *Disquisitions Relating to Matter and Spirit, to Which Is Added, The History of the Philosophical Doctrine Concerning the Origin of the Soul, and the Nature of Matter with Its Influence on Christianity, Especially with Respect to the Doctrine of the Pre-Existence of Christ* (London, 1777). The epigraph on the title page is a quotation from Charles Bonnet: "Si quelqu'un demontreroit jamais, que l'âme est materielle, loin de s'en alarmer, il faudroit admirer la puissance qui auroit donné a la matiere la capacité de penser" (from *La palingénésie philosophique, ou Idées sur l'état passé et sur l'état futur des êtres vivants: Ouvrage destiné à servir de supplément aux derniers écrits de l'auteur et qui contient principalement le précis de ses recherches sur le christianisme*, 1770).

29. Roy Porter, *Flesh in the Age of Reason: How the Enlightenment Transformed the Way We See Our Bodies and Souls* (London: Penguin, 2004), p. 351.

30. It is Casanova who reports this statement of Haller's in his account of their meeting in 1760, in his *Memoirs*. Casanova later recounted this meeting to Voltaire, on his last visit to the writer. See *Memoires de Jacques Casanova de Seingalt* (Paris: Garnier Frères, 1880), 8 vols.: vol. 4, ch. 14, p. 423.

31. Roger K. French, *Robert Whytt, the Soul, and Medicine* (London: The Wellcome Institute of the History of Medicine, 1969), p. 82.

32. Roy Porter, *Mind-Forg'd Manacles: A History of Madness in England from the Restoration to the Regency* (London: Penguin, 1987), pp. 47–52.

33. Ann Thomson, "La Mettrie et Diderot," *Recherches sur Diderot et l'Encyclopédie: Groupe de travail sur Le rêve de d'Alembert*, at http://www.sigu7.jussieu.fr/diderot/travaux/revseance2.htm (2000).

34. Georges Vigarello, *Le propre et le sale: L'hygiène du corps depuis le Moyen Age* (Paris: Éditions du Seuil, 1985).

35. This treatment was shown in the 1994 film directed by Nicholas Hytner and written by Alan Bennett, *The Madness of King George*. Ten years later, Richard Hunter and Ida Macalpine argued that the King actually suffered not from mental illness but from porphyria, a disease whose symptoms were gradually worsened by the doctors' regular administration of antimony against his supposed madness—a common purgative, similar to arsenic and as poisonous.

36. Porter, *Mind-Forg'd Manacles*, pp. 181–182.

37. Reproduced in Hunter and Macalpine, *Three Hundred Years of Psychiatry*, pp. 43–44.

38. Reproduced ibid., pp. 344–347.

39. Robert Darnton, *Mesmerism and the End of the Enlightenment in France* (Cambridge, Mass.: Harvard University Press), pp. 6–7.

40. Ibid., pp. 10–11.

41. Samuel Hahnemann, *Organon of the Medical Art*, ed. Wenda Brewster O'Reilly (Palo Alto, Calif.: Birdcage, 1996), p. 258.

42. Ibid., p. 58

43. Mesmer's method and its effects are similar to those of Reiki, an energy-based practice that originated in late-nineteenth-century Japan and was exported to the West via Hawaii by the 1930s.

44. There was a contemporary attempt to connect magnetism and Hippocratism, by a Frenchman named Élie de la Poterie, who in 1785 wrote *Examen de la doctrine d'Hippocrate sur la nature des êtres animés, sur le principe de mouvement et de la vie, sur les périodes de la vie humaine*. See Jackie Pigeaud, *Aux portes de la psychiatrie: Pinel, l'ancien et le moderne* (Paris: Aubier, 2001), p. 204.

45. Darnton, *Mesmerism*, p. 8.

46. Another member of the commission was the physician Guillotin, who, as a deputy in the new revolutionary government in 1789, would think up the so-called guillotine as a democratic means of ensuring to those condemned to death a painless end. In 1794, at the height of the Reign of Terror, it would kill Lavoisier.

47. Pigeaud, *Aux portes de la psychiatrie*, pp. 204–205.

48. Ibid., pp. 208–211.

49. Hunter and Macalpine, *Three Hundred Years of Psychiatry*, pp. 473–474.

50. Lisa Roscioni, *Il governo della follia: Ospedali, medici, e pazzi nell'età moderna* (Milan: Mondadori, 2003), pp. 10–13, 232–233. See also Mario Galzigna, "Lo

psichiatra e il libertino," in *Storia delle passioni,* ed. Silvia Vegetti Finzi (Rome and Bari: Laterza, 1995), p. 228–232.

51. Reproduced Hunter and Macalpine, *Three Hundred Years of Psychiatry,* pp. 475–478.
52. Robert Burton, *Anatomy of Melancholy,* (New York: New York Review Books, 2001), partition I, section 1, member 2, subsection 2.
53. On the political import of the debates between mechanists and vitalists, see Anne Harrington, *Reenchanted Science: Holism in German Culture from Wilhelm II to Hitler* (Princeton, N.J.: Princeton University Press, 1996).
54. F. J. Gall and J. C. Spurzheim, *Recherches sur le système nerveux en général, et sur celui du cerveau en particulier* (Paris, 1809), pp. 228–230. Cited in Edwin Clarke and L. S. Jacyna, *Nineteenth-Century Origins of Neuroscientific Concepts* (Berkeley, Los Angeles, and London: University of California Press, 1987), p. 227.
55. See the section on mesmeric phrenology in George Combe, *A System of Phrenology* (Edinburgh, first ed. 1825). One can read a part of it, from the 5th edition of 1853, at http://poyes.britishlibrary.net/phrenology/system/mesmeric.htm
56. Clarke and Jacyna, pp. 277–279, 302–307.
57. Claude Bernard, *Introduction à l'étude de la médecine expérimentale* (Paris: Garnier-Flammarion, 1966), II, 4, p. 158. (Trans. author.)
58. Ibid., p. 161.
59. Reproduced in Hunter and Macalpine, *Three Hundred Years of Psychiatry,* p. 910.
60. For an interpretation, see Georges Didi-Huberman, *Invention of Hysteria: Charcot and the Photographic Iconography of the Salpêtrière* (Cambridge, Mass.: MIT Press, 2003); and a review of the book by Carol Armstrong, "Probing Pictures," *Artforum International* (January 9, 2003).
61. "Aphasie," in *A Moment of Transition: Two Neuroscientific Articles by Sigmund Freud,* trans. and ed. Mark Solms and Michael Saling (London and New York: Karnak, 1990), pp. 35–36.
62. "Gehirn," in ibid., p. 62.
63. Gerald Edelman, *Wider Than the Sky: The Phenomenal Gift of Consciousness* (New Haven, Conn., and London: Yale University Press, 2004), p. 31.

NOTES TO CHAPTER SEVEN

1. From a sketch in the *Flying Circus* series (1972).
2. Thomas Nagel, *The View from Nowhere* (New York, Oxford: Oxford University Press, 1986), p. 108.
3. J. M. Coetzee, *Slow Man* (New York: Viking, 2005), p. 26.
4. The experiments are reported by their authors in Andreas Bartels and Semir Zeki, "The Neural Basis of Romantic Love," *NeuroReport,* 17 (11) (2000), pp. 3829–3834; "The neural correlates of maternal and romantic love," *NeuroImage,* 21 (3) (2004), pp. 1155–1166. On the role of oxytocin, see for instance T. R. Insel, L. J. Young, "The neurobiology of attachment," *National Review of Neuroscience* 2 (2) (2001), pp. 129–136. See also "The science of love: I get a kick out of you," *The Economist,* February 12th, 2004, reporting on Helen Fisher, *Why We Love: The Nature and Chemistry of Romantic Love* (New York: Henry Holt, 2004).

5. Sandra Blakeslee, "Humanity? Maybe It's in the Wiring," *New York Times* (December 9, 2003).

6. Jerry Fodor, "Making the Connection. Axioms from Axons—Why we need to think harder about thinking" (review of Joseph LeDoux's *Synaptic Self*), *Times Literary Supplement* (May 17, 2002).

7. Eric R. Kandel, "A New Intellectual Framework for Psychiatry," *American Journal of Psychiatry*, 155:4 (April 1998), p. 464.

8. Ian Hacking, *Multiple Personality and the Sciences of Memory* (Princeton, N.J.: Princeton University Press, 1995), pp. 128–134.

9. First published in Latin in 1768, and in French in 1771.

10. Peter D. Kramer, "There's Nothing Deep about Depression," *New York Times* (April 17, 2005).

11. There is a computer under development by Peter Robinson, derived from research on autism by the Cambridge psychologist Simon Baron-Cohen, which can learn how to "read" emotions from facial expressions. So the fantasy that machines will one day be able to "know" us and interpret our motions and gestures accurately might become true soon enough. But even then, what a machine could read on surfaces is bound to differ from the human mind's interpretive powers. See http://www.cl.cam.ac.uk/Research/Rainbow/emotions/ and links from there, including to press coverage.

12. Esther A. Nimchinsky, Emmanuel Gilissen, John A. Allman, Daniel P. Perl, Joseph M. Erwin, and Patrick R. Hof, "A Neurological Morphologic Type Unique to Humans and Great Apes," *Proceedings of the National Academy of Sciences* (United States), 96 (April 1999), pp. 5268–5273. See also Blakeslee, "Humanity." Whales, too, have spindle cells: see Hof & Van der Gucht, *Anatomical Record*, Nov. 2006.

13. Israel Rosenfield, *The Strange, Familiar, and Forgotten: An Anatomy of Consciousness* (New York: Random House, 1992), p. 49.

14. Paul Schilder, *The Image and Appearance of the Human Body* (New York: International Universities Press, Inc., 1950: 1970) p. 10.

15. First published in "What Is an Emotion?" *Mind*, 9 (1884), pp. 188–205. Online at http://psychclassics.yorku.ca/James/emotion.htm And see William James, *The Principles of Psychology* (Cambridge, Mass., and London: Harvard University Press, 1981), ch. 25, pp. 1058–1097.

16. Paul Ekman, "Basic Emotions," in *The Handbook of Cognition and Emotion*, eds. T. Dalgleish and T. Power (Sussex, U.K., 1999: Wiley), pp. 45–60. Also at http://www.paulekman.com/pdfs/basic_emotions.pdf.

17. Joseph LeDoux, *The Emotional Brain: The Mysterious Underpinnings of Emotional Life* (New York: Simon and Schuster, 1996).

18. Ibid., p. 29.

19. Ibid., p. 69.

20. Ibid., p. 20.

21. See Jerome Burne, "Cure for Stress?" *Independent* (London, May 28, 2006). See also http://www.heartmath.com.

22. See, for example, the work of Jon Elster, Martha Nussbaum, or Ronald de Sousa.

23. Jean-Didier Vincent, *Biologie des passions* (Paris: Odile Jacob, 1986, 2002), p. 29.

24. Antonio and Hanna Damasio, "Minding the Body," *Daedalus: Journal of the American Academy of Arts and Sciences*, Summer 2006, p. 15.

25. Jerome Kagan, *Galen's Prophecy: Temperament in Human Nature* (New York: Basic Books, 1994), p. 52.

26. Ibid., p. 53.

27. Annie Murphy Paul, *The Cult of Personality: How Personality Tests Are Leading Us to Miseducate Our Children, Mismanage Our Companies, and Misunderstand Ourselves* (New York: Free Press, 2004), p. 113. Murphy Paul writes that Ernst Kretschmer's "classifications were employed by the Third Reich to select military officers." The tradition of physiognomy was used by the 19th-century psychiatrist and criminologist Cesare Lombroso, who argued that criminals were a biological category.

28. For a full analysis of the current use of personality "typing" in America, see Murphy Paul, ibid.

29. Irving Kirsch and Guy Sapirstein, "Listening to Prozac but Hearing Placebo: A Meta-Analysis of Anti-Depressant Medication," *Prevention & Treatment*, 1 (June 26, 1998).

30. "Placebo" acquired its present meaning in the late 1700s; before then it meant a flatterer. It literally means "I shall please" in Latin—the first words of a psalm sung in medieval vespers for the dead. See *Concise Oxford English Dictionary*.

31. See http://www.gerson.org.

32. Marshall Jon Fisher, "Better Living through the Placebo Effect," *Atlantic Monthly* (October 2000).

33. Dylan Evans, *Placebo: Mind over Matter in Modern Medicine* (Oxford and New York: Oxford University Press, 2004), p. 72.

34. Ibid., p. 45.

35. Ibid., pp. 44–69, especially p. 56.

36. Daniel Moerman, *Meaning, Medicine, and the "Placebo Effect"* (Cambridge: Cambridge University Press, 2002), p. 66.

37. Arthur K. Shapiro and Elaine Shapiro, "The Placebo: Is It Much Ado about Nothing?" in *The Placebo Effect: An Interdisciplinary Exploration*, ed. Anne Harrington (Cambridge, Mass., and London: Harvard University Press, 1997).

38. Pellegrino Artusi, *La scienza in cucina e l'arte di mangiar bene* (first published 1891), ed. Piero Camporesi (Turin: Einaudi, 1970, 1995). *Science in the Kitchen and the Art of Eating Well*, trans. Murtha Baca and Stephen Sartarelli (New York: Marsilio, 1997).

39. Mary Ella Milham, *Platina, On Right Pleasure and Good Health: A Critical Edition and Translation of* De Honesta Voluptate et Valetudine (Tempe, Ariz.: Medieval and Renaissance Texts and Studies, 1998).

40. Galen's book on dieting was taken up by Oribasius and Aetius, translated into Arabic by Hunayn ibn Ishaq, into Latin by Reggio da Calabria, and first published in Paris in 1530.

41. Galen, *La dieta dimagrante*, ed. Nino Marinone (Turin: G. B. Paravia, 1973).

42. See Jack Turner's exhaustive account in *Spice: The History of a Temptation* (New York: Knopf, 2004); here, pp. 106–113. See also the erudite and useful book by Andrew Dalby, *Dangerous Tastes: The Story of Spices* (London: British Museum Press, 2000).

43. Shakespeare, *Henry V,* Act IV, Scene 3.
44. Ian Johnston, "Spice Helps to Stop the Spread of Breast Cancer," *Scotsman,* 15 (October 2005). Articles about this include Ruey-Long Hong, William H. Spohn, and Mien-Chie Hung, "Curcumin Inhibits Tyrosine Kinase Activity of p185neu Depletes p185neu1," *Clinical Cancer Research,* 5 (July 1999), pp. 1884–1891. The article Johnston refers to is Bharat B. Aggarwal, Shishir Shishodia, Yasunari Takada, Sanjeev Banerjee, Robert A. Newman, Carlos E. Bueso-Ramos, and Janet E. Price, "Curcumin Suppresses the Paclitaxel-Induced Nuclear Factor-B Pathway in Breast Cancer Cells and Inhibits Lung Metastasis of Human Breast Cancer in Nude Mice," *Clinical Cancer Research,* 11 (October 15, 2005), pp. 7490–7498.
45. There exists in America, for instance, an American Herbalists Guild, which defines itself as "a non-profit, educational organization for the furtherance of herbalism" and on whose Web site one reads: "Most pharmaceutical drugs are single chemical entities that are highly refined and purified and are often synthesized. In 1987 about 85% of modern drugs were originally derived from plants. Currently, only about 15% of drugs are derived from plants. In contrast, herbal medicines are prepared from living or dried plants and contain hundreds to thousands of interrelated compounds. Science is beginning to demonstrate that the safety and effectiveness of herbs is often related to the synergy of its many constituents." See http://www.americanherbalistsguild.com.
46. The principal books that contain the ayurveda guidelines are the *Charaka Samhita* and the *Sushrutha Samhita.*
47. Jean Filliozat, *The Classical Doctrine of Indian Medicine* (Paris: Imprimerie Nationale, 1949), pp. 196–228.
48. Ibid., pp. 229–237.
49. See Guido Majno, *The Healing Hand: Man and Wound in the Ancient World* (Cambridge, Mass., and London: Harvard University Press, 1975), p. 261. See also Filliozat, pp. 253–254.
50. See Ullmann, *Islamic Medicine* (Edinburgh: Edinburgh University Press, 1978), p. 52. See also http://www.itmonline.org/arts/unani.htm.

References

This list of references includes, chapter by chapter, the books referred to in the endnotes, as well as those consulted or of some relevance to the topics raised in this history of humours but not directly cited in the text.

CHAPTER ONE

PRIMARY

Aristotle, *Metaphysics*, trans. W. D. Ross (London, c. 1908).

Celsus, *De medicina*, I–II (London: Loeb, William Heinemann, 1971).

Galien, *L'âme et ses passions*, trans. and ed. Vincent Barras, Terpsichore Birchler, and Anne-France Morand (Paris: Les Belles Lettres, 1995).

Galen, *On the Natural Faculties*, trans. Arthur John Brock (Cambridge, Mass.: Loeb, 1916, 2000).

Galen, *On the Usefulness of the Parts of the Body (De usu partium)*, trans. Margaret Tallmadge May (Ithaca, N.Y.: Cornell, 1968).

Galen, *De placitis Hippocratis et Platonis* or *On the Doctrines of Hippocrates and Plato* (2 vols., Books I–IX), ed., trans., and comm. Phillip de Lacy (Berlin: Akademie-Verlag, 1978).

Galen, *Daß die Kräfte der Seele den Mischungen des Körpers folgen*, in arabischer Übersetzung, ed. Hans Hinrich Biesterfeldt, (Marburg: Deutsche Morgenländische Gesellschaft, Wiesbaden: Kommissions Verlag F. Steiner, 1973).

Herodotus, *The Histories*, trans. Aubrey de Sélincourt (London, 1954).

"Herophilus," in *The Art of Medicine in Early Alexandria*, ed., trans., essays Heinrich von Staden (Cambridge and New York: Cambridge University Press, 1989).

Hippocrates, *Nature of Man*, trans. W. H. S. Jones (Cambridge, Mass., and London: Loeb, 1931).

Hippocrates, *De natura hominis: La nature de l'homme*, ed., trans., comm. Jacques Jouanna (Berlin 1975).

Hippocratic Writings, ed. G. E. R. Lloyd, trans. J. Chadwyck and W. N. Mann. (London: Penguin, 1983).

Hippocrates, *L'art de la médecine,* trans. Jacques Jouanna and Caroline Magdelaine (Paris: Garnier Flammarion, 1999).

Hippocrates, Vol. IV, trans. W. H. S. Jones (Cambridge, Mass.: Loeb, 1931, 1998).

Julius Caesar, *The Civil Wars,* trans. A. G. Peskett (Cambridge, Mass.: Loeb, 1914).

Plato, *The Republic,* trans. Desmond Lee (London, 1955).

Pliny, *Natural History,* Vol. VIII, trans. W. S. H. Jones (Cambridge, Mass.: Loeb, 1956).

Plutarch, *Lives,* Vol. VII, trans. Bernadotte Perrin (Cambridge, Mass.: Loeb, 1919).

Les Présocratiques, ed. Jean-Paul Dumont, Daniel Delattre, and Jean-Louis Poirier (Paris: Gallimard/Pléiade, 1988).

SECONDARY

Bailey, James E., "Asklepios: Ancient Hero of Medical Caring," *Annals of Internal Medicine,* 124:2 (1996), pp. 257–263.

Douglas, Mary, *Leviticus as Literature* (Oxford and New York: Oxford University Press, 1999).

El-Abbadi, Mostafa, *Vie et destin de l'ancienne bibliothèque d'Alexandrie* (Paris: Organisation des Nations Unies pour l'éducation, la science et la culture, 1992).

Empereur, Jean-Yves, *Alexandria: Past, Present and Future,* trans. Jane Brenton (London: Thames & Hudson, 2002; Paris: Gallimard, 2001).

Grmek, Mirko, ed., *Histoire de la pensée médicale en Occident,* Vol. I, *Antiquité et Moyen Age* (Paris: Seuil, 1995).

Huffman, Carl, "Alcmaeon," in *Stanford Encyclopedia of Philosophy,* ed. Edward N. Zalta, at http://plato.stanford.edu/archives/sum2004/entries/alcmaeon.

Jackson, Ralph, *Doctors and Diseases in the Roman Empire* (London: British Museum Publications, 1988).

Jones, W. H. S., "Introduction" in *Hippocrates,* Vol. I, pp. ix–lv (Cambridge, Mass.: Loeb, 1931, 1998).

Jouanna, Jacques, *Hippocrates,* trans. M. B. DeBevoise (Baltimore, Md., and London: Johns Hopkins University Press, 1999).

Jouanna, Jacques, "La naissance de l'art médical occidental," in Mirko Gremk, *Histoire de la pensée médicale en Occident,* Vol. I, *Antiquité et Moyen Age* (Paris: Seuil, 1995).

Klibansky, Raymond, Erwin Panofsky, Fritz Saxl, *Saturn and Melancholy: Studies in the History of Natural Philosophy Religion and Art* (London: Nelson, 1964). Italian edition: *Saturno e la melanconia: Studi su storia della filosofia naturale, medicina, religione e arte* (Turin: Einaudi, 1983, 2002).

Lloyd, G. E. R., *The Revolutions of Wisdom: Studies in the Claims and Practice of Ancient Greek Science* (Berkeley, Los Angeles, and London: University of California Press, 1987).

Longrigg, James, "Anatomy in Alexandria in the Third Century BC," *British Journal of the History of Science,* 21, 1988, pp. 455–488.

MacLeod, Roy, ed., *The Library of Alexandria: Centre of Learning in the Ancient World* (London and New York: I. B. Tauris, 2000).

Majno, Guido, *The Healing Hand: Man and Wound in the Ancient World* (Cambridge, Mass., and London: Harvard University Press, 1975).

McEvilley, Thomas, *The Shape of Ancient Thought: Comparative Studies in Greek and Indian Philosophy* (New York: Allworth, 2002).

Manuli, Paolo, and Mario Vegetti, *Cuore, sangue e cervello: Biologia e antropologia nel pensiero antico* (Milan: Episteme, 1977).

Nussbaum, Martha, *Upheavals of Thought: The Intelligence of Emotions* (Cambridge and New York: Cambridge University Press, 2001).

Nutton, Vivian, "Galen in the Eyes of His Contemporaries," *Bulletin of the History of Medicine*, 58 (1984), pp. 315–324. Reprinted in Vivian Nutton, *From Democedes to Harvey: Studies in the History of Medicine* (London: Variorum Reprints, 1988).

Nutton, Vivian, "The Legacy of Hippocrates: Greek Medicine in the Library of the Medical Society of London," *Transactions of the Medical Society of London*, 103 (1986–1987), pp. 21–30.

Nutton, Vivian, "Healers in the Medical Market-Place: Towards a Social History of Graeco-Roman Medicine," in *Medicine in Society: Historical Essays*, ed. Andrew Wear (Cambridge, New York: Cambridge University Press, 1992), pp. 15–58.

Padel, Ruth, *In and Out of the Mind: Greek Images of the Tragic Self* (Princeton, N.J., and Chichester: Princeton University Press, 1992).

Pichot, André, *La naissance de la science*, 2 vols. (Paris: Gallimard, 1991).

Sambursky, Shmuel, *Physics of the Stoics* (Princeton: Princeton University Press, 1959).

Sarton, George, *Galen of Pergamon* (Lawrence, Kans.: University of Kansas Press, 1954).

Sassi, Maria Michela, *The Science of Man in Ancient Greece* (Chicago and London: University of Chicago Press, 2001).

Schouten, J., *The Rod and Serpent of Asklepios: Symbol of Medicine* (Amsterdam, London, and New York: Elsevier, 1967).

Staden, Heinrich von, ed., trans., essays, *The Art of Medicine in Early Alexandria* (Cambridge and New York: Cambridge University Press, 1989).

Verbeke, G., *L'évolution de la doctrine de pneuma du Stoïcisme à Saint Augustin: Étude philosophique* (Paris and Leuwen: Desclée de Brouwer, 1945).

Wright, John P., and Paul Potter, eds., *Psyche and Soma: Physicians and Metaphysicians on the Mind-Body Problem from Antiquity to Enlightenment* (Oxford: Clarendon Press, 2000), chapters 1–4.

CHAPTER TWO

PRIMARY

Agnellus of Ravenna, *Lectures on Galen's* De sectis, Latin text and translation (Buffalo, N.Y.: Seminar Classics 609, 1981).

Avicenna, *The General Principles of Avicenna's Canon of Medicine*, ed. and trans. Mazhar H. Shah (Karachi: Naveed Clinic, 1966).

Israeli, Isaac, *A Neoplatonic Philosopher of the Early Tenth Century: His Works Translated with Comments and an Outline of His Philosophy*, trans. A. Altmann and S. M. Stern (Oxford: Oxford University Press, 1958).

John of Alexandria, *Commentary on Hippocrates' Epidemics VI—Fragments; Commentary of an Anonymous Author on Hippocrates' Epidemics VI—Fragments,* ed., trans., and notes John M. Duffy (Berlin, Akademie Verlag, 1997). In the same volume: *Commentary on Hippocrates' On the Nature of the Child,* ed. and trans. T. A. Bell et al.

Maimonides, *On Asthma,* ed., trans., and annotated Gerrit Bos. Volume 1 of the complete medical works of Maimonides; parallel Arabic and English texts. (Provo, Utah: Brigham Young University Press, 2002).

Maimonides, "Treatise on Cohabitation," trans. Fred Rosner, in *Sex Ethics in the Writings of Moses Maimonides.* Includes excerpts from other writings. (Northvale, N.J., and London: J. Aronson, 1994).

Paulus Aeginata, *The Seven Books,* 3 vols., trans. and comm. Francis Adams (London: Sydenham Society, 1894).

Rhazes (Ar-Razi), *The Spiritual Physick,* trans. Arthur J. Arberry (London: Murray, 1950).

Rhazes (Ar-Razi), *La médecine spirituelle,* trans. and ed. Rémi Brague (Paris: Flammarion, 2003).

Schacht, Joseph, and Max Meyerhof, *The Medico-Philosophical Controversy between Ibn Butlan of Baghdad and Ibn Ridwan of Cairo: A Contribution to the History of Greek Learning Among the Arabs* (Cairo: Egyptian University, Faculty of Arts, Publication no. 13, 1937).

Yuhanna ibn Masawayh (Jean Mesue), *Le livre des axiomes médicaux (Aphorismi),* ed. and trans. Danielle Jacquart and Gérard Troupeau. Arabic and Latin texts translated into French. (Geneva: Droz, and Paris: Champion, 1980).

SECONDARY

Burnett, Charles, "Physics before the Physics," *Medioevo,* XXVII (2002).

Burnett, Charles, *Al-Qabisi (Alcabitius): The Introduction to Astrology* (London, 2004).

Burnett, Charles, "Antioch as a Link between Arabic and Latin Culture in the Twelfth and Thirteenth Centuries," in *Occident et Proche-Orient: Contacts scientifiques au temps des Croisades. Actes du colloque de Louvain-la-Neuve, 24 et 25 Mars 1997,* ed. Isabelle Draelants, Anne Tihon, and Baudouin van den Abeele (Brepols, 2002).

Burnett, Charles, and Danielle Jacquart, eds., *Constantine the African and 'Ali Ibn Al-Abbas Al-Magusi: The "Pantegni" and Related Texts* (Leiden, New York, and Cologne: Brill, 1994).

Cardaillac, Louis, ed., *Tolède, XIIe–XIIIe: Musulmans, chrétiens et juifs: Le savoir et la tolérance* (Paris: Autrement, 1991).

Daniel, Norman, *Islam and the West: The Making of an Image* (Edinburgh: Edinburgh University Press, 1960).

Dols, Michael W., *Medieval Islamic Medicine: Ibn Ridwan's Treatise "On the Prevention of Bodily Ills in Egypt,"* Arabic text ed. Adil S. Gamal (Berkeley, Los Angeles, and London: University of California Press, 1984).

Dols, Michael W., *Majnun: The Madman in Medieval Islamic Society,* ed. Diana E. Immisch (Oxford: Clarendon Press, New York: Oxford University Press, 1992).

Friedenwald, Harry, *The Jews and Medicine: Essays,* Vol. I (Baltimore, Md.: Johns Hopkins University Press, 1944).

García-Ballester, Luis, Roger French, Jon Arrizabalaga, and Andrew Cunningham, eds., *Practical Medicine from Salerno to the Black Death* (Cambridge and New York: Cambridge University Press, 1994).

Grmek, Mirko, ed., *Histoire de la pensée médicale en Occident,* Vol. I, *Antiquité et Moyen Age* (Paris: Seuil, 1995).

Gruner, O. Cameron, MD, *A Treatise on the Canon of Medicine of Avicenna, Incorporating a Translation of the First Book* (London: Luzac & Co., 1930).

Gutas, Dimitri, *Greek Thought, Arabic Culture: The Graeco-Arabic Translation Movement in Baghdad and Early 'Abbasic Society (2nd–4th/8th/10th Centuries)* (London and New York: Routledge, 1998).

Jarcho, Saul, *The Concept of Heart Failure, from Avicenna to Albertini* (Cambridge, Mass., and London: Harvard University Press, 1980).

Levi-Provençal, Évariste, *Histoire de l'Espagne musulmane,* Vol. 1, *La conquête et l'émirat hispano-umaiyade (710–912)* (Paris and Leiden: G. P. Maisonneuve, 1950). 3 vols.

Levi-Provençal, Évariste, *La civilisation arabe en Espagne: vue générale* (Paris and Leiden: G.P. Maisonneuve et Larose, 1961).

Majno, Guido, *The Healing Hand: Man and Wound in the Ancient World* (Cambridge, Mass., and London: Harvard University Press, 1975).

Mantra, Robert, *L'expansion musulmane (VIIe–XIe siècles)* (Paris: Presses Universitaires de France, 1969).

Meyerhof, Max, "New Light on Hunain Ibn Ishaq and His Period," *Isis,* 8 (1926), pp. 685–724.

Temkin, Owsei, *Galenism: Rise and Decline of a Medical Philosophy* (Ithaca, N.Y., and London: Cornell University Press, 1973).

Temkin, Owsei, *Hippocrates in a World of Pagans and Christians* (Baltimore, Md., and London: Johns Hopkins University Press, 1991).

Ullmann, Manfred, *Islamic Medicine* (Edinburgh: Edinburgh University Press, 1978).

Walker Bynum, Caroline, *Fragmentation and Redemption: Essays on Gender and the Human Body in Medieval Religion* (New York: Zone, 1991).

Wolfson, Harry A., "The Internal Senses in Latin, Arabic, and Hebrew Philosophic Texts," *The Harvard Theological Review,* 28:2 (1935) pp. 69–133.

CHAPTER THREE

PRIMARY

Boccaccio, Giovanni, *Decameron* (Milan, 1966). English: *The Decameron of Giovanni Boccaccio,* trans. J. M. Rigg (London, Toronto: J. M. Dent; New York: E. P. Dutton, 1930).

Chaucer, Geoffrey, *Canterbury Tales,* ed. Scott Gettman. Online at http://www.canterburytales.org/canterbury_tales.html.

Coppo Stefani, Marchione di, *Cronaca fiorentina,* Rerum Italicarum Scriptores, Vol. XXX., ed. Niccolo Rodolico (Città di Castello, 1903–1913). At http://www.iath.virginia.edu/osheim/marchione.html.

Fourth Lateran Council, 1215, at http://www.piar.hu/councils/ecum12.htm. Source: H. J. Schroeder, ed. *Disciplinary Decrees of the General Councils: Text, Trans-*

lation, and Commentary (St. Louis, Mo., and London: Harry Rothwell, 1937).

Herbarium Apulei (1481) and *Herbolario Volgare* (1522), 2 vols., intro. Erminio Caprotti (Milan: Il Polifilo, 1979).

Galen, *On the Usefulness of the Parts of the Body*, 2 vols., trans. and ed. Margaret Tallmadge May (Ithaca, N.Y.: Cornell University Press, 1968).

Galen, *Commentary on the Hippocratic Treatise Airs, Waters, Places in the Hebrew translation of Solomon ha-Me'ati*, ed., trans. and intro. Abraham Wasserstein (Jerusalem: Israel Academy of Sciences and Humanities, 1982).

Gerard, John, *The Herball or Generall Historie of Plantes, Very Much Enlarged and Amended by Thomas Johnson* (London, 1633; Facsimile: New York: Dover, 1975).

Ishaq ibn 'Imran, *Magala fi l-malihuliya (Abhandlung über die Melancholie) and Constantini Africani, Libri due de melancholia*, ed. Karl Garbers (Hamburg: H. Buske, 1977).

Mondino de' Liuzzi, *Anothomia*. At http://cis.alma.unibo.it/Mondino/liber.html

Mills, Simon, *The Essential Book of Herbal Medicine* (London: Penguin, 1991).

Oribasius, *Synopsis* (Venice: Aldus Manutius, 1554).

Paracelsus, *Four Treatises of Theophrastus von Hohenheim*, ed. Henry G. Sigerist (Baltimore, Md., and London: Johns Hopkins University Press, 1941).

Paracelsus, *De l'astrologie*, intro., trans., and ed. Lucien Braun (Strasbourg: Presses Universitaires de Strasbourg, 2002).

Platearius, *Le livre des simples médecines*, trans. and adaptation Ghislaine Malandin (Paris: Editions Ozalid et textes cardinaux: Bibliothèque Nationale, 1986). From French MS in the Bibliothèque Nationale Française.

Pseudo-Galen, *De spermate*, in Päivi Pahta, *Medieval Embryology in the Vernacular: The Case of* De spermate, Mémoires de la Société Néophilologique de Helsinki, LIII (Helsinki: Société Néophilologique, 1998).

Le Regime Tresutile et Tresproufitable pour Conserver et Garder la Santé du Corps Humain: with the commentary of Arnou de Villeneuve, corrected by the "docteurs regens" of Montpellier, 1480, Lyon, 1491, ed. Patricia Willet Cummins (Chapel Hill, N.C.: U.N.C. Dept. of Romance Languages, 1976). Also at http://www.godecookery.com/regimen/regimn07.htm

Regimen sanitatis Salernitatum, A Poem on the Preservation of Health in Rhyming Latin verse—Addressed by the School of Salerno to Robert of Normandy, Son of William the Conqueror, with an Ancient Translation, ed. and intro. Sir Alexander Croke (Oxford: D. A. Talboys, 1830).

Rule of Saint Benedict at http://www.osb.org/rb/text/toc.html#toc.

Sa'adyah Gaon on the Influence of Music, Henry George Farmer, ed. (London: Arthur Probsthain, 1943).

The Syriac Book of Medicines: Syrian Anatomy, Pathology, and Therapeutics in the Early Middle Ages, 2 vols., ed. and trans. Ernest W. Wallis Budge (London: Humphrey Milford, Oxford University Press, 1913).

Tacuinum sanitatis: Das Buch der Gesundheit, ed. Luisa Cogliati Arano, intro. Heinrich Schipperges and Wolfram Schmitt (Munich: Heimeran, 1973).

The Trotula, ed. and trans. Monica H. Green (Philadelphia: University of Pennsylvania Press, 2002).

Welsted, Robert, *De medicina mentis* (London, 1726).

SECONDARY

Benedictow, Ole J., *The Black Death 1346–1353: The Complete History* (Woodbridge: Boydell, 2004).

Brown, Peter, *The Body and Society: Men, Women, and Sexual Renunciation in Early Christianity* (New York: Columbia University Press, 1988).

Celenza, Christopher S., *The Lost Italian Renaissance: Humanists, Historians, and Latin's Legacy* (Baltimore, Md., and London: Johns Hopkins University Press, 2004).

Cipolla, Carlo M., *Cristofano e la peste* (Bologna: Il Mulino, 1976).

Cipolla, Carlo M., *Miasmi e umori* (Bologna: El Mulins, 1989).

Colapinto, John, "Bloodsuckers," *New Yorker* (July 25, 2005), pp. 72–81.

Courtenay, William J., "Curers of Body and Soul: Medical Doctors as Theologians," in *Religion and Medicine in the Middle Ages*, ed. Peter Biller and Joseph Ziegler (Woodbridge, Sufforlk; Rochester, N.Y.: York Medieval Press, 2001).

An Exploration into Medieval Medicine, at http://www.calvin.edu/academic/medieval/medicine/index.htm.

Frazier, Alison Knowles, *Possible Lives: Authors and Saints in Renaissance Italy* (New York: Columbia University Press, 2005).

Freedberg, David, *The Power of Images: Studies in the History and Theory of Response* (Chicago, Ill., and London: University of Chicago Press, 1989).

García-Ballester, Luis, Roger French, Jon Arrizabalaga, and Andrew Cunningham, eds., *Practical Medicine from Salerno to the Black Death* (Cambridge and New York: Cambridge University Press, 1994).

Gentilcore, David, *Healers and Healing in Early Modern Italy* (Manchester and New York: Manchester University Press, 1998).

Gottfried, Robert S., *The Black Death: Natural and Human Disaster in Medieval Europe* (New York: Free Press, 1983).

Grafton, Anthony, with April Shelford and Nancy Siraisi, *New Worlds, Ancient Texts: The Power of Tradition and the Shock of Discovery* (Cambridge, Mass., and London: Harvard University Press, 1992).

Grant, Edward, "Scientific Imagination in the Middle Ages," *Perspectives on Science*, 12:4 (2004), pp. 394–423.

Green, Monica H., ed. and trans., *The Trotula: An English Translation of the Medieval Compendium of Women's Medicine* (Philadelphia, Pa.: University of Pennsylvania Press, 2002).

Harrison, Peter, *The Bible, Protestantism and the Rise of Natural Science* (Cambridge and New York: Cambridge University Press).

Hunting, Penelope, *A History of the Society of Apothecaries* (London, 1998).

Jacquart, Danielle, *La science médicale occidentale entre deux renaissances (XIIe siècle–XVe siècle)* (Aldershot and Brookfield: Variorum, 1997).

Nicoud, Marilyn, "Le médecin et l'Office de santé: Le cas milanais au XVe siècle." http://www.esh.ed.ac.uk/urban_history/text/NicoudS12.doc

Nutton, Vivian, "The Reception of Fracastoro's Theory of Contagion: The Seed That Fell among Thorns?" *Osiris*, 6 (1990), pp. 196–224.

Nutton, Vivian, "God, Galen, and the Depaganization of Ancient Medicine" (The 1999 York Quodlibet Lecture), in *Religion and Medicine in the Middle Ages*, ed. Peter Biller and Joseph Ziegler (Woodbridge, Suffolk; Rochester, N.H.: York Medieval Press, 2001).

Nutton, Vivian, "Medicine in Medieval Western Europe, 1000–1500," in *The Western Medical Tradition 800 BC to AD 1800,* ed. Lawrence I. Conrad, Michael Neve, Vivian Nutton, Roy Porter, and Andrew Wear (Cambridge, New York: Cambridge University Press, 1995), pp. 139–205.

Park, Katharine, *Doctors and Medicine in Early Renaissance Florence* (Princeton, N.J.: Princeton University Press, 1985).

Park, Katharine, "Medicine and Society in Medieval Europe, 500–1500," in Andrew Wear, ed., *Medicine in Society: Historical Essays* (Cambridge: Cambridge University Press, 2002), pp. 59–90.

Pomata, Gianna, *Contracting a Cure: Patients, Healers, and the Law in Early Modern Bologna* (Baltimore, Md., and London: Johns Hopkins University Press, 1998).

Pray Bober, Phyllis, *Art, Culture, and Cuisine: Ancient and Medieval Gastronomy* (Chicago, Ill., and London: University of Chicago Press, 1999).

Roberts, Gareth, *The Mirror of Alchemy: Alchemical Ideas and Images in Manuscripts and Books from Antiquity to the Seventeenth Century* (London: British Library, 1994).

Ryan, Will F., and Charles B. Schmitt, eds., *Pseudo-Aristotle, The Secret of Secrets: Sources and Influences* (London: Warburg Institute, 1982).

Sadek, Mahmoud M., *The Arabic Materia Medica of Dioscorides* (*Saint-Jean-Chrysostome*) Quebec: Editions du Sphinx, 1983).

Sarton, George, *Galen of Pergamon* (Lawrence, Kans.: University of Kansas Press, 1954).

Silini, Giovanni, *Umori e Farmaci: Terapia medica tardo-medioevale* (Bergamo: Servitium, 2001).

Siraisi, Nancy, *Taddeo Alderotti and His Pupils: Two Generations of Italian Medical Learning* (Princeton, N.J.; Princeton University Press, 1981).

Siraisi, Nancy, *Medieval and Early Renaissance Medicine: An Introduction to Knowledge and Practice* (Chicago, Ill., and London: University of Chicago Press, 1990).

Skinner, Patricia, *Health and Medicine in Early Medieval Southern Italy* (Leiden, New York: Brill, 1997).

Stannard, Jerry, "Greco-Roman Materia Medica in Medieval Germany," *Bulletin of the History of Medicine,* 46:5 (September–October, 1977), pp. 455–468.

Starobinski, Jean, "Le passé de la passion: Textes médicaux et commentaires," *Nouvelle Revue de Psychanalyse,* 31 (1980), pp. 51–76.

Strelcyn, Stefan, *Médecine et plantes d'Ethiopie: Les traités médicaux éthiopiens* (Warsaw, 1968; Naples: Instituto Universitario Orientale, 1973).

Ullmann, Manfred, *Islamic Medicine* (Edinburgh: Edinburgh University Press, 1978).

Walker Bynum, Caroline, *The Resurrection of the Body in Western Christianity, 200–1336* (New York: Columbia University Press, 1995).

Wear, Andrew, ed. *Medicine in Society: Historical Essays* (Cambridge and New York: Cambridge University Press, 2002).

Weiss-Amer, Melitta, "Medieval Women's Guides to Food during Pregnancy: Origins, Texts, and Traditions," *Canadian Bulletin of Medical History,* 10 (1993), pp. 5–23.

CHAPTER FOUR

PRIMARY

Brandolini, Raffaele, *On Music and Poetry (De musica et poetica,* 1513), trans., intro, and notes Ann Moyer and Marc Laureys, Medieval and Renaissance Texts and Studies, Vol. 232 (Tempe, Ariz.: Arizona Center for Medieval and Renaissance Studies, 2001).

Breton, Nicholas, *Melancholicke Humour, in Verses of Diverse Natures* (London, 1600), ed. with *Essay on Elizabethan Melancholy* by G. B. Harrison (London: Scholartis, 1929).

Bright, Timothy, *Treatise of Melancholy,* 3rd ed. (London, 1612). Originally published 1586.

Burton, Robert, *The Anatomy of Melancholy* (Oxford, 1628; New York: New York Review Books, 2001).

Capella, Martianus, *Le nozze di Filologia e Mercurio,* ed. and trans. Ilaria Ramelli (Milan: Bompiani, 2001).

Cardano, Girolamo, *The Book of My Life,* trans. Jean Stoner, intro. Anthony Grafton (New York: New York Review Books, 2002).

Cellini, Benvenuto, *Autobiography,* trans. John Addington Symonds (New York: Collier Press—Kessinger Publishing, 2005).

della Porta, Giambattista, *Magiae naturalis libri XX* (Naples, 1584); translated into English as *Natural Magick* (London, 1658); facsimile reprint (New York: Basic Books, 1957).

Fernel, Jean, *De abditis rerum causis: On the Hidden Causes* (Paris, 1548), ed., trans. John Henry and John M. Forrester (Leiden: Brill, 2005).

Ferrand, Jacques, *Traité de l'essence et guérison de l'amour ou De la mélancolie érotique* (Paris, 1610, 1623). Also ed. Gérard Jacquin and Eric Foulon (Paris: Economica, 2001).

Ferrand, Jacques, *A Treatise on Lovesickness,* trans., ed., and critical intro. Donald A. Beecher and Massimo Ciavolella (Syracuse, N.Y.: Syracuse University Press, 1990).

Ficino, Marsilio, *De vita triplici: Three Books on Life,* trans. and ed. Carol V. Kaske and John R. Clark (Binghamton, N.Y.: Medieval and Renaissance Texts and Studies, 1989).

Ficino, Marsilio, *Sopra lo amore o ver'Convito di Platone: Comento di Marsilio Ficini Fiorentino sopra il Convito di Platone,* ed. G. Ottaviano (Milan: Celuc, 1973).

Fuchs, Leonhart, *New Kreüterbuch (Basel, 1543): The New Herbal of 1543,* ed. Klaus Dobat and Werner Dressendörfer (Köln: Taschen, 2001). Complete reprint.

Galen, *La dieta dimagrante,* ed. Nino Marinone (Turin: Paravia, 1973).

Galen, *De temperamentis, et de inaequali intemperie, libri tres, Thoma Linacro anglo-interprete: Galen On Temperaments,* trans. Thomas Linacre (Cambridge, 1881. Anastatic ed.; originally published Oxford, 1521.)

Galen, *On the Natural Faculties,* trans. Arthur John Brock (Cambridge, Mass.: Harvard University Press, 2000).

Galen on Food and Diet, ed. Mark Grant (London and New York: Routledge, 2000).

Galen, *De la bile noire,* ed. and trans. Vincent Barras, Terpsichore Birchler, and Anne-France Morand (Paris: Gallimard, Le Promeneur, 1998).

Paracelsus, *Selected Writings*, ed. Jolande Jacoby; trans. Norbert Guterman (Princeton and Chichester: Princeton University Press, 1951, 1988).

Paré, Ambroise, *Journeys in Divers Places*, Vol. 28:2, trans. Stephen Paget (New York, 1909, 1914). Also at http://www.bartleby.com/38/2/.

Scalona, Vincenzo della, in *Carteggio degli oratori mantovani alla corte sforzesca (1450–1500)*, ed. France Leverotti, Vol. 3: 1461 (Rome: Ministero peri beni e le attività culturali. Ufficio centrale per í beni archivistici, 2000). Letters to Barbara di Brandeburgo, May 2, 1461, pp. 195–197; and to Ludovico Gonzaga, September 20, 1461, pp. 333–335.

Servetus, Michael, *Christianismi restitutio, A Translation of His Geographical, Medical, and Astrological Writings with Introductions and Notes*, trans. Charles Donald O'Malley (Philadelphia, Pa.: Americal Philosophical Society and London: Lloyd-Luke Medical, 1953), pp. 201–208.

Vesalius, Andreas, *De fabrica corporis humani* (Basel, 1543). Translation at http://vesalius.northwestern.edu.

Walkington, Thomas, *The Optick Glass of Humours* (London, 1607, 1631), photo reprint and intro. John A. Popplestone and Marion White McPherson (Delmar, N.Y.: Scholars' Facsimiles and Reprints, 1981).

SECONDARY

Arikha, Noga, "La mélancolie et les passions humorales au début de la modernité," in *Mélancolie: Génie et folie en Occident*, ed. Jean Clair (Paris: Réunion des Musées Nationaux/Gallimard, 2005), pp. 232–240.

Babb, Lawrence, *Sanity in Bedlam: A Study of Robert Burton's Anatomy of Melancholy* (East Lansing: Michigan State University Press, 1959).

Carlino, Andrea, *Paper Bodies: A Catalogue of Anatomical Fugitive Sheets, 1538–1687*, trans. Noga Arikha (London: Wellcome Institute for the History of Medicine, 1999).

Copenhaver, Brian P., "Scholastic Philosophy and Renaissance Magic in the *De Vita* of Marsilio Ficino," *Renaissance Quarterly,* 37:4 (1984), pp. 523–554.

Cosmacini, Giorgio, "La malattia del duca Francesco," in *Carteggio degli oratori mantovani alla corte sforzesca (1450–1500)*, Vol. III, *1461*, ed. Franca Leverotti (Rome: Ministero per i beni e le attività culturali. Ufficio centrale per i beni archivistici, 2000), pp. 23–26.

Dandrey, Patrick, *Anthologie de l'humeur noire: Écrits sur la mélancolie d'Hippocrate à l'Encyclopédie* (Paris: Gallimard, Le Promeneur, 2005).

Debus, Allen, *Paracelsus and the Medical Revolution of the Renaissance: A 500th Anniversary Celebration.* At http://www.nlm.nih.gov/exhibition/paracelsus/.

Eamon, William, *Science and the Secrets of Nature: Books of Secrets in Medieval and Early Modern Culture* (Princeton, N.J.: Princeton University Press, 1994).

Gouk, Penelope, "Some English Theories of Hearing in the Seventeenth Century: Before and After Descartes," in *The Second Sense: Studies in Hearing and Musical Judgement from Antiquity to the Seventeenth Century,* ed. Charles Burnett, Michael Fend, and Penelope Gouk (London: Warburg Institute, 1991).

Grafton, Anthony, *Cardano's Cosmos: The Worlds and Works of a Renaissance Astrologer* (Cambridge, Mass., and London: Harvard University Press, 1999).

Hersant, Yves, *Mélancolies: De l'Antiquité au XXème siècle* (Paris: Robert Laffont, Bouquins, 2005).

Horden, Peregrine, ed., *Music as Medicine: The History of Music Therapy since Antiquity* (Aldershot; Brookfield, Singapore; Sydney: Ashgate, 2000).

Jarcho, Saul, *The Concept of Contagion in Medicine, Literature, and Religion* (Malabar, Fla.: Krieger, 2000).

Kagan, Richard L., *Lucrecia's Dreams: Politics and Prophecy in Sixteenth-Century Spain* (Berkeley, Los Angeles, and London: University of California Press, 1990).

Klibansky, Raymond, Erwin Panofsky, and Fritz Saxl, *Saturn and Melancholy: Studies in the History of Natural Philosophy, Religion, and Art* (London: Nelson, 1964). Italian edition: *Saturno e la melancolia: Studi en storia della filosofia naturale, medicina, religione e arte* (Turin: Einaudi, 1983, 2002).

Koyré, Alexandre, *Mystiques, spirituels, alchimistes du XVIe siècle allemand* (Paris: Gallimard, 1971).

Kümmel, Werner Friedrich, *Musik und Medizin: Ihre Wechselbeziehungen in Theorie und Praxis von 800 bis 1800* (Munich: Alber, 1977).

Lippincott, Kristen, "Two Astrological Ceilings Reconsidered: The Sala di Galatea in the Villa Farnesina and the Sala del Mappamondo at the Caprarola," *Journal of the Warburg and Courtauld Institutes,* 53 (1990), p. 185.

Mori, Elisabetta, "L'onore perduto del duca di Bracciano: Dalla corrispondenza di Paolo Giordano Orsini e Isabella de' Medici," *Dimensioni e Problemi della Ricerca Storica,* No. 2 (2004), pp. 135–174.

Newman, William R., *Promethean Ambitions: Alchemy and the Quest to Perfect Nature* (Chicago, Ill.: University of Chicago Press, 2004).

Nutton, Vivian, *Introduction to* Vesalius, *De humani corporis fabrica.* At http://vesalius.northwestern.edu/books/FA.aa.html.

Quinlan-McGrath, Mary, "The Astrological Vault of the Villa Farnesina: Agostino Chigi's Rising Sign," *Journal of the Warburg and Courtauld Institutes,* 47 (1984), pp. 91–105.

Radden, Jennifer, ed., *The Nature of Melancholy from Aristotle to Kristeva* (New York: Oxford University Press, 2000).

Roberts, K. B., and J. D. W. Tomlinson, *The Fabric of the Body: European Traditions of Anatomical Illustration* (Oxford: Clarendon, 1992).

Rosen, Charles, "The Anatomy Lesson" (on Burton's *The Anatomy of Melancholy),* *New York Review of Books* (June 9, 2005), pp. 55–59.

Sawday, Jonathan, *The Body Emblazoned: Dissection and the Human Body in Renaissance Culture* (London: Routledge, 1995).

Schmitt, Charles and Quentin Skinner, *The Cambridge History of Renaissance Philosophy* (Cambridge and New York: Cambridge University Press, 1988).

Seznec, Jean, *La survivance des dieux antiques,* Studies of the Warburg Institute, Vol. XI (London: Warburg Institute, 1940).

Shiloah, Amnon, "Jewish and Muslim Traditions of Music Therapy," in *Music as Medicine: The History of Music Therapy since Antiquity,* ed. Peregrine Horden (Aldershot; Brookfield; Singapore; Sydney: Ashgate, 2000).

Siraisi, Nancy, *The Clock and the Mirror: Girolamo Cardano and Renaissance Medicine* (Princeton, N.J.: Princeton University Press, 1997).

Starobinski, Jean, "Le passé de la passion: Textes médicaux et commentaires," *Nouvelle Revue de Psychanalyse*, XXI (1980), pp. 51–76.

Stimilli, Davide, *The Face of Immortality: Physiognomy and Criticism* (Albany, N.Y.: State University of New York Press, 2005).

Teich, Mikuláš, and Robert Young, eds., *Changing Perspectives in the History of Science: Essays in Honour of Joseph Needham* (London: Heinemann Educational, 1973).

Tomlinson, Gary, *Music in Renaissance Magic: Toward a Historiography of Others* (Chicago, Ill., London: University of Chicago Press, 1993).

Voss, Angela, "Marsilio Ficino: The Second Orpheus," in *Music as Medicine: The History of Music Therapy since Antiquity,* ed. Peregrine Horden (Aldershot; Brookfield; Singapore; Sydney: Ashgate, 2000).

Walker, D. P., *Spiritual & Demonic Magic from Ficino to Campanella* (University Park, PA.: Pennsylvania State University Press, 2000; first published as Vol. 22, Studies of the Warburg Institute, University of London, 1958).

Webster, Charles, *From Paracelsus to Newton: Magic and the Making of Modern Science,* Eddington Memorial Lectures (Cambridge, New York: Cambridge University Press, 1982).

Yates, Frances, *The Occult Philosophy in the Elizabethan Age* (London and Boston: Routledge & Kegan Paul, 1979).

CHAPTER FIVE

PRIMARY

Bacon, Francis, *The Great Instauration: The New Organon or True Directions for the Interpretation of Nature* (London, 1620), in *Francis Bacon, The Works,* 3 vols., trans. Basil Montague (Philadelphia, Pa.: 1854).

Bayle, Pierre, *Dictionnaire historique et critique,* 16 vols. (Rotterdam, 1697; Paris, 1820–1824).

Browne, Thomas, *Religio Medici* (London, 1643).

Campanella, Tommaso, "Apologia for the Opuscule on *De siderali fato vitando,*" trans. Germana Ernst and Noga Arikha, *Culture and Cosmos,* 6:1 (2002).

Campanella, Tommaso, *La città del sole: Dialogo poetico—The City of the Sun: A Poetical Dialogue,* trans. and ed. Daniel J. Donno (Berkeley, Los Angeles, and London: California University Press, 1981).

Cordemoy, Gérauld de, "Six Discours sur la distinction et l'union du corps et de l'âme" (Paris, 1666); "Discours physique de la parole" (Paris, 1668), in *Oeuvres philosophiques, avec une étude bio-bibliographique,* ed. Pierre Clair and François Girbal (Paris: Presses Universitaires de France, 1968).

Culpeper, Nicholas, *The English Physitian: Or An Astrologo-Physical Discourse of the Vulgar Herbs of this Nation* (London, 1652). Also at http://www.bibliomania.com/2/1/66/113/frameset.html.

Darmanson, Jean-Marie, *La bête transformée en machine: Divisée en deux Dissertations prononcées à Amsterdam par J. Darmanson dans ses Conférences Philosophiques* (Amsterdam, 1684).

Defoe, Daniel, *Journal of the Plague Year* (London, 1722).

Descartes, René, *Oeuvres philosophiques,* 3 vols., ed. Ferdinand Alquié (Paris: Bordas, Classiques Garnier, 1988–1989).

Digby, Kenelm, *A Discourse Made in a Solemne Assembly of Noble and Learned Men at Montpellier in France, by Sir Kenelme Digby, Knight &c., Touching the Cure of Wounds by the Powder of Sympathy* (London, 1658).

Duncan, Daniel, *Explication nouvelle et mechanique des actions animales* (Paris, 1678).

Galilei, Galileo, "The Starry Messenger" and "Letter to the Grand Duchess Christina," in *Discoveries and Opinions of Galileo,* ed. Stillman Drake (New York: Random House, 1957).

Gassendi, Pierre, "Syntagma philosophicum," in *Opera* (Lyon, 1658).

Harvey, William, "Exercitatio anatomica de motu cordis et sanguinis in animalibus (Frankfurt, 1628)," ed. and trans. Kenneth J. Franklin, in *The Circulation of the Blood and Other Writings* (London and New York: Dent, 1963).

Harvey, William, "Exercitationes duae anatomicae de circulatione sanguinis," ed. and trans. Kenneth J. Franklin, in *The Circulation of the Blood and Other Writings* (London and New York: Dent, 1963). Originally published Cambridge and Rotterdam, 1649; English ed., 1653.

Johnson, Samuel, "Hermann Boerhaave," in *Works of Samuel Johnson* (Troy, N.Y., 1903), Vol. 14, pp. 154–184. Also at http://www.samueljohnson.com/boerhaave.html.

Lower, Richard, "Diatribae Thomae Willisii de febribus vindicatio (1665)," trans. Leofranc Holford-Strevens, ed. Kenneth Dewhurt, in *Vindicatio: A Defence of the Experimental Method* (Oxford: Oxford University Press, 1983).

Malpighi, Marcello, *Correspondence* Vol. 5: 1693–1694 (Ithaca, N.Y., and London: Cornell University Press, 1975).

Mersenne, Marin, "Questions théologiques, physiques, morales, et mathematiques (Paris, 1634)," in *Questions inouyes. Questions harmoniques. Questions théologiques. Les méchaniques de Galilée. Les préludes de l'harmonie universelle* (Paris: Fayard, 1985).

Montagu, Mary Wortley, *Selected Lettters,* ed. Isobel Grundy (London: Penguin, 1997).

Paré, Ambroise, *Des monstres et des prodiges (Paris, 1575): On Monsters and Marvels,* trans. Janis L. Pallister (Chicago, Ill., and London: University of Chicago Press, 1982).

Pepys, Samuel, *Diary,* Vol. VIII, 1667 ed. Robert Latham and William Matthews (London: Bell and Hyman, 1974).

Sydenham, Thomas, "Theologia rationalis," in *Thomas Sydenham (1624–1689): His Life and Original Writings,* ed. Kenneth Dewhurst (Berkeley, Los Angeles, and London: University of California Press, 1966).

SECONDARY

Arikha, Noga, *Adam's Spectacles: Nature, Mind, and Body in the Age of Mechanism,* (unpublished doctoral thesis, Warburg Institute, London, 2001).

Bossi, Laura, *Histoire naturelle de l'âme* (Paris: Presses Universitaires de France, 2003).

Bradburne, James M., ed., *Blood: Art, Power, Politics and Pathology* (Munich, London, New York: Prestel, 2001)

Camporesi, Piero, *Il sugo della vita: Simbolismo e magia del sangue* (Milan: Garzanti, 1997).

Copenhaver, Brian P., and Charles B. Schmitt, *Renaissance Philosophy* (Oxford: Oxford University Press, 1992).

Daston, Lorraine, and Katharine Park, *Wonders and the Order of Nature, 1150–1750* (New York: Zone, 1998).

Davis, Audrey B., *Circulation Physiology and Medical Chemistry in England 1650–1680* (Lawrence, Kans.: Coronado Press, 1973).

Debus, Allen G., "Robert Fludd and the Use of Gilbert's *De magnete* in the Weapon-Salve Controversy," *Journal of the History of Medicine and Allied Sciences*, 19:4 (1964), pp. 389–417.

Debus, Allen G., *The Chemical Philosophy: Paracelsian Science and Medicine in the Sixteenth and Seventeenth Centuries* (Mineola, NY: Dover, 2002).

Dewhurst, Kenneth, ed., *Thomas Sydenham (1624–1689): His Life and Original Writings* (Berkeley, Los Angeles, and London: University of California Press, 1966).

Drake, Stillman, *Galileo: A Very Short Introduction* (Oxford and New York: Oxford University Press, 1980).

Ernst, Germana, "The Sky in a Room: Campanella's *Apologeticus* in Defence of the Pamphlet *De siderali fato vitando*," trans. Noga Arikha, *Culture and Cosmos*, 6:1 (2002).

Fara, Patricia, *Sympathetic Attractions: Magnetic Practices, Beliefs and Symbolism in Eighteenth-Century England* (Princeton, N.J.: Princeton University Press).

Findlen, Paula, *Possessing Nature: Museums, Collecting, and Scientific Culture in Early Modern Italy* (Berkeley, Los Angeles, and London: University of California Press, 1994).

Fisher, Saul, "Pierre Gassendi," *Stanford Encyclopedia of Philosophy*, ed. Edward N. Zalta. At http://plato.stanford.edu/entries/gassendi/.

Frank, Robert G., Jr., *Harvey and the Oxford Physiologists: A Study of Scientific Ideas* (Berkeley, Los Angeles, and London: University of California Press, 1980).

Freedberg, David, *The Eye of the Lynx: Galileo, His Friends, and the Beginnings of Modern Natural History* (Chicago, Ill., and London: University of Chicago Press, 2002).

French, R. K., *Robert Whytt, the Soul and Medicine* (London: Wellcome Institute of the History of Medicine, 1969).

French, Roger, *Medicine before Science: The Business of Medicine from the Middle Ages to the Enlightenment* (Cambridge: Cambridge University Press, 2003).

Gambaccini, Piero, *Mountebanks and Medicasters: A History of Italian Charlatans from the Middle Ages to the Present* (Jefferson, N.C., and London: McFarland & Co., 2004)

Giglioni, Guido, *Immaginazione e malattia: Saggio su Jan Baptiste van Helmont* (Milan: FrancoAngeli, 2000).

Grmek, Mirko D., ed., *Storia del pensiero medico occidentale*, Vol. 2, *Dal Rinascimento all'inizio dell'ottocento* (Rome and Bari: Laterza, 1996).

James, Susan, *Passion and Action: The Emotions in Seventeenth-Century Philosophy* (Oxford: Clarendon, 1957).

Kern Paster, Gail, *The Body Embarrassed: Drama and the Disciplines of Shame in Early Modern England* (Ithaca, N.Y.: Cornell University Press, 1983).

Lawrence, Christopher, and Steven Shapin, eds., *Science Incarnate: Historical Em-*

bodiments of Natural Knowledge (Chicago, Ill., and London: University of Chicago Press, 1998).

Miller, Peter, *Peiresc's Europe: Learning and Virtue in the Seventeenth Century* (New Haven, Conn., and London: Yale University Press, 2000).

Muller-Jahncke, Wolf Dieter, "Medical Magic of Paracelsus and Paracelsus Followers: Weapon Salve," *Sudhoffs Archiv: Zeitschrift für Wissenschaftsgeschichte,* 31 (1993), pp. 43–55.

Newman, William R., and Lawrence M. Principe, *Alchemy Tried in the Fire: Starkey, Boyle, and the Fate of Helmontian Chymistry* (Chicago, Ill., and London: University of Chicago Press, 2002).

Pagel, Walter, *Joan Baptista Van Helmont: Reformer of Science and Medicine* (Cambridge: Cambridge University Press, 1982).

Park, Katharine, "Bacon's 'Enchanted Glass,' " *Isis,* 75 (1984), pp. 290–302.

Pichot, André, *Histoire de la notion de vie* (Paris: Gallimard, 1993).

Schaffer, Simon, "Godly Men and Mechanical Philosophers: Souls and Spirits in Restoration Natural Philosophy," *Science in Context,* 1 (1987), pp. 55–85.

Serjeantson, Richard, "The Passions and Animal Language, 1540–1700," *Journal of the History of Ideas* (2001), pp. 425–444.

Shapin, Steven, *A Social History of Truth* (Chicago, Ill., and London: University of Chicago Press, 1994).

Shapin, Steven, *The Scientific Revolution* (Chicago, Ill., and London: University of Chicago Press, 1996).

Sutton, John, *Philosophy and Memory Traces: Descartes to Connectionism* (Cambridge, New York: Cambridge University Press, 1998).

Thomas, Keith, *Man and the Natural World: Changing Attitudes in England 1500–1800* (London: Penguin, 1983).

Todd, Dennis, *Imagining Monsters: Miscreations of the Self in Eighteenth Century England* (Chicago, Ill., and London: University of Chicago Press, 1995).

Tomalin, Claire, *Samuel Pepys: The Unequalled Self* (London: Penguin, 2002).

Trevor-Roper, Hugh, "Medicine at the Early Stuart Court," in *From Counter-Reformation to Glorious Revolution* (London, 1992).

Vegetti Finzi, Silvia, ed., *Storia delle passioni* (Rome and Bari: Laterza, 1995, 2004).

Wear, Andrew, ed. *Medicine in Society: Historical Essays* (Cambridge and New York: Cambridge University Press, 1992).

Wear, Andrew, *Knowledge and Practice in English Medicine, 1550–1680* (Cambridge and New York: Cambridge University Press, 2000).

Wilkins, Bridgit S., "The Spleen," *British Journal of Haematology,* 117 (2002), pp. 265–274.

Wittkower, Rudolf and Margot Wittkower, *Born under Saturn: The Character and Conduct of Artists—A Documented History from Antiquity to the French Revolution* (New York: Norton, 1963).

Woolley, Benjamin, *Heal Thyself: Nicholas Culpeper and the Seventeenth-Century Struggle to Bring Medicine to the People* (London and New York: HarperCollins, 2004).

CHAPTER SIX

PRIMARY

Bernard, Claude, *Introduction à l'étude de la médecine expérimentale* (Paris, 1865: Garnier Flammarion, 1966).

Descartes, René, *Oeuvres philosophiques,* 3 vols., ed. Ferdinand Alquié (Paris: Bordas Classiques Garnier, 1988–1989).

Freud, Sigmund, "Aphasie" and "Gehirn," in *A Moment of Transition: Two Neuroscientific Articles by Sigmund Freud,* ed. and trans. Mark Solms and Michael Saling (London and New York: Karnalk 1990).

Frederick II of Prussia, *Eulogy of Julien-Offroy de La Mettrie* (1748). At http://cscs. umich.edu/~crshalizi/LaMettrie/Machine/.

Hahnemann, Samuel, *Organon of the Medical Art,* ed. Wenda Brewster O'Reilly (Palo Alto, Calif.: Birdcage, 1996).

Choderlos de Laclos, *Des femmes et de leur éducation* (Paris, 1783; Editions mille et une nuits, 2000).

Choderlos de Laclos, *Les liaisons dangereuses* (Paris, 1782; Gallimard, 2003).

Lamy, Guillaume, *Discours anatomiques. Avec des Reflexions sur les objections qu'on luy a faites contre sa maniere de raisonner de la nature de l'homme, & de l'usage des parties qui le composent (Rouen, 1675); Explication méchanique et physique des fonctions de l'âme sensitive (Paris, 1677),* ed. Anna Minerbi Begrado (Paris: Universitas; Oxford: Voltaire Foundation, 1996).

La Mettrie, Julien Offray de, *L'homme-machine,* ed. Adam Vartanian (Leiden, 1748, 1751; Princeton, N.J.: Princeton University Press, 1960).

La Mettrie, Julien Offray de, *Man a Machine,* trans. and ed. Gertrude Carman Bussey (Chicago, Ill.: Open Court, 1912). With eulogy by Frederick the Great and extracts from *The Natural History of the Soul.*

Mackay, Charles, *Extraordinary Popular Delusions and the Madness of Crowds* (London, 1841, 1995).

Mavéric, Jean, *La médecine hermétique des plantes ou L'extraction des quintessences par art spagyrique* (Paris: Dorbon, 1911).

Mirville, Jules-Eudes de, *Des esprits et de leurs manifestations fluidiques* (Paris, 1853).

Owen, Richard, *The Hunterian Lectures in Comparative Anatomy, May and June 1837,* ed. Phillip Reid Sloan (Chicago, Ill.,: University of Chicago Press; London: Natural History Museum Publications, 1992).

Priestley, Joseph, *Disquisitions Relating to Matter and Spirit* (London, 1777). Available at Schoenberg Center for Electronic Text and Image, http://dewey.library. upenn.edu/sceti

Raspail, François-Vincent, *Manuel annuaire de la santé pour 1861, ou Médecine et pharmacie domestiques* (Paris and Brussels, 1861).

Steno, Nicolaus, *Discours sur l'anatomie du cerveau. À Messieurs de l'Assemblée de chez Monsieur Thevenot* (Paris, 1669).

Swift, Jonathan, "A Discourse Concerning the Mechanical Operation of the Spirit. In a Letter to a Friend. A Fragment" (London, 1704), in *A Tale of a Tub and Other Works* (Oxford: Oxford University Press, 1986).

Vives, Juan Luis, "De anima et vita beata (Bruges, 1538)," in *The Passions of the Soul: The Third Book of De Anima et Vita,* ed. Carlos G. Norena (Lewiston, NY: Edwin Mellen Press, 1990).

Whytt, Robert, *Works* (Birmingham: Gryphon Editions, 1984).

Willis, Thomas, *Two Discourses Concerning the Soul of Brutes, Which is that of the Vital and Sensitive of Man* (London, 1683).

Wright, Thomas, *The Passions of the Mind in General*, ed. William Webster Newhold (London, 1601, 1604; New York and London: Garland, 1986).

SECONDARY

Anderson, R. G. W., and Christopher Lawrence, *Science, Medicine, and Dissent: Joseph Priestley (1733–1804)* (London: Wellcome Trust/Science Museum, 1987).

Arikha, Noga, "'That I Can Study Thee': Seventeenth-Century Treatises on the Passions" (unpublished master's thesis, Warburg Institute, London, 1996).

Armstrong, Carol, "Probing Pictures," *Artforum International* (January 9, 2003).

Barker, Sheila, "Poussin, Plague, and Early Modern Medicine," *Art Bulletin* (December 2004), pp. 659–689.

Berrios, German, and Roy Porter, eds., *A History of Clinical Psychiatry: The Origin and History of Psychiatric Disorders* (London and New Brunswick, N.J.: Athlone Press, 1995).

Blom, Phillip, *Enlightening the world: Encyclopédie, the book that changed the course of history* (New York: Palgrave Macmillan, 2005).

Canguilhem, Georges, *La formation du concept de réflexe aux XVIIe et XVIIIe siècles* (Paris: Vrin, 1977).

Certeau, Michel de, *La possession de Loudun* (Paris, 1970). English: *The Possession at Loudun,* trans. Michael B. Smith (Chicago, Ill., London: University of Chicago Press, 1996).

Chang, Ku-Ming (Kevin), "*Motus Tonicus:* Georg Ernst Stahl's Formulation of Tonic Motion and Early Modern Medical Thought," *Bulletin of the History of Medicine,* 78 (2004), pp. 767–803.

Clair, Jean, ed., *L'âme au corps: Arts et sciences 1793–1993*, (Paris: Réunion des Musées Nationaux/Gallimard, 1993).

Clarke, Edwin, and L. S. Jacyna, *Nineteenth-Century Origins of Neuroscientific Concepts* (London, Berkeley and Los Angeles: University of California Press, 1987).

Cook, Harold J., "Boerhaave and the Flight from Reason in Medicine," *Bulletin of the History of Medicine,* 74 (2000), pp. 221–240.

Corsi, Pietro, ed., *La Fabrique de la pensée: De l'art de la mémoire aux neurosciences* (Milan: Electa, 1990).

Cosmacini, Giorgio, *Storia della medicina e della sanità in Italia: Dalla peste europea alla guerra mondiale, 1348–1918* (Rome: Laterza, 1987).

Cosmacini, Giorgio, *Il medico ciarlatano: Vita inimitabile di un europeo del Seicento* (Rome: Laterza, 1998).

Cosmacini, Giorgio, *Ciarlateneria e medicina: Cure, maschere, ciarle* (Milan: Cortina Raffaello, 1998).

Darnton, Robert, *Mesmerism and the End of the Enlightenment in France* (Cambridge, Mass.: Harvard University Press, 1968).

de Ceglia, Francesco Paolo, "The Blood, the Worm, the Moon, the Witch: Epilepsy in Georg Ernst Stahl's Pathological Architecture," *Perspectives on Science* 12:1 (2004), 12:1, pp. 1–28.

Didi-Huberman, Georges, *Invention of Hysteria: Charcot and the Photographic Iconography of the Salpêtrière* (Cambridge, Mass.: MIT Press, 2003).

Douthwaite, Julia V., *The Wild Girl, Natural Man, and the Monster: Dangerous Experiments in the Age of Enlightenment* (Chicago, Ill. and London: Chicago University Press, 2002).

Ehrard, Jean, *L'idée de nature en France dans la première moitié du XVIIIe siècle* (Paris: SEVPEN, 1963; Albin Michel, 1994).

Gauld, Alan, *A History of Hypnotism* (Cambridge, New York: Cambridge University Press, 1992).

Grmek, Mirko D., ed., *Storia del pensiero medico occidentale*, Vol. 2, *Dal Rinascimento all'inizio dell'ottocento* (Rome and Bari: Laterza, 1996).

Grmek, Mirko D., ed., *Storia del pensiero medico occidentale*, Vol. 3, *Dall'età romantica alla medicina moderna* (Rome and Bari: Laterza, 1998).

Haigh, Elizabeth L., "The Vital Principle of Paul-Joseph Barthez: The Clash between Monism and Dualism," *Medical History* 21 (1977), pp. 1–14.

Harrington, Anne, *Medicine, Mind, and the Double Brain* (Princeton, N. J.: Princeton University Press, 1987).

Harrington, Anne, *Reenchanted Science: Holism in German Culture from Wilhelm II to Hitler* (Princeton: Princeton University Press, 1996).

Harris, Henry, *The Birth of the Cell* (New Haven and London: Yale University Press, 1999).

Magli, Patrizia, *Il volto e l'anima: Fisiognomica e passioni* (Milan: Bompiani, 1995).

Martensen, Robert L., *The Brain Takes Shape: An Early History* (Oxford: Oxford University Press, 2004).

Paulson, Ronald, *Hogarth's Graphic Works* (New Haven and London: Yale University Press, 1965).

Picon, Antoine, *Claude Perrault ou la curiosité d'un classique* (Paris: Picard, 1988).

Pigeaud, Jackie, *Aux portes de la psychiatrie: Pinel, l'ancien et le moderne* (Paris: Aubier, 2001).

Porter, Roy, *Mind-Forg'd Manacles: A History of Madness in England from the Restoration to the Regency* (London: Penguin, 1987).

Porter, Roy, *The Greatest Benefit to Mankind: A Medical History of Humanity* (London: HarperCollins, 1997; New York: Norton, 1998).

Porter, Roy, *Quacks: Fakers and Charlatans in English Medicine* (Stroud: Tempus, 2000).

Porter, Roy, *Blood and Guts: A Short History of Medicine* (New York and London: Norton, 2002).

Porter, Roy, *Flesh in the Age of Reason: How the Enlightenment Transformed the Way We See Our Bodies and Souls* (London: Penguin, 2004).

Quinton, Anthony, " 'A Master Materialist': Review of Kathleen Wellman, *La Mettrie: Medicine, Philosophy, and Enlightenment*," *New York Review of Books* (March 25, 1993).

Roger, Jacques, *Les sciences de la vie dans la pensée française au XVIIIe siècle: La génération des animaux de Descartes à l'Encyclopédie* (Paris: Albin Michel, 1963, 1993).

Roscioni, Lisa, *Il governo della follia: Ospedali, medici e pazzi nell'età moderna* (Milan: Mondadori, 2003).

Shapin, Steven, "Descartes the Doctor: Rationalism and Its Therapies," *British Journal for the History of Science*, 33 (2000), pp. 131–154.

Starobinski, Jean, *Action et réaction: Vie et aventures d'un couple* (Paris: Seuil, 1999).

Thomson, Ann, "La Mettrie et Diderot," *Recherches sur Diderot et l'Encyclopédie: Groupe de travail sur Le rêve de d'Alembert* (2000). At http://www.sigu7.jussieu.fr/diderot/travaux/revseance2.htm.

Vigarello, Georges, *Le propre et le sale: L'hygiène du corps depuis le Moyen Age* (Paris: Seuil, 1985).

Williams, Elizabeth A., *The Physical and the Moral: Anthropology, Physiology, and Philosophical Medicine in France, 1750–1850* (Cambridge and New York: Cambridge University Press, 2002).

Young, Robert M., *Mind, Brain, and Adaptation in the Nineteenth Century: Cerebral Localization and Its Biological Context from Gall to Ferrier* (Oxford, 1990). Also at http://human-nature.com/mba/mba1.html.

Zimmer, Carl, *Soul Made Flesh: The Discovery of the Brain—and How It Changed the World* (New York: Free Press; London: Heinemann, 2004).

CHAPTER SEVEN

Achaya, K. T., *A Historical Dictionary of Indian Food* (New Delhi: Oxford University Press, 1998).

Adolphs, Ralph, and Daniel Tranel, "Impaired Judgments of Sadness but Not Happiness Following Bilateral Amygdala Damage," *Journal of Cognitive Neuroscience*, 16:3 (2004), pp. 453–462.

Albala, Ken, *Eating Right in the Renaissance* (Berkeley, Los Angeles, and London: University of California Press, 2002).

Artusi, Pellegrino, *La scienza in cucina e l'arte di mangiar bene* (first published 1891), ed. Piero Camporesi (Turin: Einaudi, 1979, 1995); *Science in the Kitchen and the Art of Eating Well*, trans. Murtha Baca and Stephen Sartarelli (New York: Marsilio, 1997).

Bartels, Andreas, and Semir Zeki, "The Neural Basis of Romantic Love," *NeuroReport*, 17:11 (2000), pp. 3829–3834.

Bartels, Andreas, and Semir Zeki, "The neural correlates of maternal and romantic love," *NeuroImage*, 21:3 (2004), pp. 1155–1166.

Bennett, M. R., and Hacker, P. M. S., *Philosophical Foundations of Neuroscience* (Oxford: Blackwell, 2003).

Benzon, William, *Beethoven's Anvil: Music in Mind and Culture* (New York: Basic Books, 2002).

Blakeslee, Sandra, "Humanity? Maybe It's in the Wiring," *New York Times* (December 9, 2003).

Boyer, Pascal, *Religion Explained: The Evolutionary Origins of Religious Thought* (London: William Heinemann, 2001).

Browne, Anthony, "Why It's All in the Mind: As More People Fill Surgeries Seeking a Cure for Ageing or Alcoholism, Doctors Are Rebelling—Perhaps We Need Less Medicine, Not More," *Observer*, London (April 14, 2002).

Burne, Jerome, "A Cure for Stress?" *Independent,* London (May 28, 2006).

Byatt, A. S., "Soul Searching," *Guardian,* London (February 14, 2004).

Canguilhem, Georges, *Le normal et le pathologique* (Paris: Presses Universitaires de France, 1966).

Dalby, Andrew, *Dangerous Tastes: The Story of Spices* (London: British Museum Press, 2000).

Damasio, Antonio R., *Descartes' Error: Emotion, Reason, and the Human Brain* (New York: Putnam, 1994; London: Picador, 1995).

Damasio, Antonio R., *The Feeling of What Happens: Body, Emotion and the Making of Consciousness* (London: William Heinemann, 1999; New York: Harvest, 2000).

Damasio, Antonio R. and Hanna, "Minding the Body," *Daedalus: Journal of the American Academy of Arts & Sciences,* Summer 2006, pp. 15–22.

Dolan, Ray, "The Body in the Brain," *Daedalus: Journal of the American Academy of Arts & Sciences,* Summer 2006, pp. 78–85.

Douglas, Mary, *Purity and Danger: An Analysis of the Concepts of Pollution and Taboo* (London and New York: Routledge, 1966).

Dusi, Elena, "Il computer che guarda nell'anima: Sviluppato per capire le emozioni umane—Aiuterà i bambini autistici," *Repubblica* (June 28, 2006), p. 35.

Edelman, Gerald M., *Wider Than the Sky: The Phenomenal Gift of Consciousness* (New Haven and London: Yale University Press, 2004).

Edelman, Gerald M., "Naturalizing Consciousness: A Theoretical Framework," *PNAS,* 100:9 (April 29, 2003), pp. 5520–5524.

Elster, Jon, *Strong Feelings: Emotion, Addiction, and Human Behaviour* (Cambridge, Mass., and London: MIT Press, 1999).

Elster, Jon, *Alchemies of the Mind: Rationality and the Emotions* (New York and Cambridge: Cambridge University Press, 1999).

Evans, Dylan, *Placebo: Mind over Matter in Modern Medicine* (Oxford and New York: Oxford University Press, 2003).

Evans, Dylan, *Emotion: The Science of Sentiment* (Oxford and New York: Oxford University Press, 2001).

Fadiman, Anne, *The Spirit Catches You and You Fall Down: A Hmong Child, Her American Doctors, and the Collision of Two Cultures* (New York: Noonday, 1997).

Filliozat, Jean, *La doctrine classique de la médecine indienne: Ses origines et ses parallèles grecs* (Paris: Imprimerie Nationale, 1949).

Fisher, Helen, *Why We Love: The Nature and Chemistry of Romantic Love* (New York: Holt, 2004).

Fodor, Jerry, "Making the Connection—Axioms from Axons: Why We Need to Think Harder about Thinking" (Review of Joseph LeDoux's *Synaptic Self*), *Times Literary Supplement* (May 17, 2002).

Fodor, Jerry, "How the mind works: what we still don't know," *Daedalus: Journal of the American Academy of Arts and Letters,* Summer 2006, pp. 86–94.

Fonagy, Peter, György Gergely, Elliot Jurist, and Mary Target, *Affect Regulation, Mentalization, and the Development of the Self* (New York: Other Press, 2002).

Galen, *La dieta dimagrante,* ed. Nino Marinone (Turin: G. B. Paravia, 1973).

Gazzaniga, Michael S., *The Mind's Past* (Berkeley, Los Angeles, and London: University of California Press, 1998).

Gawande, Atul, *Complications: A Surgeon's Notes on an Imperfect Science* (New York: Picador, 2002).

Gordon, Richard A., *Anorexia and Bulimia: Anatomy of a Social Epidemic* (Oxford and Cambridge, Mass.: Blackwell, 1990).

Hacking, Ian, *Rewriting the Soul: Multiple Personality and the Science of Memory* (Princeton, N.J.: Princeton University Press, 1995).

Hale, Sheila, *The Man Who Lost His Language* (London: Allen Lane, 2002).

Harrington, Anne, ed., *The Placebo Effect: An Interdisciplinary Exploration* (Cambridge, Mass., and London: Harvard University Press, 1997).

Herper, Matthew, "The Science of Love," *Forbes* (June 24, 2004).

Jacob, François, *La logique du vivant: Une histoire de l'hérédité* (Paris: Gallimard, 1970).

James, William, *The Principles of Psychology* (Cambridge, Mass., and London: Harvard University Press, 1890, 1981).

Kagan, Jerome, *Galen's Prophecy: Temperament in Human Nature* (London: Free Association, 1994).

Kagan, Jerome, *Three Seductive Ideas* (Cambridge, Mass., and London: Harvard University Press, 1998).

Kagan, Jerome, *An Argument for Mind* (New Haven, Conn., and London: Yale University Press, 2006).

Kandel, Eric R., "A New Intellectual Framework for Psychiatry," *American Journal of Psychiatry,* 155:4 (April 1998), pp. 457–469.

Kandel, Eric R., *In Search of Memory: The Emergence of a New Science of Mind* (New York: Norton, 2006).

Kramer, Peter D., *Listening to Prozac* (New York: Penguin, 1993, 1997).

Kramer, Peter D., "There's Nothing Deep about Depression," *New York Times* (April 17, 2005).

Lad, Vasant, *Ayurveda: The Science of Self-Healing—A Practical Guide* (Twin Lakes, Wis., 1984).

Lears, Jackson, "The Resurrection of the Body: Medicine and the Pursuit of Happiness in America," *New Republic* (April 26, 2004).

Leary, David E., *Metaphors in the History of Psychology* (Cambridge and New York: Cambridge University Press, 1990).

LeDoux, Joseph, *The Emotional Brain: The Mysterious Underpinnings of Emotional Life* (New York: Simon and Schuster, 1996).

Maestro Martino di Como, *The Art of Cooking,* ed. Luigi Ballerini, trans. Jeremy Parzen (Berkeley, Los Angeles, and London: University of California Press, 2005).

Miller, Jonathan, ed., *States of Mind: Conversations with Psychological Investigators* (London: BBC, 1983).

Modell, Arnold H., *Imagination and the Meaningful Brain* (Cambridge, Mass., and London: MIT Press, 2003).

Moerman, Daniel, *Meaning, Medicine, and the "Placebo Effect"* (Cambridge: Cambridge University Press, 2002).

Morris, Amy S., *Brainquake: In the Grip of Epilepsy* (Xlibris, 2003).

Müller, August, "Arabische Quellen zur Geschichte der indischen Medizin," *Zeitschrift des Deutschen Morgenländischen Gesellschaft,* 34 (1880), pp. 465–558.

Murphy Paul, Annie, *The Cult of Personality: How Personality Tests Are Leading Us to Miseducate Our Children, Mismanage Our Companies, and Misunderstand Ourselves* (New York: Free Press, 2004).

"A Mystical Union," *Economist* (May 4, 2004).

Nagel, Thomas, "What Is It Like to Be a Bat?" *Philosophical Review,* 83 (1974), pp. 435–450.

Nagel, Thomas, *The View From Nowhere* (New York and Oxford: Oxford University Press, 1986).

Nimchinsky, Esther A., et al., "A Neuronal Morphologic Type Unique to Humans and Great Apes," *Proceedings of the American Academy of Sciences,* 96 (April 1999), pp. 5268–5273.

Nussbaum, Martha, *Upheavals of Thought: The Intelligence of Emotions* (Cambridge and New York: Cambridge University Press, 2001).

Oksenberg Rorty, Amélie, ed., *The Identities of Persons* (Berkeley, Los Angeles, and London: University of California Press, 1976).

Origgi, Gloria, "What Does It Mean to Trust in Epistemic Authority?", in P. Pasquino and P. Harris (eds.), *The Concept of Authority: From Epistemology to Social Sciences* (Rome: Edizioni Fondazione Olivetti, 2006).

Platina, *On Right Pleasure and Good Health,* trans. and ed. Mary Ella Milham (Tempe, Ariz.: Medieval and Renaissance Texts and Studies, 1998).

Prinz, Jesse J., *Gut Reactions: A Perceptual Theory of Emotion* (New York and Oxford: Oxford University Press, 2004).

Quirk, Gregory J., and Donald R. Gehlert, "Inhibition of the Amygdala: Key to Pathological States?" *Annals of the New York Academy of Sciences,* 985 (2003), pp. 263–272.

Redfield Jamieson, Kay, *Touched with Fire: Manic-Depressive Illness and the Artistic Temperament* (New York: Free Press, 1993).

Rosenfield, Israel, *The Strange, Familiar, and Forgotten: An Anatomy of Consciousness* (New York: Random House, 1992).

Rothbart, Mary K., and Michael I. Posner, "Temperament, Attention, and Development Psychopathology," in *Developmental Psychopathology,* ed. D. Cicchetti, (Hoboken, N.J.: Wiley, 2006).

Sample, Ian, "Tests of Faith," *Guardian,* London (February 24, 2005).

Schilder, Paul: *The Image and Appearance of the Human Body* (New York: International Universities Press, 1950).

"The Science of Love: I Get a Kick Out of You," *Economist* (February 12, 2004).

Searle, John: *Minds, Brains and Science* (Cambridge, Mass., and London: Harvard University Press, 1986).

Sicard, Didier, *Hippocrate et le scanner: Réflexions sur la médecine contemporaine* (Paris: Descleé de Brouwer, 1999).

Solomon, Andrew, *The Noonday Demon: An Anatomy of Depression* (London: Chalto & Windus, 2001); with subtitle *An Atlas of Depression* (New York: Simon and Schuster, 2001).

de Sousa, Ronald, *The Rationality of Emotion* (Cambridge, Mass., and London: MIT Press).

Sperber, Dan, "Conceptual Tools for a Natural Science of Society and Culture," *Proceedings of the British Academy,* 111 (2001), pp. 297–317. Also at http://www.dan.sperber.com/Rad-Brow.htm.

Sperber, Dan, "Seedless Grapes: Nature and Culture," *Creations of the Mind: Theories of Artifacts and their Representations,* ed. Stephen Laurence and Eric

Margolis (Oxford, 2005). Also at http://www.dan.sperber.com/Seedless%20Grapes.pdf.

Sperber, Dan (ed), *Metarepresentations: A Multidisciplinary Perspective* (New York and Oxford: Oxford University Press, 2000).

Veyne, Paul, *Les Grecs ont-ils cru à leurs mythes?* (Paris: Seuil, 1983).

Vincent, Jean-Didier, *Biologie des passions* (Paris: Odile Jacob, 1986).

Weinberger, Norman M., "Music and the Brain," *Scientific American* (October 2004).

Widmer, David A. J., "Black Bile and Psychomotor Retardation: Shades of Melancholia in Dante's *Inferno*," *Journal of the History of the Neurosciences* 13:1 (2004), pp. 91–101.

Wozniak, Robert H., *Mind and Body: René Descartes to William James* (1992). At http://serendip.brynmawr.edu/Mind/Table.html.

Young-Bruehl, Elisabeth, *Where Do We Fall When We Fall in Love?* (New York: Other Press, 2003).

Picture Credits

Index

Page numbers of illustrations are set in italics.